5th International Scientific and Practical Seminar Mobility of Transport and Technological Vehicles (MTTV 2020)

IOP Conference Series: Materials Science and Engineering
Volume 1086

Nizhny Novgorod, Russia
12 - 13 November 2020

ISBN: 978-1-7138-3349-9
ISSN: 1757-8981

Printed from e-media with permission by:

Curran Associates, Inc.
57 Morehouse Lane
Red Hook, NY 12571

Some format issues inherent in the e-media version may also appear in this print version.

Institute of Physics
Dirac House, Temple Back
Bristol BS1 6BE UK

Phone: 44 1 17 929 7481
Fax: 44 1 17 920 0979

techtracking@iop.org

Additional copies of this publication are available from:

Curran Associates, Inc.
57 Morehouse Lane
Red Hook, NY 12571 USA
Phone: 845-758-0400
Fax: 845-758-2633
Email: curran@proceedings.com
Web: www.proceedings.com

TABLE OF CONTENTS

MOBILITY OF GROUND TRANSPORT AND TECHNOLOGICAL VEHICLES 1
V V Belyakov, L N Mazunova, A I Markovnina

A CONTRIBUTION TO THE ISSUE OF WHEEL BASE CHANGE OF A MOTOR VEHICLE DURING ITS MODIFICATION FOR INCREASED AUTOMOTIVE TRAFFIC SAFETY 6
Y I Molev, M.G. Cherevastov

A METHOD FOR CONSTRUCTING A GLOBAL MOTION PATH AND PLANNING A ROUTE FOR A SELF-DRIVING VEHICLE 11
A A Sinodkin, T S Evdokimova, M I Tiurikov

A METHOD TO IMPROVE FORWARDERS' ENERGY PERFORMANCE AND ENVIRONMENTAL COMPATIBILITY 20
G O Kotiev, E E Klubnichkin

A METHODOLOGY FOR EVALUATING THE EFFICIENCY OF DRIVING A TRUCK 30
A M Umirzokov, K T Mambetalin, S S Saidullozoda, A L Berdiev

ANALYSIS OF ACCELERATED METHODS TESTING PISTON ENGINE FOR RELIABILITY 36
A N Gots, Sh A Amirseyidov

ANALYSIS OF THE DEVELOPED HINGED JOINT ASSEMBLY OF THE WHEELED HARVESTER 44
V E Klubnichkin, E E Klubnichkin, A B Kartashov, K V Vasilyeva

ASSESSING LOAD ON SUPPORT ROLLERS OF THE TRACKED LOAD-TRANSPORT TIMBER HARVESTING VEHICLE THROUGH SIMULATED MATHEMATICAL MODELING 53
V E Klubnichkin, G O Kotiev, V S Makarov

COMPARATIVE STUDY OF IMPRESSION AND CUTTING INTO ICE BY HELICAL BLADE 59
U Sh Vakhidov, I A Yerasov, Y I Molev, D S Mokerov

CONCEPTUAL MODEL OF THE AUTOMATED PEDESTRIAN SIMULATION SYSTEM FOR UNMANNED VEHICLES AND ADAS SYSTEMS TESTING 65
A A Tumanov, A V Zolotov, L O Fedosova, A Y Sizov

DAMPING OF PRESSURE PULSATIONS IN THE FUEL SYSTEM OF THE MOTOR 74
A A Lepeshkin, L A Berdnikov, M G Korzhachkin, V A Panov

DETERMINATION OF THE MOMENT OF RESISTANCE TO TURNING OF A FORESTRY ARTICULATED MACHINE 80
D Rogachev, A Alyabiev, V Latysheva

DETERMINING OF OPTIMUM COMPOSITION OF WATER-FUEL MIXTURE FOR A SPARK-IGNITION INTERNAL COMBUSTION ENGINE WITH ELECTRONIC CONTROL 86
A P Kartoshkin, A L Birykov, Sh V Busikov, D G Sergeev, S.A. Plotnikov

DETERMINING OF OPTIMUM OPERATION MODES OF A DIESEL ENGINE WITH A MULTI-COMPONENT BIO-FUEL COMPOSITION 94
S A Plotnikov, A N Kartashevich, M V Simonov, M N Glushkov

DEVELOPING AN UNMANNED VEHICLE LOCAL TRAJECTORY USING A REINFORCEMENT LEARNING ALGORITHM 100
T S Evdokimova, A A Sinodkin, M I Tiurikov, L O Fedosova, A V Kalyashina

DEVELOPMENT AND STUDY OF WAYS TO EXTEND THE MULTIFUEL CAPABILITY OF ICE 106
S A Plotnikov, A N Kartashevich, P Ya Kantor, M N Vtyurina

DEVELOPMENT OF KORSAK 8X8 SMALL-SIZE ALL-TERRAIN VEHICLE DESIGN PROJECT 116
A G Kapustin, A V Papunin, A M Belyaev

EFFICIENCY OF THE SIMULATION MODEL OF A LIGHT COMMERCIAL VEHICLE 121
D A Butin, A V Tumasov, A S Vashurin, M E Gnenik

ELASTOMER ELASTIC ELEMENT VISCO-ELASTIC PROPERTIES' IMPACT ON DAMPING IN A VEHICLE SUSPENSION 128
E V Stepanov, S M Ogorodnov, L N Orlov, M K Chegurov

ELECTRONIC COMPONENT BASE FOR CONTROLLING THE ENGINE, MECHATRONIC TRANSMISSION AND ADAPTIVE BRAKE FORCE LIMITATION SYSTEM AS PART OF THE ELECTRONIC PLATFORM OF THE FORMULA SAE CLASS SPORTS VEHICLE 134
A Kulagin, V Khozerov

ENSURING RELIABILITY OF ELECTRONIC CONTROL OF ENGINE FUEL SUPPLY SUBSYSTEM IN OPERATION 140
Y Bazhenov, M Bazhenov, V Kalyonov

EVOLUTION OF REQUIREMENTS TO STATE OF ORDER OF AUTOMOTIVE TIRES FOR TRAFFIC SAFETY 147
I A Yerasov, D N Proshin, D V Glazunov

EXPERIMENTAL AND CALCULATION METHOD FOR DETERMINING THE PARAMETERS OF A CAR SUSPENSION WITH AN ELASTIC ELASTOMERIC ELEMENT 153
E V Stepanov, S I Maleev, R A Musarsky, S M Ogorodnov

FEATURES OF MACROMODELING OF SMALL TOWNS WITH A MONOSTRUCTURED ECONOMY 161
K O Goncharov, M E Eliseev, K V Laschuk, A O Chikaev

FORMATION OF PERFORMANCE CHARACTERISTICS OF A FORMULA SAE SPORTS VEHICLE BASED ON CALCULATIONS AND SIMULATIONS OF SYSTEMIC ELEMENTS 167
A Kulagin, K Goncharov, V Kuzmichev, M Golubev, M Gulin, V Khozerov

MODELING OF THE TOTAL DISTURBING EFFECTS ON THE TRANSMISSION ELEMENTS FROM MOVEMENT ON DIFFERENT SUPPORT BASES AND THE ENGINE 177
K Y Leliovsky

MODERN METHODS OF FIELD TESTS OF DRIVER ASSIST SYSTEMS 183
Ye I Toropov, Yu P Trusov, A S Vashurin, P S Moshkov, V I Filatov, M E Gnenik

NATURAL NANOMATERIALS AS A BASIC FOR IMPROVING THE ENVIRONMENTAL SAFETY OF TRANSPORT SYSTEMS 192
D O Chernyshev, L V Denezhko, E E Bazhenov

OPTIMIZATION OF TRAFFIC FLOW SPEED IN URBAN CONDITIONS 197
E A Kindeyev

POSSIBLE SOLUTIONS TO THE PROBLEM OF RECYCLING ELECTRIC VEHICLE BATTERIES ON THE EXAMPLE OF GLOBAL MANUFACTURERS 202
D A Kolesnichenko, D N Pechalinov, R V Vyatkin, A V Erofeeva

PREDICTING MODELING OF THE SERVICE PROPERTIES OF CARRYING SYSTEMS IN THE ARTICULATED-VEHICLE-BASED SELF-PROPELLED ARTILLERY SYSTEMS 206
E E Bazhenov, A N Kustovsky, M L Usupov

PREDICTING THE DURABILITY OF PISTON ENGINE CRANKSHAFTS 211
A N Gots, Sh A Amirseyidov

SCALING AND DYNAMICS OF SPECIAL MACHINES 218
I G Kuklina

SIMULATION OF CONDUCTED EMISSIONS IN LOW VOLTAGE SWITCHING CONVERTERS 223
A V Vasin, A A Novoselov, E E Samokhvalov, S V Lazarevich

SIMULATION OF THE PASSIVE PART OF AN ACTIVE CAR MUFFLER 229
A Okunev, A Kocherov, S Mantserov, M Kolobov, D Bushmanov

SIMULATION OF THE PROCESS OF INTERACTION BETWEEN THE VEHICLE AND SNOW IN THE ENVIRONMENT OF DEM AND MBD-SYSTEMS 235
D S Teslenko, V V Belyakov

TRACK SNOW AND SWAMP-GOING VEHICLE WITH 16 T GVW 248
S E Manianin, E V Lisagov, P E Dmitriev, A A Eremin, U Sh Vakhidov, V S Makarov

USE OF ACTIVE RELATED TRANSPORTATION SYSTEMS IN THE EXTRACTIVE INDUSTRIES OF THE NATIONAL ECONOMY AND PROTECTION COMPLEX 253
E E Bazhenov, L A Novopashin, D O Chernyshev

WAYS TO REDUCE FUEL CONSUMPTION OF A LIGHT COMMERCIAL VEHICLE FOR WLTC AND NEDC CYCLES 259
A A Kolin, S E Silantyev, P S Rogov

Author Index

IOP Conf. Series: Materials Science and Engineering 1086 (2021) 011001 doi:10.1088/1757-899X/1086/1/011001

Preface

On November 12-13, 2020, the 5th International scientific and practical seminar Mobility of Transport and Technological Vehicles (MTTV) was held at the Nizhny Novgorod State Technical University named after R. E. Alekseev, dedicated to the 85th anniversary of the Department Automobiles and Tractors.

The conference was organized by Nizhny Novgorod state technical university n.a. R. E. Alekseev, LLC InTech and Nizhny Novgorod scientific and practical school of transport snow science.

Doctor of technical sciences, Professor Vladimir Belyakov opened the seminar MTTV: "Dear participants of the fifth seminar Mobility of transport and technological vehicles! I am glad to welcome you and wish you fruitful work. This seminar is being held at a difficult time. Due to the COVID-19 pandemic, we are forced to conduct the workshop remotely. I hope this will not interfere with our fruitful work."

- Dear colleagues and participants of the seminar! I congratulate you with the start of seminar! The seminar is held in the year of the 85th department Automobiles and tractors, - the head of the department, Candidate of Technical Sciences, Associate Professor Anton Tumasov made a welcoming speech. - Mobility of vehicles and tractors is one of the primary tasks in the design and exploitation of transport systems. Scientists of the Soviet, and today, and Russian science related to automotive construction, have always paid great attention to the mobility of machines, as it provides a safe and stable movement of automotive equipment. Let me wish the participants of the seminar fruitful work.

The head of the Department Building and Road Machines, Doctor of Technical Sciences, Professor Umar Vahidov congratulated the Department of AIT and all its employees, students and graduates on the 85th anniversary.

The seminar was attended by scientists, researchers and designers in the field of land transport, as well as teachers, postgraduates and students from all over the country, including from NNSTU n.a. R. E. Alekseev, Bauman Moscow State Technical University, NAMI, Moscow Polytechnic University, Vladimir State University, Ural State Agrarian University, Ural State Forestry University, Vyatka State University, Nizhny Novgorod State University of Engineering and Economics, Mininsky University and other universities.

More than 50 reports were selected for the seminar. Participants from Nizhny Novgorod presented their works in full-time format, colleagues from other cities took part in the on-line format.

As before, the main topic of the seminar was the mobility of transport and technological vehicles (TTV) - an integral exploitation characteristic of the TTV that determines its ability to perform the task with optimal adaptability to the operating conditions and technical condition of the machine itself, that is, the ability of the vehicle to resist external and internal factors that hinder the performance of the task. The seminar also discussed topics related to the design and creation of wheeled and tracked vehicles, road, construction vehicles, machines for agriculture and forestry, the features of the operation of heat engines, the issues of modeling the dynamics and testing of vehicles, including driverless vehicles, simulation of traffic flows, the issues of ecology, traffic safety and smart city systems.

After all the reports, an exchange of views was held and ways for further cooperation were outlined. Made suggestions, wishes and congratulations 5-year anniversary seminar on "Mobility",

IOP Conf. Series: Materials Science and Engineering 1086 (2021) 011001 doi:10.1088/1757-899X/1086/1/011001

professor Evgeny Bazhenov (USAU), professor Umar Vahidov (NNSTU), professor Vladislav Dygalo (VSTU), associate professor Vladislav Klubnichkin (BMSTU).

In conclusion, the organizer and secretary of the seminar MTTV, doctor of technical sciences, professor Vladimir Makarov thanked everyone for their participation and reminded that the materials of the seminar will be recommended for publication in the journals Proceedings of the NNSTU n. a. R. E. Alekseev, Transport Systems and in the journal IOP Conference Series: Materials Science and Engineering (MSE).

With Best wishes,

Prof. Vladimir Belyakov Chairman of MTTV 2020

Prof. Vladimir Makarov Secretary of MTTV 2020

NIZHNY NOVGOROD STATE TECHNICAL UNIVERSITY N.A. R.E. ALEKSEEV

НИЖЕГОРОДСКАЯ НАУЧНО-ПРАКТИЧЕСКАЯ ШКОЛА ТРАНСПОРТНОГО СНЕГОВЕДЕНИЯ

InTech

АВТОМОБИЛИ И ТРАКТОРЫ ИТС НГТУ

IOP Conf. Series: Materials Science and Engineering 1086 (2021) 011001 doi:10.1088/1757-899X/1086/1/011001

Committees of MTTV

Conference Chair

Prof. Vladimir Belyakov (Nizhny Novgorod State Technical University n.a. R. E. Alekseev)

Conference Secretary

Prof. Vladimir Makarov (Nizhny Novgorod State Technical University n.a. R. E. Alekseev)

Program Committee

Prof. Evgeny Bazhenov (Ural State Agrarian University)

Prof. Sergey Plotnikov (Vyatka State University)

Prof. Umar Vahidov (Nizhny Novgorod State Technical University n.a. R. E. Alekseev)

Prof. Yuri Molev (Nizhny Novgorod State Technical University n.a. R. E. Alekseev)

Dr. Evgeny Klubnichkin (Bauman Moscow State Technical University)

Dr. Vladislav Klubnichkin (Bauman Moscow State Technical University)

Dr. Anton Tumasov (Nizhny Novgorod State Technical University n.a. R. E. Alekseev)

Dr. Kirill Goncharov (Nizhny Novgorod State Technical University n.a. R. E. Alekseev)

Peer review declaration

All papers published in this volume of *IOP Conference Series: Materials Science and Engineering* have been peer reviewed through processes administered by the Editors. Reviews were conducted by expert referees to the professional and scientific standards expected of a proceedings journal published by IOP Publishing.

- **Type of peer review:** Single-blind
- **Conference submission management system:** articles were sent directly to secretary on seminar e-mail vladimir.makarov@nntu.ru
- **Number of submissions received:** 70
- **Number of submissions sent for review:** 70
- **Number of submissions accepted:**39
- **Acceptance Rate (Number of Submissions Accepted / Number of Submissions Received X 100):** 56%
- **Average number of reviews per paper:** 1
- **Total number of reviewers involved:**10
- **Any additional info on review process:**

  The main criteria for the review were:
  1. Is the topic in the conference scope?
  2. Is subject is worthy the publication?
  3. Is information new?
  4. Is paper written in acceptable English?

  For each question, the reviewer used scale from 1 to 5 to answer. The reviewer also checked that article was formatted according the Template, that all references, figures and tables were cited in the text and provided other comments.

- **Contact person for queries:**

  Secretary of the seminar: Vladimir Makarov, Doctor of Technical Sciences, Professor of the Department «Construction and road machines», NNSTU n.a. R.E. Alekseev, tel. +7-905-012-44-03, e-mail: vladimir.makarov@nntu.ru
  Coordinator: Kirill Goncharov, Candidate of Technical Sciences, Associate Professor, Deputy director of the Institute of Transport Systems, NNSTU n.a. R.E. Alekseev, tel. +7-910-143-33-30, e-mail: goncharov@nntu.ru

IOP Conf. Series: Materials Science and Engineering 1086 (2021) 012001 doi:10.1088/1757-899X/1086/1/012001

Mobility of ground transport and technological vehicles

V V Belyakov[1], L N Mazunova and A I Markovnina

Nizhny Novgorod State Technical University n.a. R.E. Alekseev, Minin str., 24, 603950, Nizhny Novgorod, Russia

E-mail: [1]belyakov@nntu.ru

Abstract. The article provides a definition of the mobility of ground transport vehicles and transport and technological complexes. Mobility is defined as an integral characteristic of structural and operational, economic, ergonomic and environmental properties of ground mobile systems and complexes. Also, mobility characterizes the competitiveness of objects as a product. A structural diagram of the hierarchy of control tasks for transport and technological vehicles is given containing four tasks related to the field of design parameters: maintaining mobility, maintaining directional orientation, maintaining speed, eliminating critical situations when moving a transport and technological system; and also the task: to ensure environmental safety; ensuring nominal economic efficiency and ensuring the comfort of the transport and technological system.

1. Introduction

Mobility is an integral operational property of transport and technological vehicles (TTV), which determines its ability to perform the assigned task with optimal adaptability to the operating conditions and the technical condition of the vehicle itself, that is, the ability of the machine to resist external and internal factors that impede the performance of the task [1-6]. Thus, the process of maintaining mobility is key in solving the problem of ensuring the stable and safe functioning of mobile land transport vehicles and transport and technological complexes. Also, during operation, operational and structural mobility must be ensured.

2. Components of TTV mobility

Operational mobility (OM) determines the possibility of fulfilling the task under the operating conditions, including both terrain characteristics and vehicle loading modes.

Structural mobility (SM) is associated with the technical characteristics of the vehicle, which during operation can change significantly and lead to both partial loss of mobility and its complete loss. Changes in structural mobility are associated, on the one hand, with the wear of units, systems and machine parts, which leads to functional failures, and on the other hand, with structural failures caused by the "aggressiveness" of the external environment. The aggressiveness of the external environment can manifest itself in very different forms, but most often these are mechanical influences: wear and shock. Deterioration determines the failure reliability of a machine, and shock mechanical interactions can be natural due to contact with the terrain and artificial in collisions with other machines and / or objects of the external environment, as well as damage received during hostilities and terrorist acts from the effects of weapons enemy side.

Mobility (M) of a vehicle is a generalized function of operational and structural mobility.

$$M = f(OM; SM) \tag{1}$$

Active and passive safety encompasses a wide range of tasks related to maintaining stable and safe traffic and do not have clear boundaries. More acceptable is the division of mobility into agility and mobility in terms of survivability.

Agility (dynamic adaptive mobility). Essentially, this term refers to the concept of "movement" or more broadly defines "readiness for a quick reaction, quick engagement in activity." All these concepts are combined into the ability to move in any conditions and are solved in three main tasks of vehicle control: maintaining the speed of movement, ensuring directional orientation, and eliminating critical situations.

The assessment of mobility and the construction of control algorithms for vehicle are performed on the basis of the following criteria: 1) by the stock of traction force; 2) power balance; 3) on course orientation (handling and maneuverability). In this case, the following solutions to the problem of maintaining mobility by mobility take place: 1) the concept of motion control ($\lambda_{\text{р}} = var, \lambda_{\text{к}} = const,$ $\lambda_{\text{э}} = const$) - for the given operating conditions and the given structural configuration of the vehicle, the optimal modes of motion control are determined; 2) the vehicle design concept ($\lambda_{\text{р}} = const, \lambda_{\text{к}} =$ $var,\ \lambda_{\text{э}} = const$) - for given operating conditions and selected motion control modes, a rational structural configuration of the vehicle is determined; 3) the concept of operating conditions ($\lambda_{\text{р}} =$ $const, \lambda_{\text{к}} = const,\ \lambda_{\text{э}} = var$) - for a given structural configuration of the machine and the selected motion control modes, the critical characteristics of the operating conditions are determined. Here $\lambda_{\text{к}}$ – vehicle parameters, including mover parameters; $\lambda_{\text{э}}$ – characteristics of operating conditions, including properties and track parameters; $\lambda_{\text{р}}$ – parameters characterizing kinematic and power modes of motion. Moreover, these parameters can be considered both separately and in the form of any complex characteristics $\lambda \equiv \xi(\lambda_{\text{к}}, \lambda_{\text{э}},\ \lambda_{\text{р}})$.

Mobility in terms of survivability- ensuring control of the state of the vehicle aimed at maintaining the operability of ground transport systems and transport-technological complexes to the current state of the vehicle and the terrain. In this case, one should distinguish between partial and complete loss of performance.

The survivability of vehicles can be divided into failure, associated with wear and tear, and operational, which is determined by the functioning of the machine in the natural and climatic environment and the socio-technical sphere of activity. Thus, the viability can be associated not only with the properties of the terrain, but with the impacts from human activities, for example, man-made disasters, military operations, road traffic accidents, super-heavy (irregular) operating modes.

Survivability can be divided into structural and functional components. If the study of the structural component of survivability is mainly reduced to identifying vulnerabilities in the topology of the system and determining the degree of their influence on the integrity of the system, then the study of the functional component of survivability is reduced to determining the ability of the system to solve the problems facing it with the changing capabilities of its elements.

Evaluation of the vehicle's survivability for failure reliability is carried out according to the criterion of the probability of failure-free operation, and in terms of operational reliability - the probability of serviceability in a post-accident or damaged state.

3. Calculation of mobility TTV

A comprehensive assessment of the mobility of land transport vehicles and transport and technological complexes is based on a system of criteria and limiting conditions:

by agility

$$_{\Delta}P_{\varphi}\left(\Phi_{\varphi}, \Phi_{f}, \lambda\right) \to \max_{\lambda \in \Lambda} \text{ at } _{\Delta}P_{\varphi}\left(\Phi_{\varphi}, \Phi_{f}, \lambda\right) \geq 0, \tag{2}$$

$$W_{\varphi}\left(\Phi_{\varphi}, \lambda\right) \to \min_{\lambda \in \Lambda} \text{ at } W_{\varphi}\left(\Phi_{\varphi}, \lambda\right) \geq \left[W_{f}\left(\Phi_{f}, \lambda\right) + _{\Delta}W\left(\Phi_{\varphi}, \Phi_{f}, \lambda\right)\right], \tag{3}$$

$$\Phi_R\left(\Phi_\varphi,\Phi_f,\lambda\right)\to\min_{\lambda\in\Lambda}\text{ at }\Phi_R\left(\Phi_\varphi,\Phi_f,\lambda\right)\le 0;\tag{4}$$

by survivability

$$R_{\text{он}}\left(\Phi_\varphi,\Phi_f,\lambda,t\right)\to\max_{\lambda\in\Lambda}\text{ at }R_{\text{он}}\left(\Phi_\varphi,\Phi_f,\lambda,t\right)\ge R_\eta\left(\Phi_\varphi,\Phi_f,\lambda,t\right);\tag{5}$$

$$R_{\text{эн}}\left(\Phi_{\text{ив}},\lambda,t\right)\to\max_{\lambda\in\Lambda}\text{ at }R_{\text{эн}}\left(\Phi_{\text{ив}},\lambda,t\right)\ge R_\gamma\left(\Phi_{\text{ив}},\lambda,t\right).\tag{6}$$

The criteria are built depending on the parameters of the interaction of the mover with the track Φ_φ,Φ_f and taking into account other characteristics $\lambda\equiv\xi\left(\lambda_\text{к},\lambda_\text{э},\lambda_\text{р}\right)$. Here ΔP_φ – stock of traction force, W_φ – traction power, $\Delta W = W_{\text{эу}} - W_f$ – stock of engine power, $W_{\text{эу}}$ – engine power, W_f – resistance power, Φ_R – generalized function of the radius of curvature of the trajectory, $R_{\text{он}}$ – failure-free probability, R_η – limiting probability of failure-free operation of equipment under optimal operating conditions and vehicle manufacturing technology; $R_{\text{эн}}$ – performance probabilities in a post-accident or affected state, $\Phi_{\text{ив}}$ – generalized function of the intensity of emergency or damaging impact, R_γ - the limiting probability of equipment operability after a given (in calculations) intensity of an emergency or damaging effect, t – operating time or exposure time according to the considered type of reliability.

The presented system of criteria (2) - (6) with limiting conditions can be reduced to a more substantiated system of objective functions:

$$\Phi_\varphi\left(\lambda\right)\to\max_{\lambda\in\Lambda},\ \Phi_f\left(\lambda\right)\to\min_{\lambda\in\Lambda},\ v\left(\lambda\right)\to\max_{\lambda\in\Lambda},\ \rho\left(\lambda\right)\to\min_{\lambda\in\Lambda},\ R\left(\lambda\right)\to\max_{\lambda\in\Lambda}.\tag{7}$$

Where Φ_φ – generalized function of adhesion of the vehicle mover to the support base; Φ_f – generalized function of resistance to movement of the vehicle; v – speed; $\rho = R_\text{п}/B$ – relative turning radius, where B –track, R – probability of failure-free operation of the vehicle as a function $R = (R_{\text{он}}, R_{\text{эн}})$ failure and operational reliability.

The solution to the problem of multicriteria optimization by the presented system of equations (7) is described in detail in [5, 6].

The solutions are directly related only to the limited tasks of maintaining the mobility of automotive vehicles and transport and technological machines created on its basis. However, there are other options for solving the problem of maintaining mobility, which are determined by the function «var-const» in relation to parameters $\lambda_\text{к}$, $\lambda_\text{э}$, $\lambda_\text{р}$. In this case, the most interesting is the complex solution of the mobility problem, when $\lambda_\text{к}$=var, $\lambda_\text{э}$=var, $\lambda_\text{р}$=var, that is, an assessment of the competitiveness of existing, modified or newly created vehicles.

For a qualitative and quantitative assessment of the competitiveness of automotive vehicles, a number of methods can be proposed: building an expert system, finding regression, qualimetry and multi-criteria optimization [6]. At the same time, the latter method is aimed not so much at assessing the quality of the existing structure as at the choice of rational technical, technological, operational and consumer parameters of the designed vehicle. This method, unlike the first three, can be combined with an automatic design system, and the quality of an existing vehicle from others can be assessed as their deviation from the optimal standard, which is generally also done in other methods. However, other methods do not allow making a mathematically accurate choice of rational parameters of the designed automotive equipment.

The competitiveness of automotive vehicles is understood as such a complex property of a particular machine, which determines its quality in accordance with the actual values of technical, technological, operational and consumer indicators, which characterizes its ability to compete with similar models of vehicles produced (developed) by competing firms.

The methodology for assessing the competitiveness of ground TTV is as follows: 1) to split the estimated characteristics to numerical indicators; 2) express the estimated indicators in dimensionless form; 3) accept the condition that the growth of the indicator determines the increase in efficiency, as a result of which a number of indicators should be either taken as a reciprocal value, or as a result of subtracting from one; 4) repeated values should be counted as many times as they occur, which will determine their ranking (weight significance); 5) all assessments should be taken modulo; 6) the number of estimated characteristics for compared different types of TTV should be the same. The greatest difficulty is the assessment of aesthetic indicators. It can be carried out on the basis of probabilistic estimates based on the results of consumer surveys or experts. Probabilistic assessments can be used as the basis for the scoring characteristics of aesthetics of the machine. However, one should not neglect the well-known statement that "tastes differ", and this is the most difficult task. From this point of view, the most suitable for assessing the competitiveness of automotive vehicles is the method of constructing expert systems [6].

Then an express model of the competitiveness of a vehicle, broken down by structural blocks, in general form can be represented by the following system of equations:

$$a_{\text{к}i}(\lambda)\to\underset{\lambda\in\Lambda}{\text{extr}};\ a_{\text{э}i}(\lambda)\to\underset{\lambda\in\Lambda}{\text{extr}};\ a_{\text{р}i}(\lambda)\to\underset{\lambda\in\Lambda}{\text{extr}};$$
$$a_{\text{т}i}(\lambda)\to\underset{\lambda\in\Lambda}{\text{extr}};\ a_{Ri}(\lambda)\to\underset{\lambda\in\Lambda}{\max};\ a_{Ci}(\lambda)\to\underset{\lambda\in\Lambda}{\min}; \quad (8)$$
$$_{\Delta}P_{\varphi}(\lambda)\to\underset{\lambda\in\Lambda}{\max};\ v(\lambda)\to\underset{\lambda\in\Lambda}{\max};\ Э(\lambda)\to\underset{\lambda\in\Lambda}{\text{extr}};\ П(\lambda)\to\underset{\lambda\in\Lambda}{\text{extr}}. \quad (9)$$

Where $a_i(\lambda)\in A$, $(i=1,2,\ldots,11)$; $\lambda=\xi(\lambda_{\text{к}},\lambda_{\text{э}},\lambda_{\text{р}},\lambda_{\text{т}},\lambda_{\text{п}})$ at $\lambda_{\text{к}}=\text{var}$, $\lambda_{\text{э}}=\text{var}$, $\lambda_{\text{р}}=\text{var}$, $\lambda_{\text{т}}=\text{var}$, $\lambda_{\text{п}}=\text{var}$. Objective functions (9) refer to the vehicle as a whole.

Examples of solving the problem of optimization of objective functions (7-9) in order to find a rational design of a vehicle in given operating conditions on the basis of multi-criteria optimization are considered in the works [5, 6].

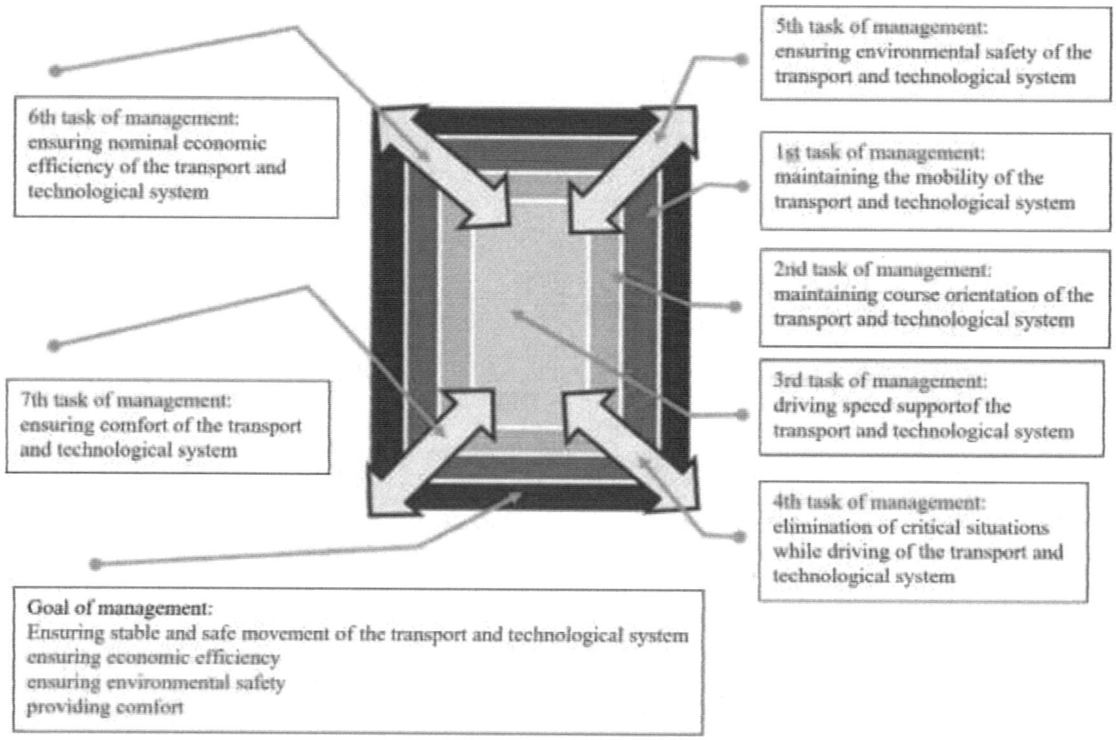

Figure 1. Hierarchy of TTV management tasks

4. Conclusion

As a result, it can be concluded that the conceptual diagram of the hierarchy of control tasks to ensure the mobility of the TTV, considered in [5], can be presented in the form of a diagram shown in Figure 1. It includes not only four tasks related to the field of design parameters (maintaining mobility, maintaining directional orientation, maintaining speed, eliminating critical situations when moving the transport and technological system), but also tasks: ensuring environmental safety; ensuring nominal economic efficiency and ensuring the comfort of the transport and technological system.

Acknowledgment

This study was conducted in continuation of the research conducted by "Nizhny Novgorod scientific and practical school of snow vehicles" and with financial support of the grant of the President of the Russian Federation No. MD-226.2020.8.

References

[1] Belyakov V V and Kulyashov A P 2004 *All-terrain transport and technological vehicles Foundations of the theory of motion.* N. Novgorod. TALAM p 961

[2] Makarov V, Zeziulin D and Belyakov V 2014 Prediction of all-terrain vehicles mobility in snowscape scenes *18th International Conference of the ISTVS*

[3] Papunin A Vet al. 2018 A dynamic model of unsupported pit traversal by a vehicle with 6×6 wheel arrangement *IOP Conference Series: Materials Science and Engineering* **386(1)** 01200 doi:10.1088/1757-899X/386/1/012001

[4] Porubov D et al. 2017 Study of efficiency of using all-terrain vehicles during the winter period *19th International and 14th European-African Regional Conference of the ISTVS*

[5] Belyakov V V et al. 2014 The conception of mobility of ground transport and technological vehicles *Transactions of NNSTU n.a. R.E. Alekseev* **3(100)** pp 145-174

[6] Belyakov V V, Bushueva M E and Sagunov V I 2001 *Multicriteria optimization in problems of mobility, competitiveness of automotive vehicles and diagnostics of complex technical systems* N.Novgorod. NNSTU p 271

IOP Conf. Series: Materials Science and Engineering 1086 (2021) 012002 doi:10.1088/1757-899X/1086/1/012002

A contribution to the issue of wheel base change of a motor vehicle during its modification for increased automotive traffic safety

Y I Molev[1], M.G. Cherevastov[2]

[1]Nizhny Novgorod State Technical University n.a. R.E. Alekseev, Minin st., 24, 603950, Nizhny Novgorod, Russia.
[2]Nizhny Novgrod State Pedagogical University n.a. K. Minin, Uyanova str., 1, 603950, Nizhny Novgorod, Russia

E-mail: [1]moleff@yandex.ru

Abstract. The paper presents a study of the change of the transient duration, of the control sensitivity of the vehicle, of the gain and the response delay of N1 Cat. cargo vehicle at instant and harmonic steering control action. The changes are caused by vehicle modification via wheel base length extension. The wheel vehicle is represented by a flat single-mass calculation model. Based on the conducted calculation, additional influence of the driving speed as a risk factor has been revealed. In the summary of the paper, a conclusion is made on the presence of the influence of the vehicle wheel base length change on the automotive traffic safety.

1. Introduction and relevance

The relevance of the mortality and the injury rate reduction among motor traffic participants can scarcely be overestimated. The World Health Organization reports annually over a million casualties on the motor roads of our planet, whereas several dozens of million people get injuries of different severity within the same period. For Russia, this situation remains in the high level of tension, being of high socio-economic importance. For the time being, widely used are motor vehicles with design modifications undertaken by their owners which, in day-to-day intended operation, exert their influence on the automotive traffic safety along with the standard motor vehicles. One of the main risk factors considerably influencing the consequences of traffic accidents is the driving speed of wheel type vehicles which is limited both by the traffic rules, and by the vehicle manufacturer. Usually, design modifications of a vehicle leave unchanged its speed limit, whereas the modification proper is not considered as a reason for conscious speed reduction.

In the course of time, the popularity of modifications to vehicle designs is growing every year. So, in the Russian Federation, within the three years from 2015 to 2017, the number of modifications has grown more than by 10 times, from 29432 to 307882 cases [1]. Thereby, this trend is very dynamical. The absolute leader among the vehicle modification types is currently the replacement of the normally carried fuel system of the internal combustion engine. During the three indicated years, the number of such modifications has grown more than thrice, whereas the percentage has grown from 23.4% to 77.6%. Along with that, as experience shows, widely used is also such type of vehicle modification at which the wheel base length of the vehicle is changed (mostly extended). Based on authors' data, the number of such modifications makes up to 15% of the total number of modifications. Hereinafter, we study the influence of the modified wheel base length on the controllability of the vehicle, that is, on the safety of the automotive traffic.

2. Theoretical research

For a preliminary analysis, we take the well-studied flat single-mass linear calculation motor vehicle model [2]. This model is limited in its use, nevertheless, it has a big advantage of simplicity and possibility of an analytical solution. For the evaluation of the influence of the undertaken design modification of the vehicle, we consider two cases of the control action (instant steering wheel turning by a certain angle and harmonic steering wheel manipulation). During modeling of the locomotion processes of the vehicle, the following was assumed: the vehicle has understeer, the driving speed is constant, the wheel slipper is linear. The applicability of the model was monitored by measurement of the wheel slipper angles, the turning angle of the steering wheel, and their non-exceeding of the admissible limits, as well as by calculation of the specific lateral load and its comparison with the admissible value. During the calculations, specifications of a N1 Category cargo motor vehicle were used for the case of the gross laden weight.

"Instant turn of steering wheel". In this case, we have evaluated the duration of the transient *(t_p)*, caused by the vehicle steering control action of the driver and the dynamic response of the vehicle to the control *(μ_d)*. The specified parameters characterize the controllability of the vehicle as control object.

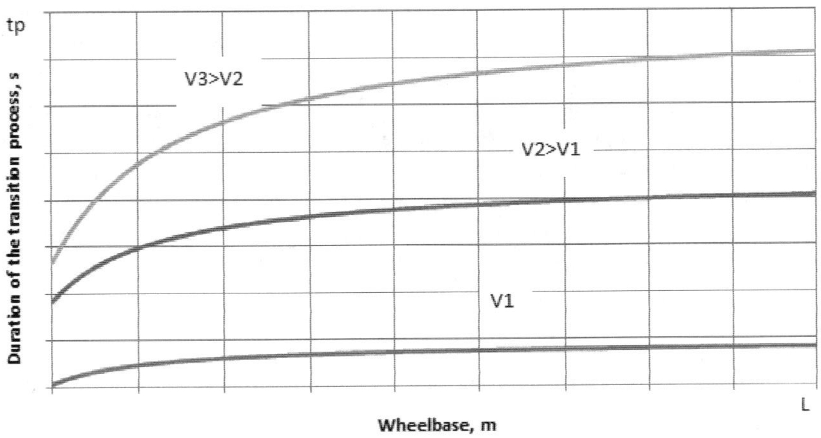

Figure 1. Influence of vehicle wheel base on transient duration at different driving speeds

It is evident from the graphs in Figure 1 that a longer wheel base will lead to the transient duration growth affecting the vehicle controllability. The increasing of the driving speed also leads to the transient duration growth which is more intensively displayed at long wheel bases.

The dynamic steering sensitivity (see figure 2) characterizing the vehicle locomotion curvature change determined with the classic method [3], gets back in a non-linear manner with the growing wheel base also affecting negatively the controllability of the vehicle. The highest reduction of the sensitivity is observed at high speeds.

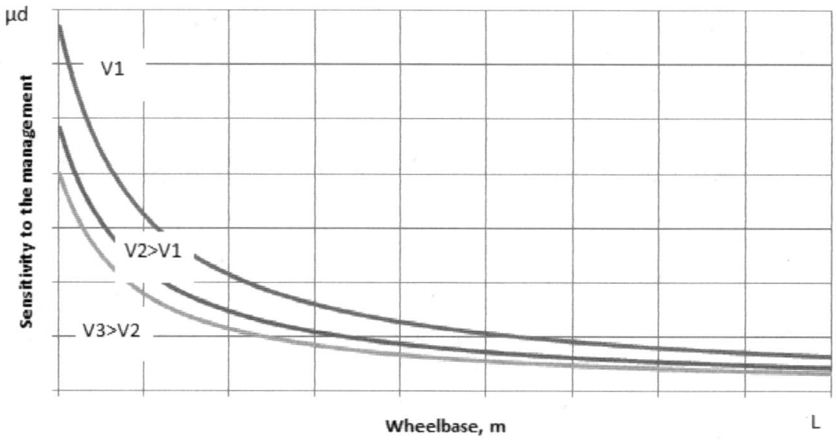

Figure 2. Influence of vehicle wheel base on dynamic steering sensitivity at different driving speeds

Further, we conducted a calculation of the dynamic sensitivity change at extension of the vehicle wheel base by 20%, 40%, 60%, 80%, and 100% which is vastly popular in the modifications of N1 Cat. motor vehicles. The calculation results are depicted graphically in Figure 3. It is evident in the Figure that the dependency is non-linear, thereby, the extension of the wheel base by 100%, that is, when it is doubled, the steering sensitivity of the vehicle is reduced by a half. The driving speed of the wheel type vehicle affects the steering sensitivity only insignificantly. This feature is a consequence of the fact that the center of gravity of the vehicle at full weight is located near the neutral steer point, at vehicle's outfit weight, the configuration will change slightly, and the speed factor will get more distinct. As for the transient duration, the doubled wheel base length will lead to a 10% longer transient process. Thus, for typical control action, the instant steering wheel turn by a certain angle one could conclude that the extension of the wheel base of the vehicle during its modification affects the controllability of the vehicle, thereby, this influence also depends on the driving speed.

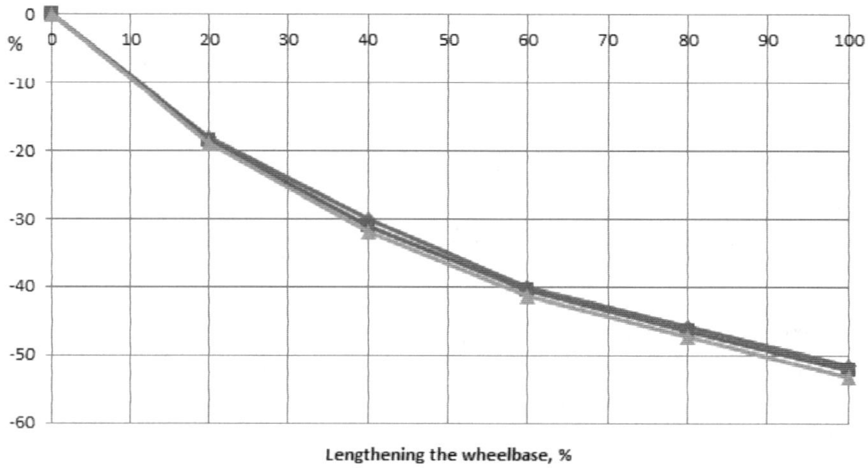

Figure 3. Reduction of dynamic steering sensitivity of vehicles versus wheel base extension

"Harmonic steering action". Specific for such steering control action are amplitude vs. frequency and phase vs. frequency curves [4] determining the ratio of the output value amplitude (angular speed of the vehicle) to the amplitude of the input value (turn angle of the steered wheels) and, respectively, the delay of the output signal versus the input signal in the control frequency. The most typical frequency is 0.75 Hz. That is why we conducted calculations of the changes of such curves with the change of the wheel base of a motor vehicle in 0.75 Hz frequency. Fig. 4 shows the dependency of the change rate of the output signal to the input signal (vehicle gain) in the specified frequency versus wheel base length.

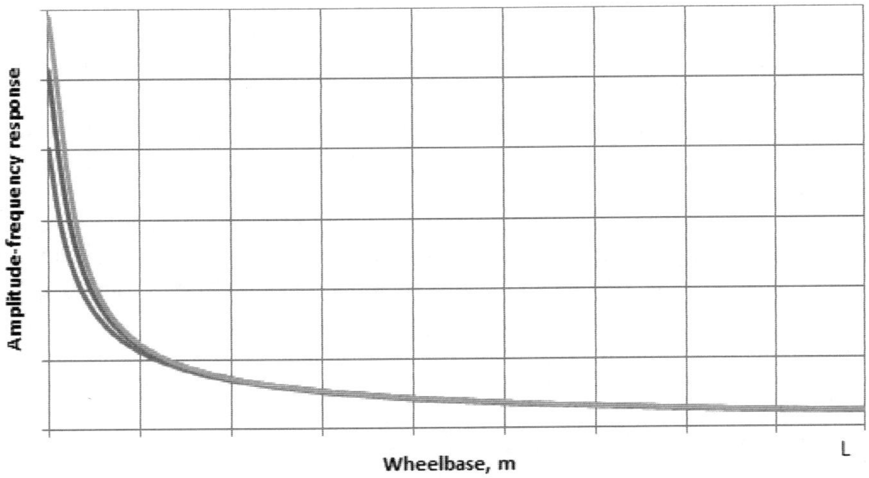

Figure 4. Vehicle gain change in fixed control frequency versus wheel base length

It is evident in the figure that the extension of the wheel base length is followed by a monotonous decrease of the vehicle gain. Thus, in identical driving conditions, a longer wheel base will be followed by a less angular speed of vehicle turning, consequently, by a less course angle affecting the maneuverability of the modified vehicle. The influence of the driving speed is only feebly marked within practically reasonable extension of the wheel base length. At decreased control frequency, this influence grows, and at increased frequency it goes down.

Figure 5 shows the dependency of the angular turning speed lag of the vehicle versus turning angle of the steered wheels in a typical control frequency. In the present case, for a practical modification range, the extension of the wheel base length will also lead to the growth of the vehicle lag. It also affects to a certain extent the maneuverability of the modified vehicle

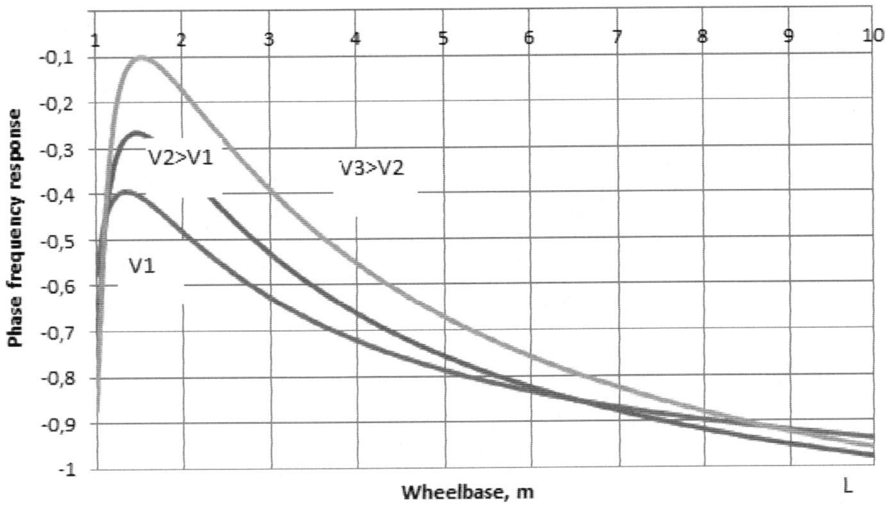

Figure 5. Vehicle lag change in fixed control frequency versus wheel base length

IOP Conf. Series: Materials Science and Engineering 1086 (2021) 012002 doi:10.1088/1757-899X/1086/1/012002

3. Results and conclusions

The conducted preliminary analysis of the influence of the wheel base length extension during the modification of the vehicle design demonstrates the following results. Firstly, the duration of the transient process gets longer, thus affecting the response of the vehicle as a control object, consequently, the control behavior of the motor vehicle worsens. Secondly, the dynamic steering sensitivity of the motor vehicle goes back also affecting the controllability of the vehicle. Thirdly, the vehicle control gain is decreased, and the response lag of the vehicle grows which also scarcely improves the controllability. Thereby, in many cases, the driving speed as a risk factor completes the overall worsening of the controllability of the vehicle.

Thus, for adequate automotive traffic safety, it is important to know the acceptable modifying range of vehicle's controllability due to the intended modification, as well as to additionally limit the admissible driving speed of the vehicle as required, to neutralize the pertaining risk factor.

References

[1] Kirillov K A 2020 *Method of safety assurance of wheel motor vehicles upon design modifications* PhD thesis. Moscow p 177

[2] Milliken U F and Whitecom D U 1963 Application of the general theory of stability and controllability in the design of motor vehicles *Controllability and stability of motor vehicles. Selection of articles* pp 145-204

[3] Bryanskiy Yu A 1983 *Controllability of heavy-duty trucks* M. Mashinostroyeniye p176

[4] Voronov A A 1977 *Basics of the theory of automatic regulation and control* M. Higher School Publishing House p 519

IOP Conf. Series: Materials Science and Engineering 1086 (2021) 012003 doi:10.1088/1757-899X/1086/1/012003

A method for constructing a global motion path and planning a route for a self-driving vehicle

A A Sinodkin [1], T S Evdokimova and M I Tiurikov

Nizhny Novgorod State Technical University n.a. R.E. Alekseev, Nizhny Novgorod, Russian Federation

E-mail: [1] alexey.sinodkin@gmail.com

Abstract. The article presents methods for constructing an unmanned vehicle (UV) motion path in an environment with obstacles to solve the problem of finding a path for a non-holonomic vehicle in an environment with obstacles. A combined approach is described and individual components are considered that allow solving the problem of planning a path for a non-holonomic vehicle. The practical use example of the combined approach for solving the problem of planning the UV path is given.

1. Introduction.

In the past few years, there has been a growing interest in research and experimentation on autonomous vehicles, especially self-driving ones. One of the key components of the unmanned vehicle development is planning and searchinga path to a specific goal. When driving autonomous vehicles, the peculiarity of their movement as non-holonomic systems should be taken into account. A non-holonomic vehicle's state can be described in three dimensions: the vehicle's two-dimensional position and its driving direction. The 2D position defines the vehicle coordinates in the environment and the driving path indicates the direction the vehicle is heading. A simple way to describe this non-holonomic nature is to say that a vehicle cannot spin on the spot: it cannot change its direction without changing its position. Classical methods based on cellular decomposition have a number of drawbacks associated with developing a pathway for vehicles with a vehicle kinematics, and require improvement [1]. In order to function properly, modern trajectory planning techniques must be able to detect safe trajectories that correctly account for the limitations inherent in vehicle dynamic performance.

Another problem with route planning for non-holonomic vehicles is that the UV moves along a continuous path, i.e. exists in a continuous environment. Standard methods of path planning based on cellular decomposition cannot ensure correct path planning in the real world without significantly increasing computational time. Additionally, the vehicle cannot navigate the discrete path that the cellular decomposition path planner would detect. As a result, it is required to consider several methods to form a non-standard (combined) approach, which should take into account the running time algorithm for constructing the UV trajectory movement and develop a safe movement trajectory.

2. Analytical review of modern methods for traffic planning in an environment with obstacles

Path planning is an important task in the field of mobile robot navigation. When solving the path planning problem, the following aspects must be considered: the developed path must pass from the starting point to the given finishing point, the path must ensure the vehicle (robot) movement as far as possible from potential obstacles. Thirdly, the path must be optimal in terms of target parameters among all possible paths.

The problem described above is solved by methods that use the environment map or describe it using graphs: methods based on a grid basis [2-4]; potential field methods [2, 5-6, 10-11]; optimization methods [2, 7-9]. In many of these methods, the result is a chain of control points connecting the starting and finishing points, followed by the problem of smoothing the resulting path.

To implement algorithms that take into account complex and heterogeneous constraints, one algorithm or method is not enough; it is necessary to form a combined approach from various methods of planning the movement path. This approach will allow taking into account various restrictions when planning the motion route.

3. Combined path planning approach of a self-driving car

When planning the path of an autonomous vehicle, the position is usually represented as a three-dimensional vector *(x, y, θ),* where *x* and *y* describe the center location of the vehicle's rear axle on the global map, and θ is the angle characterizing the vehicle's movement direction relative to the x-axis headed along the vehicle axis symmetry, ϕ is the wheels rotation angle. This 3D state is shown in (Figure 1).

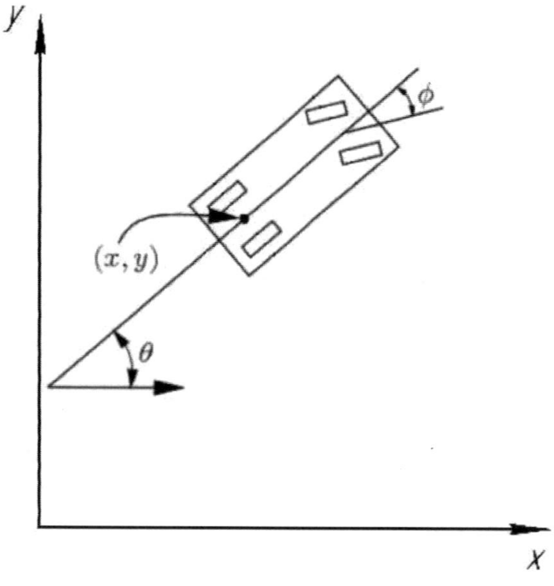

Figure 1. Three-dimensional state of the vehicle.

The path planner will take the starting position as input $S_0 = (x_0, y_0, \theta_0)$ and the final position $S_f = (x_f, yf, \theta f)$ and a sequence of vehicle positions at the output $S_0, S_1, ..., S_f$ is obtained. The movement trajectory from S_0 to S_f can be developed on this basis. This trajectory is then smoothed using gradient descent, since the trajectory obtained from the calculated positions (points) will be a polyline, not realizable for the vehicle.

Based on the above, the combined approach algorithm of constructing UV movement path (Figure. 2) consists of the following stages:

1. filling the occupancy grid;
2. making a distances map to obstacles and drawing a Voronoy diagram;
3. UV movement search for the way to the set goal;
4. smoothing the path built by the Hybrid A * algorithm;
5. UV traveling along the constructed path.

IOP Conf. Series: Materials Science and Engineering 1086 (2021) 012003 doi:10.1088/1757-899X/1086/1/012003

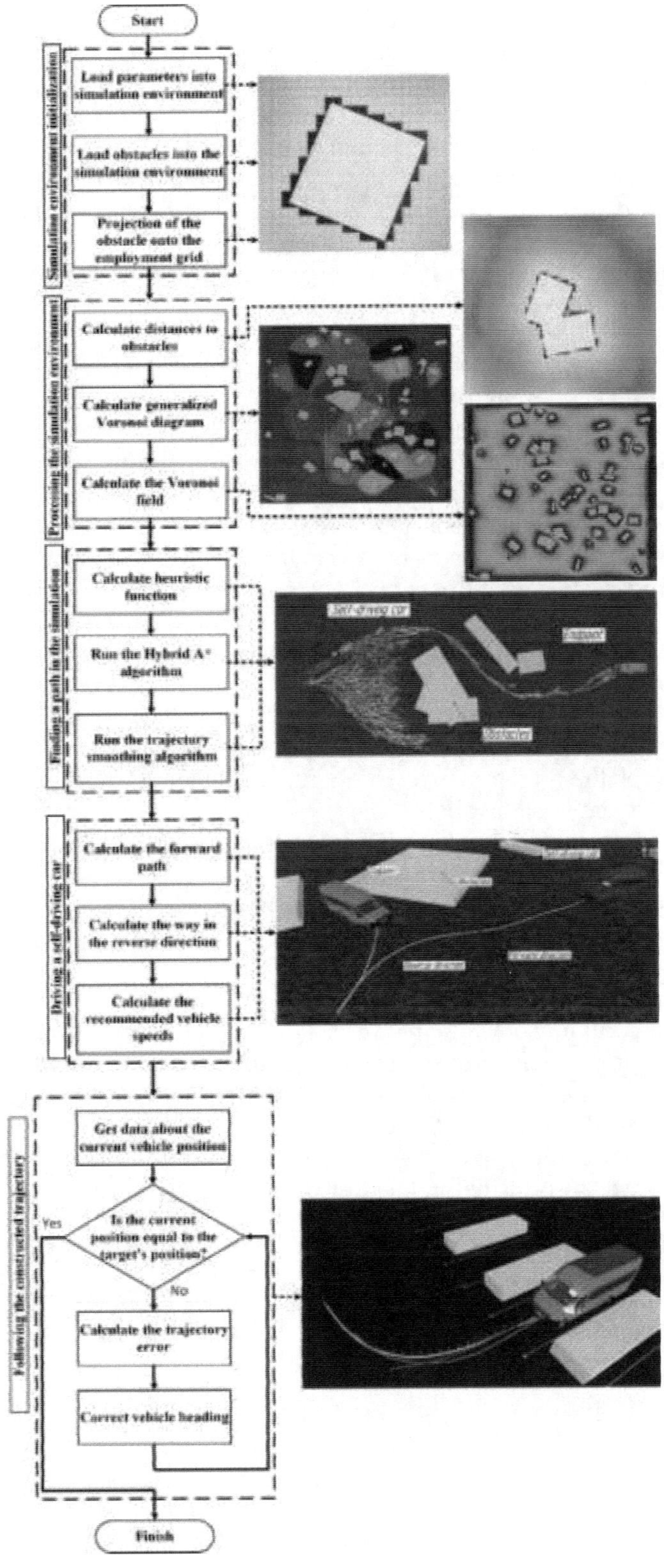

Figure 2. Flowchart of the combined approach operation algorithm

3.1. Filling in the occupancy grid

At the beginning of the path planning procedure, the environment is indexed in the occupancy grid. This occupancy grid determines which cells (nodes) in the environment are occupied by obstacles and which are free space. To simulate the occupancy grid, obstacles are projected onto the grid using the Bresenham algorithm. According to this algorithm, at each step, one of the coordinates - either x or y (depending on the angle of segment inclination) changes by one, the change in the other coordinate (by zero or one) is determined during operation and depends on the error value, the distance at which actually the installed cell is from the line. Thus, the pixel closer to the ideal segment is selected. Since when implementing the algorithm it is more convenient to analyze not the error value, but its sign, the true error value is shifted by -0.5.

3.2. Making a map of distances to obstacles and drawing a Voronoi diagram

A simple occupancy grid is not enough to find a safe path to avoid obstacles. The disadvantage of the grid is that the path planner will prefer paths that are dangerously close to obstacles as has been shown above in order to minimize path length. One of the ways to determine the trade-off between path length and proximity to obstacles is to use the potential field. The potential field can be used to increase the occupancy grid so that each cell has a calculated cost based on the distance to the obstacle. Cells that are closer to obstacles have a higher cost of passing; cells that are occupied by obstacles have an infinite cost. Costs of passing a potential field are used to bypass obstacles remotely when searching for the best route.

While the standard potential fields define the cost of passing a cell as a function of the distance to the nearest obstacle, in this paper an algorithm that uses a combination of the potential fields' method and Voronoi diagram to develope the Voronoi field will be used. It determines the cost of passing as a function of the distance to the nearest obstacle and the distance to the nearest Voronoi edge.

The key advantage of the Voronoi field over ordinary potential fields is that the field value is scaled in proportion to the total available distance between obstacles as a result of this even bottlenecks remain passable, which is not always possible with standard potential field methods.

3.3. UV movement search for the way to the set goal

Pathway planners use cell-based decomposition techniques, such as the A * algorithm commonly used in robotic pathfinding. The search algorithm applies a grid to the local vehicle environment and associates the vehicle position with each grid cell when searching.

However, the A * algorithm is not suitable for solving the problem of path planning for a vehicle, because the kinematics of the vehicle does not allow it to make turns around its axis. To solve the described problem, there is a special algorithm developed on the basis of A *. This algorithm was named Hybrid A *, which allows to take into account the vehicle kinematics.

Hybrid A * is a variant of the standard A * algorithm based on the cell decomposition method, which has been redesigned to solve the problem of finding a path at nodes, where path estimation is carried out without spinning the object around its axis.

The path planner searches in space using the state vector (x, y, θ, r), where $(x \in R, y \in R)$ is the global position of the vehicle rear axle center, $(\theta \in [-\pi, \pi])$ is the vehicle direction relative to the x-axis, and $(r \in \{0,1\})$ - represents the current direction of vehicle movement (moving forward - 0 or moving backward - 1).

All nodes found during a Hybrid A * lookup are added to the cell cost priority queue, where the node cost is calculated using (1):

$$f(n) = g(n) + h(n), \quad (1)$$

where n is the next node on the path, $g(n)$ is the cost of the path from the starting node to the node n, and $h(n)$ is a heuristic function that estimates the cost of the cheapest path from node n to the target. The cost of the path $g(j)$ at node j is calculated using the formula (2):

$$g(j)E = g(i) + d_{ij}\left(1 + v \cdot vor_j + \rho \cdot rev_j\right) + \sigma \cdot switch_{ij}, \quad (2)$$

where i is the parent node in relation to j; j - child node in relation to i ; d_{ij} - Euclidean distance between nodes i and j; rev$j$ - direction of movement of the vehicle at node j (when $rev_j = 1$, reverse movement is performed, $rev_j = 0$ - forward direction of movement); vor_j - cost at node j of the Voronoi field; $switch_{ij}$ is a change in the direction of movement (for $switch_{ij} = 1$, the directions of the vehicle's movements at node i and j are different, for $switch_{ij} = 0$ the directions of the vehicle's movements at node i and j are the same); v, ρ, σ - constant penalty coefficients determining how much each term affects the cost of the path.

The cost at the vor_j node pushes the construction of a path from obstacles to the edges of the Voronoi diagram (i.e., to transport corridors where the vehicle can safely pass). The reversing value rev is the total cost that penalizes the reversing paths, since reversing is a priori slower and more dangerous than forward. The switch shift value penalizes frequent changes in direction of travel in order to account for the time it takes to stop completely and change gear [8].

The second term $h(n)$ of $f(n)$ is an admissible heuristic function that estimates the cost of the path from node n to the final cell of the path. The heuristic function is calculated using the following formula (3):

$$h(n) = \sqrt{(x_i - x_g)^2 + (y_i - y_g)^2} \tag{3}$$

where, x_i, y_i - coordinates of the parent cell; x_g, y_g - coordinates of the target cell.

To smooth the path made by the Hybrid A * algorithm, gradient descent is used to minimize the following objective function (4) [1,12]:

$$\begin{aligned}(X_i) = \omega_\rho \sum_{i=1}^{N} \rho_V (x_i, y_i) + \omega_o \sum_{i=1}^{N} \sigma_o (|X_i - o_i| - d_{max}) + \\ +\omega_k \sum_{i=1}^{N-1} \sigma_k \left(\frac{\Delta\phi_i}{|\Delta X_i|} - k_{max}\right) + \omega_s \sum_{i=1}^{N-1} (\Delta X_{i+1} - \Delta X_i)^2\end{aligned} \tag{4}$$

where $X_i = (xi, y_i)$, $i \in [1, N]$ is the sequence of vertices, o_i is the location of the obstacle closest to the cell;$\Delta X_i = X_i$-X_{i-1} is the vector of displacement into the cell, $\Delta\phi_i = \left|arctan\left(\frac{\Delta y_{i+1}}{\Delta x_{i+1}}\right) - arctan\left(\frac{\Delta y_i}{\Delta x_i}\right)\right|$- change in the tangential angle in the cell, ρ_V - the value of the Voronoi field in the cell; k_{max} - maximum permissible curvature, limited by the turning radius of the vehicle; d_{max} - maximum distance of the collision field with an obstacle; and also ω_ρ, ω_o, ω_k, ω_s are weight constants.

3.4 UV motion algorithm along the constructed path

At this stage, segments of the movement trajectory in the forward and reverse directions are constructed, based on the calculated data obtained by the Hybrid A * algorithm operation and the movement speeds on these path segments are also calculated. Then the algorithm checks whether the vehicle has reached the final point of the path and, at the same time, as the vehicle moves, calculates the deviation of the vehicle from the nominal trajectory developed by the algorithm, and corrects the direction.

To control the vehicle along the movement trajectory, a PID controller is used to determine the desired angle of rotation of the wheels δ, due to the received error e, to correct the trajectory. The wheel turning angle δ is determined by the formula (5):

$$\delta(t) = P + I + D = K_p e(t) + K_i \int_0^t e(\tau) d\tau + K_d \frac{d_e}{dt}, \tag{5}$$

where, e is the closest distance between the nominal track and the vehicle.

To implement the PID controller in vehicle control, the maximum steering wheel angle must be observed. It means that $\delta(t) \in [\delta min, \delta max]$, the minimum and maximum values of wheel rotation are set based on the vehicle characteristics.

To ensure the vehicle smooth running, the movement speed in each cell (node) of the track is calculated. Each cell in the constructed path has a corresponding recommended speed.

The recommended vehicle speed at each path point, during the simulation, was determined using the speed controller based on three parameters: the global speed limit, the instantaneous curvature of the path in a specific cell, and the deceleration limit based on the speed of subsequent states in the constructed path.

The global speed limit exists in all cells of the path, regardless of the peculiarities of each cell and is set during the simulation by the operator.

The recommended curvature speed for each waypoint can be calculated using equation (6):

$$v = \sqrt{a_{max} \frac{\|\Delta x_i\|}{\left|\arctan\left(\frac{\Delta y_{i+1}}{\Delta x_{i+1}}\right) - \arctan\left(\frac{\Delta y_i}{\Delta x_i}\right)\right|}}, \quad (6)$$

where, $\Delta x_i = x_i - x_{i-1}$ is the displacement vector at the i point of the trajectory; a_{max} is the maximum lateral acceleration that the vehicle can achieve when driving on a curve.

Because of this lateral acceleration, the vehicle is forced to decelerate during tight bends on the trajectory. A maximum lateral acceleration of 0.5 m/s ^{2} was set in the simulation.

It is not enough to limit speed on steep curves; the vehicle must slow down before it enters a bend in order to track the correct speed in the right place. To decelerate the vehicle while driving on slow sections, a deceleration limitation is imposed on the path. The following kinematic equation (7) shows how the deceleration limiting speed for node i is calculated:

$$v_{\text{зам}}(i) = \sqrt{v_{i-1}^2 + 2a_{\text{зам}}\Delta x}, \quad (7)$$

where, v_{i-1} - speed in the current cell; $v_{\text{зам}}$- recommended deceleration speed; $a_{\text{зам}}$ - acceleration of deceleration, specified during simulation by the operator; Δx - distance between the current and the next cell.

4. Practical implementation of the combined approach for constructing the self-driving vehicle movement trajectory

Unity, a cross-platform computer game development environment, was chosen to test the operation of the algorithm. The tests were carried out in a full-scale virtual environment, replicating real-life conditions to complete the tasks of finding the safe shortest path to the goal.

As a vehicle for testing the combined approach, a full-scale model of an all-metal GAZ-A65R32 bus with sensors for obstacle detection and orientation in the environment (Figure 3) was used.

Figure 3. Bus model GAZ-A65R32 with sensors

Several virtual experiments have been carried out to check for finding paths of trajectory construction accuracy. For the tests, two scenes were selected: with a random environment and parking spaces shown in (Figure 4) and (Figure 5). These environments have been used to represent collision-free driving through obstacles and accurate parking in parking lots.

To navigate successfully in high density obstacle environments, the vehicle must find a safe, controlled path, and the controller must follow the constructed path as closely as possible. The comfort of vehicle passengers is also important so as the vehicle path must be smooth and free from frequent changes in direction. The speed of calculating the search for a suitable route should not be high in order to provide only a small delay between the moment when the target is selected and the vehicle starts moving towards it.

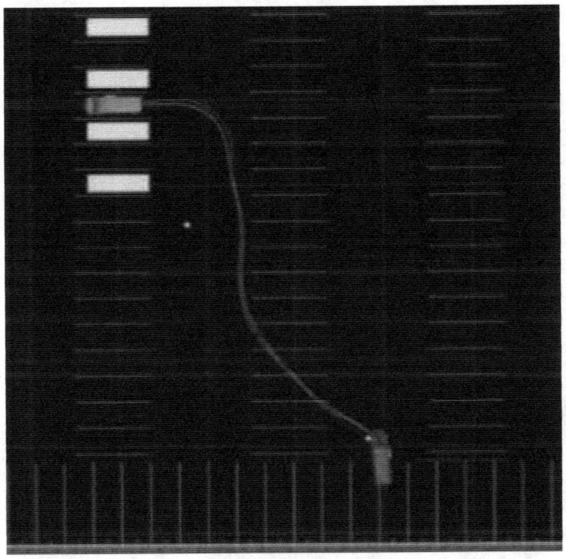

Figure 4. Simulation environment for testing automatic parking

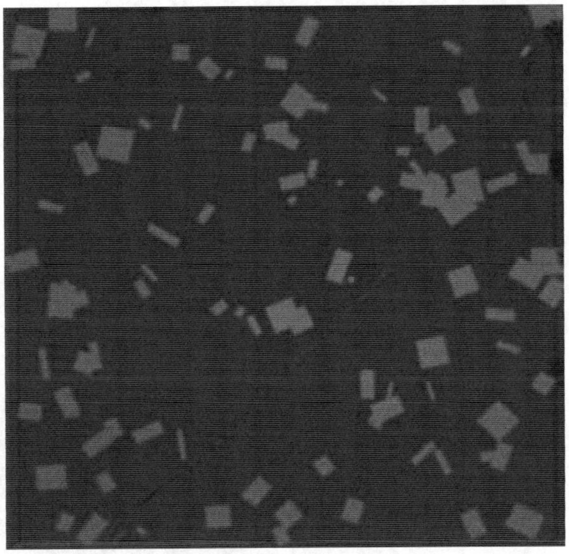

Figure 5. Simulation environment for testing in a random environment

The start and position of each environment was the same; only the positions, sizes and orientations of 100 obstacles were random. Table 1 shows the runtime results for each step in the path planning process. Generalized Voronoi's chart summarizing is the time required to load all obstacles into the obstacle grid, summarize a generalized Voronoi's chart and calculate Voronoi's field. The heuristic clculation is the time required to compute the "holonomic with obstacles" heuristic. The Hybrid A * runtime is the time it takes to find a path in the environment. The number of nodes extended for each A * search is also shown. Path smoothing time measures how long it took to smooth the path found by the Hybrid A *.

Table 1. Runtime of the path planning algorithm for a simulation environment with a random environment

Stages of the algorithm	Simulation environment with random environment (sec)	Simulation environment with parking spaces (sec)
GVC summarizing	0,391	0,375
Heuristic calculation	0,016	0,004
Hybrid A * operation	0,516	0,328
Smoothing path	0,344	0,031
Total time	1,267	0,738

In order to successfully navigate around densely spaced obstacles with minimal discomfort for passengers, the desired path must be properly smoothed. Table 2 shows the results for the paths developed by the Hybrid A * algorithm, and subsequently smoothed, as well as the results if the constructed vehicle path is polyline.

Table 2. Statistics of overcoming the path along the smoothed and polyline path.

Motion parameters	Simulation environment with random environment		Simulation environment with parking spaces	
	Smoothed trajectory	Polyline trajectory	Smoothed trajectory	Polyline trajectory
Travel time	64 sec.	77 sec.	22 sec.	24 sec.
Average travel speed	16 km/h	11 km/h	8 km/h	7 km/h
Number of direction corrections on the trajectory	8 pieces.	27 pieces.	3 pieces.	12 pieces.

The positioning accuracy for the vehicle in free terrain was from 50 mm to 250 mm. In order to perform the task at hand in parking conditions, the vehicle's control controller must be able to track the path accurately, especially in the reverse direction. Since this experiment involves reversing entering a parking space (Figure 6), the effectiveness of tracking the reverse path directly affects the target error. As well as in the past testing, data on the Hybrid A * algorithm operation speed presented in Table 1, and the efficiency of its operation when smoothing the broken path developed by the Hybrid A * algorithm (table 2) were obtained.

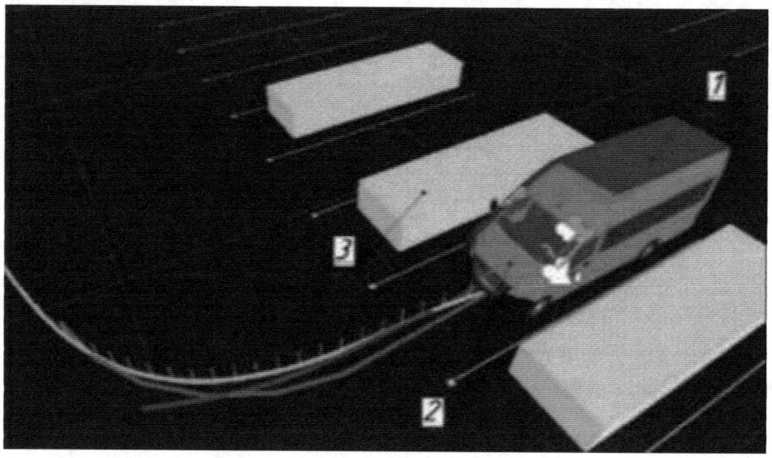

Figure 6. Illustration of the algorithm operation when parking in reverse: 1 - a vehicle parked at a given point, 2 - the end point of the path, marked by a static orange vehicle, 3 - an obstacle

The results shown in Table 1 represent an approximate order of magnitude for the time it takes to plan a path through a moderate-sized environment. This shows that the presented algorithms can be executed fast enough to provide only a small delay between the moment when the target is selected and the vehicle starts moving towards it.

Table 2 shows how track smoothing can affect driving performance as well as travel comfort. First, when driving on a flat path, the vehicle can smoothly enter corners, which allows increasing the speed on them. The smoothed trajectory gave a significant increase in the average vehicle speed compared to the polyline trajectory. Even if the distance from the start to the target was the same in every randomly generated environment, the vehicle can reach the target in less time when the path is smoothed. Second, flattening the desired trajectory provides a much smoother ride for the vehicle's passengers.

The accuracy in the studies can be considered satisfactory for driving safely in real conditions and performing an automatic parking maneuver.

5. Conclusions and Future work

Based on the simulation results, the vehicle positioning accuracy in an environ with a random environment was from 50 to 250 mm and in an environment with parking spaces from 50 to 105 mm. The simulation accuracy can be considered satisfactory for safe driving in real conditions, and automatic parking. The described combined approach is universal in terms of its application to vehicles with different technical parameters (such as overall dimensions; length and width of the wheelbase; engine power, etc.). The technical characteristics of the vehicle are set at the beginning of the simulation by the operator. In the future, it was decided to implement a combined approach on a real unmanned vehicle based on the GAZ A65R32 vehicle as part of the annual field tests of unmanned robotic systems "RoboCross".

6. Recommendations

As a recommendation for further work, it is worth noting the need to finalize the terrain maps in the Unity simulation environment for specific tasks for testing a self-driving vehicle: vehicle movement in an urban environment, on highways, and over rough terrain.

References

[1] Izotova T Yu 2016 Review of algorithms for finding the shortest path in the column *New information technologies in automated systems* **19** pp 341-344

[2] Liu V 2018 Path planning methods in an environment with obstacles (review) *Mathematics and mathematical modeling* **1** pp15–58.

[3] Sleumer N H and Tschichold–Gurman N 1999 Exact cell decomposition of arrangements used for path planning in robotics *Zurich: Inst. of Theoretical Computer Science*

[4] LaValle S 1998 Rapidly-exploring random trees : a new tool for path planning *The annual research report*

[5] Ge S S and Cui Y J 2000 New potential functions for mobile robot path planning *IEEE Transactions on Robotics and Automation* **16-5** pp 615-620 doi: 10.1109/70.880813

[6] Ren J, McIsaac K A and Patel R V 2006 Modified Newton's method applied to potential field-based navigation for mobile robots *IEEE Transactions on Robotics* **222** pp 384-391 doi: 10.1109/TRO.2006.870668

[7] Ross I M and Fahroo F A 2002 Perspective on methods for trajectory optimization. AIAA/AAS *Astrodynamics specialist conf. and exhibit* pp 1–7.

[8] Fliess M, Levine J, Martin P and Rouchon P 1995 Flatness and defect of non–linear systems: introductory theory and examples *Intern. J. of Control* **61-6** pp 1327–1361

[9] Culligan K, Valenti M, Kuwata Y and How J P 2007 Three-Dimensional Flight Experiments Using On-Line Mixed-Integer Linear Programming Trajectory Optimization *American Control Conference* pp 5322-5327 doi: 10.1109/ACC.2007.4283101

[10] Khatib O 1985 Real-time obstacle avoidance for manipulators and mobile robots *Proceedings. 1985 IEEE International Conference on Robotics and Automation* pp 500-505 doi: 10.1109/ROBOT.1985.1087247

[11] Bounini F, Gingras D, Pollart H and Gruyer D 2017 Modified artificial potential field method for online path planning applications *2017 IEEE Intelligent Vehicles Symposium (IV)* pp 180-185 doi: 10.1109/IVS.2017.7995717

[12] Yang K and Sukkarieh S 2008 3D smooth path planning for a UAV in cluttered natural environments *2008 IEEE/RSJ International Conference on Intelligent Robots and Systems*pp. 794-800 doi: 10.1109/IROS.2008.4650637

IOP Conf. Series: Materials Science and Engineering 1086 (2021) 012004 doi:10.1088/1757-899X/1086/1/012004

A method to improve forwarders' energy performance and environmental compatibility

G O Kotiev[1] and E E Klubnichkin[1,2]

[1]Mytishi branch Bauman Moscow State Technical University, Department of Transport and technological means and equipment of the forest complex, 1st Institutskaya street, 1, 141005, Mytischi, Moscow region, Russian Federation

E-mail: [2]klubnichkin@mgul.ac.ru

Abstract. The article presents the results of theoretical justifications of a method improving energy performance and environmental compatibility of the load-haul-dump vehicles (forwarders) used during timber cut-to-length operations.

1. Vehicles for timber cut-to-length operation

The basis of the modern mechanized cut-to-length technology are feller-delimber-bunchers (harvesters) and load-haul-dump vehicles (forwarders). The harvesters and forwarders are manufactured both wheel-based and track-based.

There is a big number of manufactures of this type equipment on the market, the harvesters and forwarders are divided into three classes: light, middle and heavy.

The most popular vehicle system for shortwood-at-the stump operation in case clear-cut harvesting is the one comprising an 18-20-ton heavy harvester, an 8-10 m long manipulator boom equipped with a harvester end able to cut trees up to 70 cm in dia. The employed forwarders which come as a set feature a weight of around 20 tons and working load of 17-18 tons.

The performance of such vehicle systems depending on the operation conditions may be up to 350 m^3 per day (20 – 25 m^3 per hour). Today there is a tendency to increase power and weight characteristics of the vehicles.

During timber logging, a transportation (skidding) operation is the most complex and heavy one in terms of energy performance, production capacity and environmental compatibility. In this regard we have focused on a forwarder (skidder) since the harvester's energy performance (if we consider a power plant and a need in energy) is mainly dependent on operation of the processing equipment. The harvester does not require slow speed of movement over cutting area, high production capacity during transportation operations. It needs high production capacity during processing operations including felling, knotting, crosscutting: a vehicle remains stationary during those processing operations. It is the opposite with the forwarder: its energy performance depends on speed of loading-hauling (skidding) operations.

The leading scholars note that the design concept of vehicles for cut-to-length technology of timber harvesting is different for a harvester and a forwarder [1-7].

They believe that in case of cut-to-length timber harvesting a forwarder is the vehicle which expends on more energy for movement.

We offer a method to improve energy performance and environmental compatibility of the forwarders as follows.

2. A method to determine output of the forwarder power plant

Wood moisture content and timber species affect the amount of timber taken for transportation (skidding) since the weight impacts the utilization of the forwarder's lifting capacity. Round wood is calculated as a solid volume of timber by determining the volume of each assortment or stacked volume. To carry out the calculations related to determination of output of the forwarder power plant during loading operations, it is necessary to determine an average mass of 1 м^3 of timber each working shift, which can be measured as a quotient obtained when a quantity of loaded tons is divided by a number of cubic meters of the loaded timber.

For example: weight of 1 m^3 of timber at moisture of 12% stands for birch at 630-650 kg; for European aspen at 470-490 kg; for oak at 680-720 kg; for fir at 440-460 kg; for pine at 370-600 kg; for larch at 650-800 kg. For easy calculation we use tons instead of m^3. Let us determine the harvester performance (P_{SMH}).

$$P_{SMH} = M_{GR}/T_X \tag{1}$$

where: $M_{GR}$is mass of assortment cut by the harvester; T_X is time of approach, felling, knotting and crosscutting during harvesting of assortment of mass M_{GR} ; Let us determine performance of the forwarder or its load flow (P_{SMF}).

$$P_{SMF} = M_{GRF}/T_F \tag{2}$$

where: $M_{GRF}$is mass of assortment hauled on the log deck at the storage place; T_F is time for loading and hauling of assortment of mass M_{GRF}; Condition $P_H = P_F$.

Given that S_H is a path passed by the harvester during cutting assortment of mass $M_{GR}$for a time of P_{GR}. Let's determine the forwarder performance (P_{GR}) during loading or unloading of assortment of mass M_{GRF} for a time of

$$T_{pogr/razgr}.\, P_{GR} = M_{GRF}/T_{pogr/razgr} \tag{3}$$

Let us determine a coefficient of utilization of the forwarder mass (k_{mf})

$$k_{mf} = M_{grf}/M_f \tag{4}$$

where: $M_f$is the forwarder mass.

Assumptions:

1) The forwarder follows the harvester's path.
2) The forwarder's resistance-to-movement coefficient does not change depending on the number of passages and forwarder mass.
3) Transmission efficiency is equal to 1; no slipping.

The mass M_{GR} of assortment cut by the harvester is equal to the mass M_{GRF} of assortment hauled by the forwarder to the log deck.

Solution:

Time T_F of forwarder one trip includes: time of empty forwarder movement from a loading site to an assortment load collection site, time to pick up a load, time of loaded forwarder movement to a loading site, time to unload assortment from the forwarder to a stack, min.

$$t_n = M_{gr}/P_{gr} \tag{5}$$

$$t_{gri} = S_i/V_{gri} \tag{6}$$

$t_{gri}$is time of loaded forwarder movement over path S_i at speed V_{gri}.

The speed of movement of the forwarder loaded with assortment to the loading site can be determined as follows

$$V_{gri} = N_{dvf}/(M_f + M_{grf}) \cdot f \cdot g \quad (7)$$

$$T_{dvig} = \sum_{i=1}^{n} t_{gri} \quad (8)$$

where: f is a summing coefficient of resistance to movement; n is number of passages from the harvester cutting site to the log deck; N_{dvf} is forwarder engine power;

$$t_{b/gri} = \frac{S_i}{V_{b/gri}} \quad (9)$$

$t_{b/gri}$is time of empty movement over path S_i at speed $V_{b/gri}$.

$$V_{b/gri} = N_{dv}/M_f \cdot f \cdot g \quad (10)$$

$$T_{dvig\ b\ gru} = \sum_{i=1}^{n} t_{b/gri} \quad (11)$$

where: n is number of skiddings (passages) from the harvester cutting site to the log deck.

$$T_f = 2\frac{M_{gr}}{P_{gr}} + \sum_{i=1}^{n} \frac{S_i(M_f+M_{gri})\cdot f\cdot g}{N_{dv}} + \sum_{i=1}^{n} \frac{S_i M_f \cdot f \cdot g}{N_{dv}} \quad (12)$$

$$T_f = \frac{2M_{gr}}{P_{gr}} + \frac{2M_f \cdot f \cdot g}{N_{dv}} \sum_{i=1}^{n} + \frac{f\cdot g}{N_{dv}} \sum_{i=1}^{n} + \frac{f\cdot g}{N_{dv}} \sum_{i=1}^{n} S_i M_{grf} \quad (13)$$

The specificity of processing operation of the mechanized cut-to-length technology is that the forwarder always follows the harvester. The forwarder reaches the harvester, loads the assortment and moves back to the log deck or hauling road and this repeats several time per shift. If we know the harvester's performance, performance of the loading-unloading operations of the forwarder, then we can determine the general level of the designed structure of the forwarder (k_{mf}). Then we determine or select the harvester's performance. It can be different but for the sake of calculation we shall specify high performance figure. High performance figures shall also be specified for loading-unloading equipment installed on the forwarder. S_X is a path passed by the harvester but not the vehicle's movement: it is related to the frequency of trees location on the cutting area.

$$N_{dv} = \frac{f\cdot g\cdot S_H(\frac{1}{k_{mf}}+\frac{1}{2})}{(\frac{1}{P_H}-\frac{2}{P_{gr}})} \quad (14)$$

This equation is convenient for use, if we specify the performance, then we can obtain the engine power, if we divide the engine power by f and g and specify movement speed, average speed (for example, 10 km/h), then we obtain average mass, since a speed in one direction and the other differs, the power curve is hyperbolic. We obtain a mass, then using a coefficient of forwarder mass utilization we find the amount of assortment to be carried by the forwarder and then go to transportation operation.

This formula can be developed as follows:

- first development, the forwarder always follows the harvester and repeats the same path, another model of the forwarder movement can be selected depending on how it moves between the cutting areas and collects assortment loads;

- second development, it is necessary to create different resistance to movement of the loaded and empty forwarder. It is clear that the coefficient of resistance to movement does not depend on mass, but soil pressure varies depending on whether the forwarder is empty of loaded and thus depends on the damaged soil;

- third development, a coefficient of resistance to movement depends on the number of passages. But the model operating principle is more understandable in this form with due consideration of the taken assumptions. It is necessary to check the obtained conclusions through modelling in the applied program packages. Further development of the proposed formula is as follows: for example, resistance to movement decreases, the consumed engine power drops, we alternatively increase and reduce it and then determine the number of forwarders required to have time to complete handling operations after the harvester depending on the cutting area type, wood density and volume, as well as on the path passed the harvester and its performance. The proposed model skips the economic issues as other researchers in this field do [8-12].

The principle of this model consists in that the performance of the felling-knotting-crosscutting operations is equal to performance of the handling operations, so neither harvester nor forwarder is idle. We obtained a dependency which helped us to find out that very fast riding over cutting area is possible but it leads to transportation of small load (assortment volume) since the power output is pre-determined: if small load needs to be carried then it requires fast riding, if slow riding is required then heavy loads shall be carried.

The balance rational design of the forwarder's construction for the specific conditions of timber harvesting will be specified with due consideration for the harvester's performance, forwarder's resistance to movement f and number of its passages. This model allows for the following main conclusion – the harvester's performance shall be equal to the forwarder's load flow (everything cut by the harvester shall be picked up by the forwarder). If the harvester's performance rises, the path passed increases, then the forwarder will not be able to keep pace with the harvester, so the forwarder requires a running gear with high off-road performance, travel speed and lifting capacity, and more productive processing equipment needs to be employed. Should the proposed measures fail to improve the forwarder's situation, then it will be necessary to use two forwarders for timber cutting. However this is a different model. As an example, f varies depending on whether the forwarder which moves over a forest swath is loaded or empty, as well as on a number of its passages, etc.

The scientific novelty is as follows. The forwarder engine power has been obtained taking into account the principles of equality of the forwarder load flow, harvester's performance, that helps to specify the design of the developed vehicle basing on the resistance to movement and the path to be passed by the harvester depending on the stand density.

T_H is time of the harvester felling-knotting-crosscutting operations and movement to the next tree, use is made of the function which links the stand density and performance of the processing equipment of the manipulator and harvester end.

In order to use this formula

$$N_{dv} = \frac{f \cdot g \cdot S_H(\frac{1}{k_{mf}} + \frac{1}{2})}{(\frac{1}{P_H} - \frac{2}{P_{gr}})} \tag{15}$$

Coefficients P_H and P_{gr} if unknown shall be experimentally found through work measurements of the harvester and forwarder operation.

P_H and P_{gr} are taken as kg per sec for easy calculation.

The proposed calculation procedure allows for defining a propulsion unit type and design of the forwarder running gear at the forwarder construction designing stage.

The forwarder's path changes as the harvester constantly moves further.

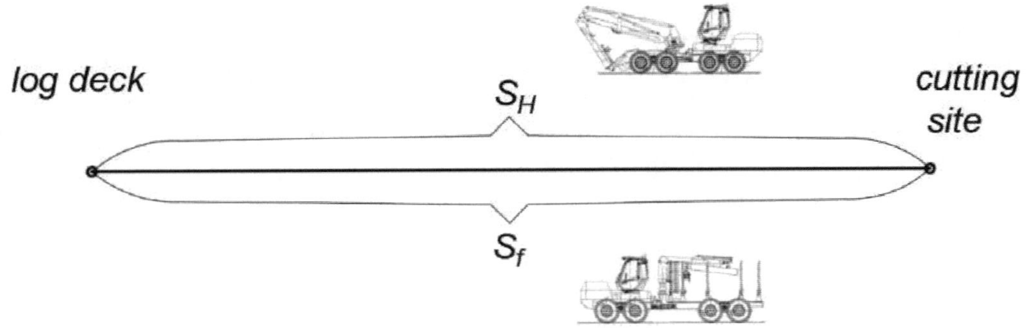

Figure 1. Assortment taken away in one passage

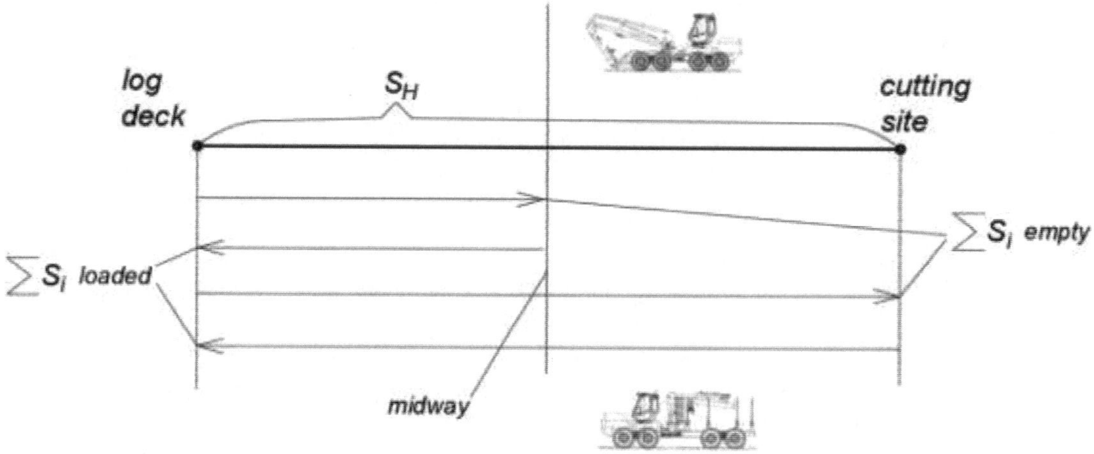

Figure 2. Assortment taken away in two passages

Figure 1 shows the first case when a number of passages is equal to one, the harvester has cut wood, the forwarder has quickly arrived and picked up all ready assortment and taken it on the log deck.

Definition

$$\sum_{i=1}^{n} S_i \tag{16}$$

a) Option n = 1

$$\sum_{i=1}^{n} S_i = S_H \tag{17}$$

where: S_H – harvester path; S_f – forwarder path.

Figure 2 shows the second case when the harvester continues moving over the forest swath, the forwarder covers a path n=2 to the midway, then it leaves for unloading and arrives to the loading site.

b) Option n = 2

$$\sum_{i=1}^{2} S_i = \frac{S_H}{2} + S_H \frac{2}{2} \tag{18}$$

The same is for path n=3 and so on for each and every trip till the end of the shift.

c) Option n = 3

$$\sum_{i=1}^{3} S_i = \frac{S_H}{3} + \frac{2}{3} S_H + S_H \frac{3}{3} \tag{19}$$

$$\sum_{i=1}^{n} S_i = \frac{S_H \sum_{i=1}^{n} i}{n} = \frac{S_n}{n} \frac{n(n+1)}{2} \tag{20}$$

$$\sum_{i=1}^{n} S_i = \frac{S_X(n+1)}{2} \tag{21}$$

Next presented is a very interesting case when n=2, i.e. the forwarder has once covered half-distance and once till the end. S_i is the forwarder's path in one direction. The more the harvester operates S_H, the more precise is calculation, since S_H is a path passed by the harvester while cutting assortment of mass M_{gr} for the time t_H per shift. The longer is S_H, the easier to neutralize the power oscillations. S_H depends on stand density at the developed cutting area.

Prior to calculations it is necessary to obtain data on the stand characteristics, species composition, diameter, etc.

Definition:

$$M_{gr} = \sum_{i=1}^{n} M_{grf} = P_f M_{grf} \tag{22}$$

$$T_f = \frac{2M_{gr}}{P_{gr}} + \frac{f \cdot g}{N_{dv}} \left(2M_f + M_{grf}\right) \frac{S_H(n+1)}{2} = \frac{M_{gr}}{P_H} \tag{23}$$

$$\frac{1}{P_H} = \frac{2}{P_{gr}} + \frac{f \cdot g}{N_{dv}} \frac{S_H(n+1)}{2} \left(2\frac{M_f}{M_{grf}} + \frac{M_{grf}}{M_{gr}}\right) \tag{24}$$

$$\frac{1}{n} \to (\frac{M_{grf}}{M_{gr}}) \tag{25}$$

$$\frac{1}{P_H} = \frac{2}{P_{gr}} + \frac{f \cdot g \cdot S_H (n+1)}{N_{dv} \; 2} (\frac{2}{n \cdot k_{mf}} + \frac{1}{n}) \tag{26}$$

assuming that $n > 20$

$$n + 1 \approx n \tag{27}$$

$$\frac{n+1}{n} \approx 1 \tag{28}$$

$$\frac{1}{P_H} = \frac{2}{P_{gr}} + \frac{f \cdot g \cdot S_H}{N_{dv}} (\frac{1}{k_{mf}} + \frac{1}{2}) \tag{29}$$

$$N_{dv} = \frac{f \cdot g \cdot S_H (\frac{1}{k_{mf}} + \frac{1}{2})}{(\frac{1}{P_H} - \frac{2}{P_{gr}})} \tag{30}$$

This method is applicable to the harvesting vehicles with wheeled, tracked and combined running gear. The engine power calculation formula (30) includes f as a total coefficient of resistance to movement which impacts selection of the running gear type. For example, wheel running gear is employed in case of minor resistance, if resistance rises then wheeled-tracked gear is used, high resistance requires tracked running gear. The total coefficient of resistance to movement f depends on soil category and its bearing capacity. The method will further focus on slipping of the harvesting vehicle running gear.

3. Coefficient of resistance to movement

Our model of forwarder selection shall be added with required power, specific characteristics of the cutting area, its length and width, stand volume and density, estimated harvester capacity per shift, coefficient of utilization of the forwarder mass.

Main stipulations and assumptions

1. Forwarder capacity (per shift) is a lifting capacity is equal to harvester capacity (per shift);
2. Resistance-to-movement along the forwarder travel changes after it takes load (assortment) only;
3. Coefficient f is an initial value of resistance-to-movement.

Coefficient Δf stands for variation of resistance-to-movement per each trip.

$$T_f = 2\frac{M_{gr}}{P_{gr}} + \sum_{i=1}^{n} \frac{S_i (M_f + M_{grf}) t_i g}{N_{dv}} + \sum_{i=1}^{n} \frac{S_i M_f t_i g}{N_{dv}} \tag{31}$$

where: $2\frac{M_{gr}}{P_{gr}}$ is time of forwarder loading and unloading; $\sum_{i=1}^{n} \frac{S_i (M_f + M_{grf}) t_i g}{N_{dv}}$ is time of loaded forwarder movement; $\sum_{i=1}^{n} \frac{S_i M_f t_i g}{N_{dv}}$ is time of empty forwarder movement; n is number of passages (during movement of both empty and loaded vehicle).

$$T_f = 2\frac{M_{gr}}{P_{rg}} + \frac{g}{N_{dv}} (\sum_{i=1}^{n} S_i t_i)(2M_f + M_{grf}) \tag{32}$$

Let us find

$$\sum_{i=1}^{n} S_i t_i$$

IOP Conf. Series: Materials Science and Engineering 1086 (2021) 012004 doi:10.1088/1757-899X/1086/1/012004

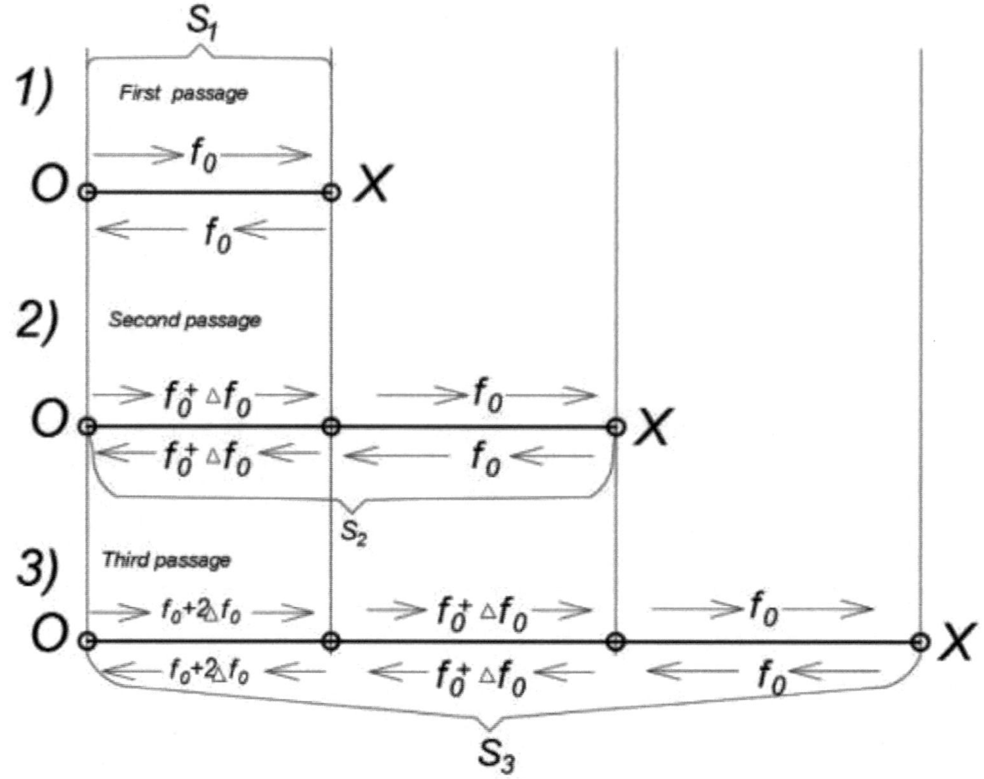

Figure 3.

Third passage

$$(S_3 - S_2)f_0 + (S_2 - S_1)(f_0 + \Delta f) + S_1(f_0 + 2\Delta f) = S_3 \cdot f_0 + S_2\Delta f + S_1\Delta f = S_3 f_0 + \sum_{i=1}^{2} S_i \Delta f \quad (33)$$

n–passage

$$S_n f_0 + \sum_{j=1}^{n-1} S_i \Delta f \quad (34)$$

$$\sum_{i=1}^{n} S_i f_i = \sum_{i}^{n}\left(S_i f_0 + \sum_{j=1}^{i-1} S_j \Delta f\right) = f_0 \sum_{i=1}^{n} S_i + \Delta f \cdot \sum_{i=1}^{n} \sum_{j=1}^{i-1} S_j \quad (35)$$

The forwarder has made n-number of passages, for example 20, resistance changes, that is quite a total resistance which is finally found through approximation.

$$S_i = S_j = \frac{S_H}{P} \quad (36)$$

$$\sum_{i=1}^{n} S_i t_i = \frac{f_0 S_H}{n} \sum_{i=1}^{n} i + \frac{\Delta f S_H}{n} \sum_{i=1}^{n} \sum_{j=1}^{i-1} j \quad (37)$$

$$\sum_{i=1}^{n} i\} \frac{n(n+1)}{2} \quad (38)$$

$$\sum_{i=1}^{n} \sum_{j=1}^{i-1} j\} \sum_{i=1}^{n} \frac{(i-1)\cdot i}{2}\} \frac{1}{2}\left(\sum_{i=1}^{n} i^2 - \sum_{i=1}^{n} i\right) \quad (39)$$

$$\sum_{i=1}^{n} i^2\} \frac{n(n+1)(2n+1)}{6} \rightarrow$$

$$\sum_{i=1}^{n} i\} \frac{n(n+1)}{2} \rightarrow \frac{n(n+1)}{2\cdot 2}\left(\frac{2n+1}{3} - 3\right) \rightarrow \frac{n(n+1)\cdot(n-1)\cdot 2}{2\cdot 2\cdot 3} \rightarrow \frac{n(n+1)\cdot(n-1)}{2\cdot 3} \quad (40)$$

$$\sum_{i=1}^{n} S_i t_i = \frac{f_0 S_H}{2}(n+1) + \frac{\Delta f S_H (n+1)(n+1)}{6} \quad (41)$$

$$T_f = 2\frac{M_{gr}}{P_{gr}} + \frac{g(2M_{gr} + M_{grf}) S_H (f_0 + \Delta f(n-1))(n+1)}{N_{dv}\, 2} \quad (42)$$

$$\frac{1}{P_H} = \frac{2}{P_{gr}} + \frac{g S_H \left(f_0 + \frac{\Delta f(n-1)}{3}\right)}{N_{dv}} \frac{\left(\frac{1}{k_{mf}} + \frac{1}{2}\right)(n+1)}{n} \quad (43)$$

With due regards for:

$$M_{gr} = n \cdot M_{grf} \tag{44}$$

Then multiply Δf by n minus one plus f_0 – this is quite a final total resistance. In other words, there is certain resistance prior to movement in the field, after a vehicle has made n-number of passages, for example 20, resistance changes. That is quite a total resistance which is finally found through approximation.

$$N_{dv} = \frac{gS_H}{\frac{1}{P_H} - \frac{2}{P_{gr}}}\left(\frac{1}{k_{mf}} + \frac{1}{2}\right)\frac{n+1}{n}\left(f_0 + \frac{\Delta f(n-1)}{3}\right) \tag{45}$$

Let's set the expression to A

$$\frac{gS_H}{\frac{1}{P_H} - \frac{2}{P_{gr}}}\left(\frac{1}{k_{mf}} + \frac{1}{2}\right)\frac{n+1}{n} \rightarrow A \tag{46}$$

$$N_{dv} = \frac{gS_H}{\frac{1}{P_H} - \frac{2}{P_{gr}}}\left(\frac{1}{k_{mf}} + \frac{1}{2}\right)\frac{n+1}{n}\left(f_0 + \frac{\Delta f(n-1)}{3}\right) = A(\frac{n+1}{n} f_0 + \frac{\Delta f}{3n} n^2 - 1) \tag{47}$$

$$\frac{\partial N_{dv}}{\partial n} = A\left(f_0 \frac{n-n-1}{n^2} + \frac{\Delta f}{3}\frac{2n \cdot n - n^2 + 1}{n^2}\right) = \frac{A}{n^2}(-f_0 + \frac{\Delta f}{3}(n^2+1)) \tag{48}$$

$$\frac{\partial N}{\partial n} = 0 \rightarrow n^* = \sqrt{\frac{3f_0}{\Delta f} - 1} - minN_{dv} \tag{49}$$

where: n^*- number of passages per shift corresponds to the minimum estimated engine power $minN_{dv}$

$$\frac{\partial^2 N_{dv}}{\partial n^2} = A\left(\frac{2f_0}{n^3} - \frac{\Delta f}{3} \cdot \frac{2}{n^3}\right) = \frac{2A}{n^3}(f_0 - \frac{\Delta f}{3}) \tag{50}$$

For example, if we take the forwarder movement speed equal to 10 km/h, skidding distance of 300 m.

Calculation example at f_0=0.1; Δf=0,008 number of passages per shift is equal to six.

$$n^* = \sqrt{\frac{3 \cdot 0{,}1}{0{,}008} - 1} = 6{,}123724 \approx 6$$

4. Running gear slipping correction

The following relationship has been obtained to find required power for the forwarder running gear drive with due consideration for a change in resistance to movement

$$\frac{1}{P_{SMH}} = \frac{2}{P_{gr}} + \frac{g \cdot S_H (f_0 + \frac{\Delta f(n-1)}{3})}{N_{dvig}} \cdot \frac{(\frac{1}{k_{mf}} + \frac{1}{2})(n+1)}{n} \tag{51}$$

where n is number of the loaded forwarder passages; Δf is increase in resistance to movement per one pass of the loaded forwarder; N_{dvig} is dependence on change in the forwarder resistance to movement.

The relationship is obtained with due consideration for the running gear slipping

1) Suppose $n \gg 1 \rightarrow \frac{n+1}{n} \approx 1$;

where: n – number of passages

2) $f_0 + \Delta f(n-1) = f_{max}$

where: f_0 is initial resistance to movement; f_{max} is forwarder resistance to movement after n-passages

For example:

Dry earth road → bumpy earth road; Suppose G_0 is running gear slipping at f_0;

$$G = \frac{\omega z_0 - V}{\omega z_0},$$

where: Z_0 is radius of the driving wheel; G_{max} is running ear slipping at f_{max}.

$$N_{dv} = \frac{N_{dvig}}{1-G},$$

$$N_{dv} = fmgV + R_X(\omega z_0 - V) = fmg\omega z_0 = \frac{fmgV}{1-G},$$

$$\frac{1}{3}\left(3f_0 + \Delta f(n-1)\right),$$

$$\frac{1}{P_{SMH}} = \frac{2}{P_{gr}} + \frac{gS_H(2f_0+f_{max})(\frac{1}{k_{mf}}+\frac{1}{2})}{3N_{dvig}},$$

$$\frac{1}{P_{SMH}} = \frac{2}{P_{gr}} + (\frac{1}{k_{mf}} + \frac{1}{2})\frac{gS_X(2\frac{f_0}{1-G_0}+\frac{f_{max}}{1-G_{max}})}{3N_{dvig}}$$

Specific energy consumption per unit of the passed path during movement for loading. $f_w = \frac{N_{dvig}}{mgV}$, coefficient of free thrust $K_T = \frac{P_T}{mg}$

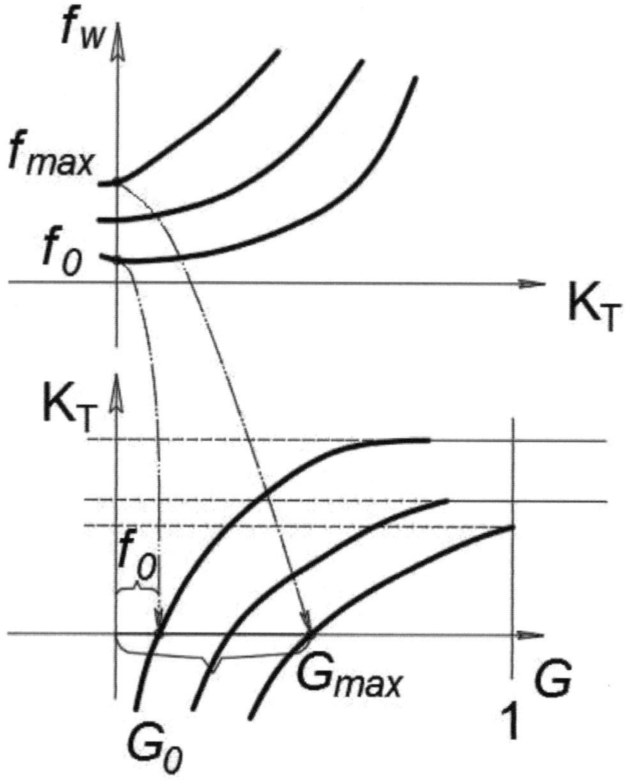

Figure 4.

Resistance to movement varies depending on the number of passages, as the skidding track (road/path) gets broken, the movement becomes more difficult, then we abandon the wheeled running gear in favor of the tracked one, and if the skidding track is not broken we continue using the wheeled gear.

The proposed method is for every occasion. For example, if we have ten harvesters operating, how many forwarders do we need and what shall be their power? Will it be dry or wet? Will the snow be deep? What shall we drive, how much shall we carry? What type of vehicle is required? Etc. When we consider the running gear, we need to know its specific capacity. It can be assumed that transmission is continuously variable. We do not consider speeds but shifts and passed path.

5. Conclusion

Based on the presented calculations, the engine power does not depend on the amount of assortment the forwarder carries (skids) but on the technical level of the harvester's performance, stand (forest) density and loading-unloading operations. This conclusion needs to be proved, to do so it is necessary to simulate the proposed equation.

The proposed calculation method allows for determining a type of running gear and design of the undercarriage of the forwarder at the forwarder designing stage.

The harvester's performance shall be equal to the forwarder's load flow (everything cut by the harvester shall be picked up by the forwarder).

It is necessary to draw a table of a coefficient of utilization of the forwarder mass which is based on the amount of assortment carried and the forwarder's own mass at the moment.

Acknowledgments
The work has been accomplished at the Bauman Moscow State Technical University with financial support from the Ministry of Science and Higher Education of the Russian Federation under the agreement No. 075-11-2019-030 as of 22 November 2019.

References

[1] Klubnichkin V et al. 2019 Researching rolling resistance of the wheeled forestry tractor at skidding *IOP Conference Series: Materials Science and Engineering* **695(1)** 012003 doi: 10.1088/1757-899X/695/1/012003
[2] Makarov V S et al. 2019 Study of uneven surfaces distribution on forestry roads *Journal of Physics: Conference Series* **1177(1)** 012041 doi: 10.1088/1742-6596/1177/1/012041
[3] Klubnichkin V E et al. 2018 Interaction between elements of the track ground contacting area with the soil at curvilinear motion of the timber harvesting machine *IOP Conference Series: Materials Science and Engineering* **386** 012016 doi: 10.1088 / 1757-899X / 386/1/012016
[4] Klubnichkin E and Klubnichkin V 2017 Theoretical foundations of preliminary assessment of the impact of the timber harvesting vehicles on forest soils in the course of logging. *Proceedings of the 19th International & 14th European-African Regional Conference of the ISTVS*
[5] Cambi M, Certini G, Neri F and Marchi E 2015 The impact of heavy traffic on forest soils. *Forest Ecology and Management* **338** pp 124-138
[6] Ringdahl O, Hellström T and Lindroos O 2012 Potentials of Possible Machine Systems for Directly Loading Logs in Cut-to-Length Harvesting *Canadian Journal of Forest Research* **42 - 5** pp 970-985
[7] Anisimov G M and Kochnev A M 2008 Testing of cutting machines *Textbook "SPbGLTA". St.Petersburg* p 488
[8] Goberman L A, Goberman V A and Ksenevich I P 2003 Ground trailer transport systems Technical and economic bases of designing machines and processes Methodological aspects of project activities and decision making *Moscow Encyclopedia* **3**
[9] Balyakov V V et al. S 2013 Mobility of Ground Transport-Technological Vehicles *Transactions of NNSTU n.a. R.E. Alekseev* **4** pp 72-77
[10] Proto A R, Macrì G, Visser R, Harrill H, Russo D, Zimbalatti G 2018 Factors Affecting Forwarder Productivity *European Journal of Forest Research* **137 - 2** pp 143-151
[11] Allman M, Jankovský M, Messingerová V, Allmanová Z and Ferenčík M 2015 Soil compaction of various central european forest soils caused by traffic of forestry machines with various *Forest Systems* **24(3)**
[12] Klubnichkin E E, Klubnichkin V E, Kotiev G O 2018 Theoretical research of soil packing by timber harvester running gear *IOP Conference Series: Materials Science and Engineering* **386** 012025 doi: 10.1088 / 1757-899X / 386/1/012025

IOP Conf. Series: Materials Science and Engineering 1086 (2021) 012005 doi:10.1088/1757-899X/1086/1/012005

A methodology for evaluating the efficiency of driving a truck

A M Umirzokov[1], K T Mambetalin[2], S S Saidullozoda[1,2,3] and A L Berdiev[1]

[1]Tajik Technical University named after M.S. Osimi, 11, academicians of the Radjabovs, Dushanbe 734042, The Republic of Tajikistan

[2]Polytechnic Institute, South Ural State University, 76, Lenin Avenue, Chelyabinsk 454080, Russian Federation

E-mail: [3]saivali.saidullo@mail.ru

Abstract. The article considers the issues of evaluating the efficiency of converting thermal energy into mechanical work when operating pit vehicles - BELAZ-7540B dump trucks in mountain conditions and the role of the driving skills in this process. The article also defines the efficiency of driving a vehicle, based on which it presents a methodology and mathematical models for evaluating the driver's efficiency. At the same time, the basis is formed by several operational indicators, out of which the most important ones for mountain operating conditions of heavy-duty trucks - dump trucks are energy, economic and reliability indicators. The driver's efficiency is evaluated by the energy and economic indicators based on the trip fuel consumption and the performance of the vehicle - dump truck per shift. The efficiency of driving a heavy-duty truck - a dump truck by the material costs indicator is evaluated by the consumption of spare parts and repair and maintenance materials. The driver's efficiency is evaluated by the reliability indicators based on the specific time, labor, and money costs spent on eliminating operational and resource failures.

1. Introduction

A vehicle is a kind of an energy machine. In a vehicle, thermal (as usual) energy is converted into mechanical work used to transport passengers and cargo. Unlike a passenger car, a truck typically converts more energy into mechanical work, which is used to perform a large number of useful operations. In both cases, the efficiency of the conversion of thermal energy into mechanical work used to perform some useful operations depends on a large number of parameters (criteria). Along with the design parameters and characteristics of the vehicle, an important role is given to the driver's efficiency. The driver acts as a key factor in the formation and regulation of the efficiency of converting thermal energy into mechanical work, which is used to perform useful operations (transport operations). Consequently, the driver of a heavy-duty truck is actively involved in the conversion of much (significantly) more thermal energy into useful operations, thus forming the operating efficiency of the vehicle and the Driver-Vehicle-Road-Environment system in general.

It is one thing when the driver participates in the conversion of thermal energy into useful work in normal vehicle operating conditions, but it is absolutely different to drive a heavy-duty vehicle, or rather to manage the conversion of thermal energy to perform a large volume of useful operations according to a two-shift working schedule, in severe mountain and high-altitude conditions with a complex road geometric pattern in plan and profile with a gravel coating, with inevitably frequent and sharp maneuvers.

2. Nature of the problem

The truck driver's efficiency can be defined as a ratio of the achieved result (in its energy, reliability, economic, environmental, and material terms) in the process of converting thermal energy into mechanical work, which is used to transport passengers and cargo involving the driver to the resources used.

The essence of the concept of the driver's efficiency is much broader than the concept of the driver's reliability. If the driver's reliability is the ability to faultlessly drive a vehicle in various road and weather conditions during working hours, the efficiency is the driver's reliability achieved with the least energy, economic and material costs, as well as with the least environmental damage, which can be generally expressed by the ratio:

$$E_D = \frac{R_L}{R_A}, \tag{1}$$

where R_L is the result achieved by the driver during lean driving; R_A is the result achieved by the driver during aggressive driving.

For a comprehensive evaluation of the driver's efficiency E_D when building hydraulic structures in mountain conditions, we use the following expression

$$E_D = E_{en} \cdot E_{ec} \cdot E_r \cdot E_m, \tag{2}$$

where E_{en}, E_{ec}, E_r, E_m are, respectively, the components of the vehicle driving efficiency in terms of the energy, economic, reliability indicators, as well as in terms of material costs.

The achieved result of the truck driver's efficiency consists of a large number of factors: energy, economic, reliability, material, environmental, and efficiency by ensuring traffic safety.

Not all of the above factors are significant for mountain and high-altitude operating conditions for trucks [1]. The achieved environmental result of the driver's efficiency in the conditions of building hydraulic structures is negligible [2]. The fact is that these conditions are characterized by good ventilation. Pit roads are characterized by low traffic density and intensity, and the truck fleet is equipped with diesel engines [3-4].

3. Experimental study

Over the entire period of building Rogunskaya HPP, no traffic accidents with serious consequences were recorded.

Given the insignificance of the influence of the achieved results of the environmental efficiency, as well as the efficiency due to ensuring traffic safety in the above conditions, they may not be taken into account in practical calculations of evaluating the driving efficiency [5-6].

The energy component of the driver's efficiency E_{en} is mainly formed due to the specific fuel consumption per unit of the performed work or per unit of the operating time (Figure 1).

The mountain and high-altitude conditions of operating heavy-duty dump trucks are characterized by a relatively low energy efficiency of the driver, and it depends on a large number of factors, out of which the driver's skills, knowledge, and experience of driving in difficult mountain conditions are important [7-8].

The efficiency of driving a heavy-duty truck - a dump truck by the economic indicator E_{ec} can be indirectly evaluated by the performance per shift, which varies widely depending on the driving style (Figure 1). In the conditions of building Rogunskaya HPP, the best indicators of the performance per shift are observed, starting from the driver's three-year to sixteen-year length of service, reaching the maximum value by the eighth year of the driver's length of service.

The efficiency of driving a heavy-duty truck - a dump truck by the material costs indicator E_m can be evaluated by the consumption of spare parts and repair and maintenance materials (Figure 2).

The economic criteria for the vehicle's failure-free operation and durability are taken as a parameter to evaluate the efficiency of driving a heavy-duty truck - a dump truck by the reliability indicator E_r, in particular the specific time, labor and money costs spent on eliminating operational and resource failures (Figure 3).

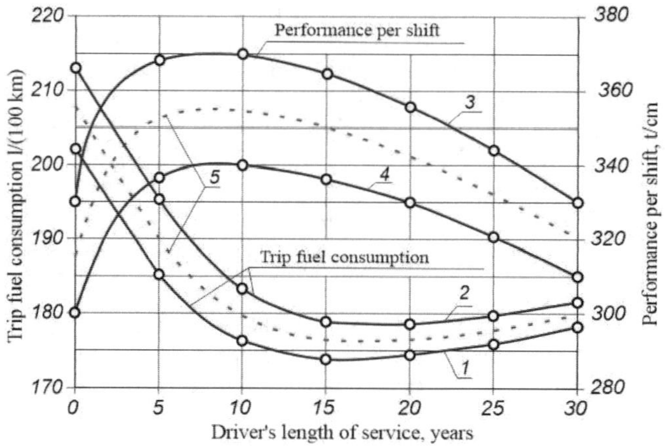

Figure 1. The dependence of the trip fuel consumption and performance per shift on the driver's length of service: 1 and 3 - lean driving; 2 and 4 - aggressive driving; 5 – the average values of the indicators.

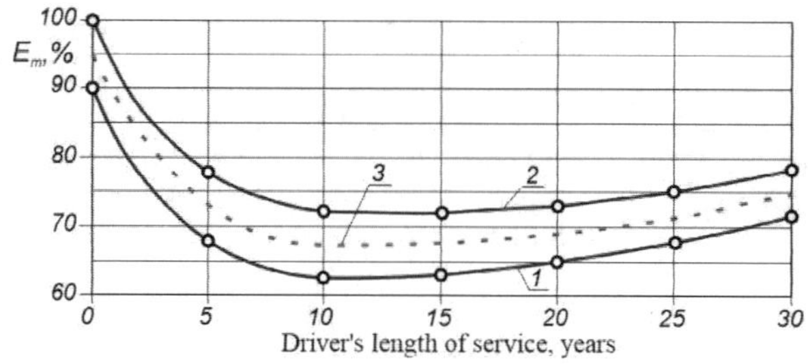

Figure 2. The dependence of the consumption of spare parts and materials on the driver's length of service: 1 - lean driving; 2 - aggressive driving; 3 - the average value of the indicator.

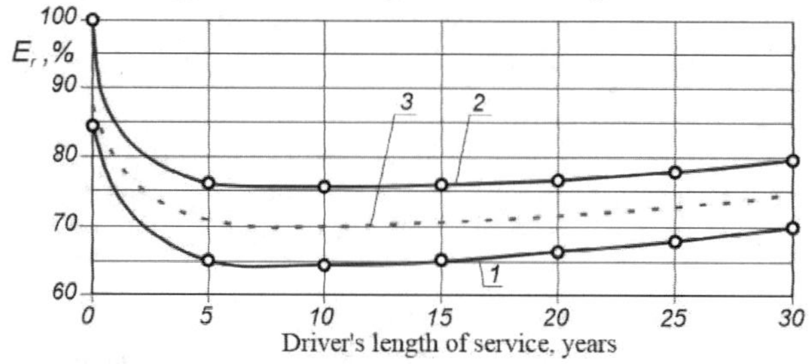

Figure 3. The dependence of the vehicle reliability indicators on driver's length of service: 1 - lean driving; 2 - aggressive driving; 3 - the average value of the indicator.

The energy efficiency of driving can be determined from the ratio

$$E_{en} = \frac{Q_{\min}}{Q_i}, \tag{3}$$

where Q_{min} and Q_i are the minimum and current values of the trip fuel consumption, respectively.

The economic efficiency of driving be determined from the ratio

$$E_{ec} = \frac{P_{max}}{P_i}, \tag{4}$$

where P_{max} and P_i are the maximum and current values of the vehicle performance per shift, respectively.

The driving efficiency by the reliability indicators is expressed by the ratio

$$E_r = \frac{R_{max}}{R_i}, \tag{5}$$

where R_{max} and R_i are the maximum and current values of the vehicle's estimated reliability indicators, respectively.

The driving efficiency by the material costs indicator can be determined from the ratio

$$E_m = \frac{M_{min}}{M_i}, \tag{6}$$

where M_{min} and M_i are the minimum and current values of material costs, respectively.

Taking into account expressions (3 – 6), expression (2) can be rewritten as follows

$$E_D = \frac{Q_{min}}{Q_i} \cdot \frac{P_{max}}{P_i} \cdot \frac{R_{max}}{R_i} \cdot \frac{M_{min}}{M_i}. \tag{7}$$

An empirical formula in the form of the Lagrange interpolation polynomial is proposed to calculate the efficiency of driving BelAZ-7540B heavy-duty pit dump trucks used in the construction of Rogunskaya HPP [9-10]

$$E_D = 0.008857 \cdot t^3 + 0.5824 \cdot t^2 + 10.63 \cdot t + 41.31, \tag{8}$$

where t is the driver's length of service, years.

4. Results

Based on the long-term observations, we made calculations for BelAZ-7540B heavy-duty pit dump trucks in the operating conditions when building Rogunskata HPP to evaluate the driving efficiency, using expression (7). The calculation results are given in Table 1.

Table 1. Results of calculating the driving efficiency

Efficiency evaluation indicator	Length of service						
	0	5	10	15	20	25	30
Trip fuel consumption **Q_i,** l/(100 km)	208	190	179.5	176.5	176.5	178	180
Q_{min} / Q_i	0.8486	0.9289	0.9833	1.0	1.0	0.9916	0.9806
Performance per shift **P_i,** t/cm	315	353	354	350	343	333	320
P_{max} / P_i	0.8873	0.9994	0.9972	0.9859	0.9662	0.9380	0.9014
Material costs, **M_i**	95.0	80.0	67.0	68.0	39.0	71.0	75.0
M_{min} / M_i	0.7053	0.8375	1.0	0.9853	0.9710	0.9437	0.8933
Vehicle reliability indicators, **R_i**	90.0	71.0	70.0	71.0	72.0	73.0	75.0
R_{max} / R_i	0.7778	0.9859	1.0	0.9859	0.9722	0.9589	0.9333

The calculation results by formula (7) using the data of the long-term observations are presented in the form of a graph of the dependence of the efficiency of driving a heavy-duty pit truck on the driver's length of service (Figure 4).

The graphs presented in the work (Figure 1-3) clearly show the efficiency of driving a vehicle depending on the driving style. The discrepancies between the efficiencies of the lean and aggressive driving styles by individual efficiency indicators vary over a wide range depending on the driver's length of service.

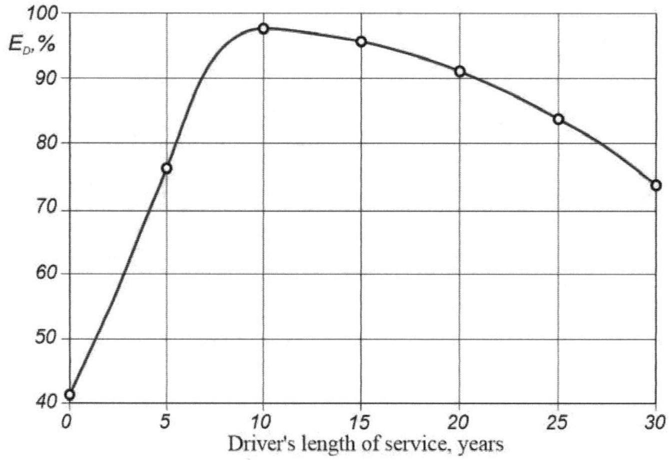

Figure 4. The dependence of the efficiency of driving a BelAZ-7540B dump truck on the driver's length of service in mountain conditions .

5. Conclusions

1. We proposed a methodology for evaluating the efficiency of driving a vehicle. We experimentally established the dependences of the efficiency of driving a vehicle on the driver's length of service, as well as the limits of changes in the efficiency indicators using the lean and aggressive driving styles.

2. We developed a mathematical model of the dependence of the driving efficiency on the driver's length of service, which is congruent with the results of the experimental studies for the mountain conditions of the Republic of Tajikistan.

References

[1] Vakhlamov V K 2005 *Cars; Serviceability* Moscow: Akademia p 240

[2] Ismonzoda S M 2019 Problems of ensuring traffic safety in the Republic of Tajikistan *Federal State Government Educational Institution of Higher Education Orel Law Institute of the Ministry of Internal Affairs of the Russian Federation named after V.V. Lukyanov* 1(2)

[3] Tursunov A A and Yusupov Ya 1999 Reliability of a car driver in mountain conditions. *Proceedings of TTU, Series: Transport and Road Facilities* **2**

[4] Nicholas P Patorniti, Nicholas J Stevens and Paul M Salmon 2017 A systems approach to city design: Exploring the compatibility of sociotechnical systems. *Habitat International* **66**

[5] Ryabchinsky A I, Trofimenko Yu V and Shelmakov S V 2000 *Environmental car safety.* Moscow Automobile and Road Construction Technical University. Moscow p 95

[6] Grafkina M V, Mikhailov V A and Ivanov K S 2016 *Ecology and environmental car safety: textbook* Moscow. FORUM. INFRA-M p 320

[7] Umirzokov A M, Mambetalin K T, Saidullozoda S S and Saibov A A 2019 Conceptual Model of the Estimation of the Efficiency of the Driver-Car-Road-Environment System *Bulletin of the*

South Ural State University. Ser. Mechanical Engineering Industry **19(1)**
[8] Umirzokov A M, Mallaboev U M, Saidullozoda S S and Khabibullozoda Kh Kh 2020 Classification of factors influencing the reliability of the driver-vehicle-road-environment (DVRE) system in the conditions of mountain quarries. *IOP Conference Series: Materials Science and Engineering* **817**
[9] Korn G and Korn T 1977 *Mathematical Handbook for Scientists and Engineers* Moscow. Nauka Publishing house p 832
[10] Gnedenko B V, Belyaev Yu K and Soloviev A D 2010 *Mathematical methods in the theory of reliability* Moscow. Nauka p 524

Analysis of accelerated methods testing piston engine for reliability

A N Gots and Sh A Amirseyidov[1]

Vladimir State University n.a. Alexander and Nikolay Stoletovs, Vladimir, Gorky street 87, Russia, 600000

E-mail: [1] shihamir@mail.ru

Abstract. The analysis of accelerated bench methods of testing piston engines for reliability is carried out. For accelerated testing, it is necessary to reduce the engine operating time before failures occur. Although the loads on the engine parts during accelerated tests differ from the loads in real operation, it is necessary to ensure a qualitative similarity of real and simulated loading processes when organizing them. To assess the quality of accelerated tests, use the acceleration coefficients for operating time or acceleration over time. Various methods of accelerated testing are analyzed due to forcing the conditions for using a special working cycle of the internal combustion engine, for example, for such failures as coking of diesel injectors. For an accelerated assessment of the fatigue strength of the parts surrounding the combustion chamber, the firing surface is heated and then cooled with water. More severe external conditions can also be achieved by increasing the amount of dust supplied with air and oil, increasing or decreasing the ambient temperature. In such tests, after comparing failures during accelerated (model) tests, as well as in operation, the test methodology is adjusted.

1. Introduction

Maintenance of required reliability – one of the basic problems at creation of the engine of internal combustion. Its feature will be, that the estimation of reliability and decision-making on its sufficiency can be carried out only after the expiration of rather long time of operation of the machine on which the engine will be established. In case of revealing insufficient reliability internal combustion engine (ICE) it is required not only to remove the reason of refusals with development of new constructive or technological decisions, but also again to carry out a cycle of long tests [1,2,3]. This process named operational development of a design, as a rule, happens multi-cyclic and frequently long time proceeds. Only acceleration of tests is the most effective way of creation of new perspective engines or modernizations serial [4,5,6].

The concept "reliability" – integrated property of object also includes durability, non-failure operation, maintainability and a retention. Maintainability and a retention at the majority of designs ICE does not render determining influence on complex parameters of reliability [7,8,9,10]. Therefore on a design stage and technological preparation of manufacture the basic attention should be directed on achievement of required parameters of durability and non-failure operation. Accelerated definitions of the specified parameters and a comparative estimation of various variants of a design it is possible only at an intensification of the reasons of refusals, decrease(reduction) in a resource by increase of speed and quantity loads on a time unit, increase in influence of an environment, reduction of breaks in work, etc [10].

The accelerated tests essentially differ from real operation, but at their organization it is necessary to provide, first of all, qualitative similarity of the real and simulated processes, there should not be the refusals which are not meeting in operation.

At the accelerated tests for durability and non-failure operation widely use factors of acceleration on operating time K_1 and acceleration on time K_2 [10,11,12].

At an estimation of a level of non-failure operation factor of acceleration on an operating time define as the attitude of an operating time of products in operation to an operating time of the same products at tests before occurrence of the first refusals, refusals of the given kind or identical number of characteristic refusals of the given group of complexity. Similarly define factor of acceleration on time.

Depending on features of a specific target the factor of acceleration can be attributed to the certain parameter of reliability estimated at tests (to durability, to an average time between failures, to an operating time to first failure, etc.).

Factor of acceleration - size statistical, dependent on dispersion of parameters of durability of a tested product. Accuracy of definition of factors of acceleration, as well as accuracy of forecasts of operational durability and non-failure operation by results of the accelerated tests, depends on volume of the available information on reliability ICE in operation and at tests.

The occurrence of identical damages, failures or signs of a limit state is a clear criterion when a specific type of failure or a specified limit state of the engine is implied. Usually, during accelerated tests of the internal combustion engine, it is considered that it reaches the limit state if during the tests or during the final examination, failures classified as resource were detected, as well as if the wear of parts is such that its operation is disrupted [10, 13, 14].

Various details of the tested engine, as a rule, have the features, differ on bearing ability, on miscellaneous are loaded, in this connection, at tests in one mode it various factors of acceleration are peculiar.

Different parts of the tested engine design, as a rule, have their own characteristics, differ in bearing capacity, are loaded differently, and therefore, when tested in the same mode, they are characterized by different acceleration coefficients.

For different types of damage to the product elements, various methods and test modes can be used to assess their reliability, obtaining different values of the acceleration coefficients K_n, respectively.

The value of K_n depends on the differences between the operating mode of the product and its test mode. If these modes coincide K_n =1, then acceleration is achieved only by using the calendar time more fully during testing in comparison with its use in operation. If the processes of destruction of the product during testing are intensified, K_n increases. In such cases, a reduction in the duration of tests can be achieved both by better use of calendar time and by speeding up the destruction processes [10, 13, 14].

2. Purpose of research

To analyze modern methods of accelerated testing of internal combustion engines for reliability with an assessment of the results obtained.

3. Materials and methods of research

Let's performed analysis modern ways of accelerated tests ICE on reliability. Method for increasing the use of time and frequency of internal combustion engine loading without changing the loading level and operating conditions. Such engine tests are performed in the start-up and warm-up modes. Similarly, launch units are tested when they work not for a few seconds as when starting an internal combustion engine, but for dozens of hours and continuously, for example, in dusty conditions. The dust is fed to the engine with air or oil. Similarly, round-the-clock engine testing, when it tested near 500 ... 800 operating hours per month, allows you to intensify failures. The method allows you to significantly reduce the time to obtain indicators of reliability and durability.

Method of the accelerated tests without change of operational modes. A principle of formation of tests - condensation of the most loaded modes due to full use of calendar time. If the engine under operating conditions works basically on partial modes duration of its tests for non-failure operation and durability can be reduced by change of character of casual dispersion loads in test specifications in the same range loads and operating conditions, characteristic for operation of the engine of concrete purpose. This method is used at the accelerated tests of details of cylinder-piston group (CPG) []1,3,4].

Method of the accelerated tests of the engine and his(its) details at toughening modes loads in comparison with operational. It is reached by increase in average effective pressure, frequency of rotation of a cranked shaft, to increase of pressure in outlet, decrease or rise in temperature of cooling of water, increase in backlashes, change of speed loads, etc. So, for the accelerated check of durability of some elements of pistons (edges of the chamber of combustion or ring and partition be-tween valve are increased with rigidity of a running cycle of the engine , the maximal pressure of combustion . For speeding up cavitation sleeves of cylinders increase a backlash in connection "piston - sleeve". The method is distributed enough, known firms Cummins (USA), Leyland (England), Comutzu (Japan) use it [5,9].

Method of the accelerated tests on heavy modes of tests that change character loads of ICE or their details. For example, for the accelerated check of thermal strength head of cylinders of tractor diesel engines of liquid cooling use a method of heat strokes due to serial submission in system of cooling of cold or hot water. For the accelerated check of fatigue durability of details or elements of the chamber of combustion use cycles with the increased values p_z and $\Delta p / \Delta \varphi$. For the accelerated estimation of fatigue durability of pistons fire a surface heat up, and then cool water. It is possible to provide more rigid external conditions also due to increase in quantity of the dust submitted with air and oil, increase or downturn of an ambient temperature. At such tests it is especially important to compare in details with refusals at accelerate tests, and also in operation and on this basis to correct a technique of tests.

Method of use of dependence of wear resistance of details from speed of their deterioration. Reduction of time of tests is provided due to use of a known method of alignment of speeds of deterioration.

Let's assume, that deterioration U of a detail and time t its wear processes are connected by dependence: $U = At^{\alpha}$, where A - casual factor; α - not casual and constant for the certain type of details an exponent determined experimentally.

The size α can be defined by results of operational or bench tests for deterioration (at corresponding operating time).

Forcing the conditions for using a special working cycle of the internal combustion engine. Acceleration of the process of coking the atomizers of diesel fuel injectors can be provided by 70 ... 80 times by increasing the cyclic fuel supply at the expansion stroke. Additional fuel is injected from an Autonomous fuel supply system. At the same time, the total cycle feed is 10...20% higher than the maximum cycle feed in operation.

Leading engine-building companies pay great attention to the organization and conduct of accelerated reliability tests, and each company, as a rule, independently selects accelerating effects and loading cycles.

For example, firm Leyland (England) test of diesel engines carries out(spends) in three stages for 500 hours everyone half-hour cycles loads at what from a stage to a stage of loading (Table 1) grow. At last stage of tests, the engine has forced on 5 % on cyclic submission of fuel and on 10 % on frequency of rotation of a cranked shaft at the maximal capacity [15,16].

Fiat (Italy) tests for non-failure operation and durability automobile petrol engines two methods. For research of details CPG 100-hour tests by 10-minute cycles in conditions of sharp change of thermal, high-speed and loading operating modes of engines (Table 2) are carried out. Duration of tests on each mode - 1 minutes. Reliability of a gas joint is simultaneously checked also.

At test of engines for "heatstroke" firm Fiat spends 100-hour tests by 10-minute cycles, thus the temperature of a cooling liquid is constantly supported on maximal for operation a level 90…95° C, and the temperature of oil - a constant during each cycle, but in each new cycle toil changes: 110 - 120

IOP Conf. Series: Materials Science and Engineering 1086 (2021) 012006 doi:10.1088/1757-899X/1086/1/012006

- 130ºC. It is more temperature then in exploitation.

Table 1. Cycles loads diesel engines at the accelerated tests of firm Leyland

Duration of a mode, mines	Operating modes of a diesel engine		
	I a stage (500 h)	II stage (500 h)	III stage (500 h)
20	N_{en}	$M_{e.max}$	$M_{e.for.}$
10	stop	stop	stop
20	N_{en}	M_{emax}	$M_{e.for.}$
10	Idling	Idling	Idling

Table 2. Cycles loads automobile petrol engines at accelerated tests CPG of firm Fiat

Operating modes	Frequency of rotation of a cranked shaft, mines^{-1}	Temperature of a cooling liquid, ºC
Idling	1000	30
Full load	5000 (0,9 n_{n}.)	30
Idling	1000	90
Full load	5000	90
Idling	1000	90
Full load	5000	30
Idling	1000	30
Full load	5000	90
Idling	1000	90
Full load	5000	90

Firm Comatzu (Japan) with the purpose of definition of non-failure operation and durability of the engine carries out the bench accelerated tests of various types: the first - with the purpose of check of an opportunity of occurrence of cracks owing to weariness, seize or scratch surfaces of friction, occurrence of other refusals under loading from gas forces, and also intensive deterioration of pairs friction (general duration of tests makes 1000 hours).

Before the beginning of tests the engine force on average effective pressure upon 15 % more, and during tests it works as 6-minute cycles on variable modes from the minimal frequency of rotation of idling up to nominal and on the contrary (Table 3). The temperature of a cooling liquid is supported at the maximal level for operation 90 … 95º C.

To obtain the maximum temperatures of the internal combustion engine parts, the engine is forced and it forced to work most of the time under load at increased values of gas, inertial and thermal loads, which allows you to bring the acceleration coefficient K_{1n} to 6...9 for reliability and to 1.2...4.7– to assess the durability of the friction surface wear.

The second test is carried out on a constant engine operation mode in order to assess the properties of the oil and burning its consumption.

Table 3. Cycles loads of firms Comatzu on variable modes before fail-safe test

Duration of cycles, secund	Frequency of rotation of a cranked shaft	Values of the effective twisting moment
30	1,05 $n_{idling.min}$	0
300	1,05 n_{n}	1,05 M_{emax}
30	1,05 $n_{idling.min}$	0

In Ukraine (Kharkov) for an estimation thermal fail sharp edges of the chamber of combustion of pistons of tractor diesel engines the method accelerated cyclic load is developed. Feature of work of a diesel engine is put in its basis during heavy agricultural works. For bench tests the cycle of variables load duration 44 with (dump of loading – 22 is chosen with; increase – 22 second). For this time capacity varies from zero up to the maximal value (N_{emax}), and frequency of rotation of a cranked shaft – from the minimal frequency of rotation idling up to frequency of rotation on a nominal mode and on the contrary. The analysis of results of operational and bench tests has shown, that the resource of the engine 6000 engine hours is equivalent 105 cycles loads, that provides destruction of edges of pistons owing to weariness till 125 o'clock, i.e. K_1 =6000/125 ≈ 50 [7, 17].

Widely uses methods of the accelerated tests association "Autodiesel engine" (Yaroslavl). By development of a technique of the accelerated tests for wear resistance of piston rings and sleeves of cylinders the problem was put to provide similarity outline deterioration in operation and at bench tests. As reduction of temperature of a cooling liquid influenced deterioration to its increase have applied the abrasive particles providing concentration of a dust corresponding work of the engine in conditions of a dusty storm a little. Maximum permissible deterioration of cylinders of engine reached after 90 … 160 h. tests. The form outline deterioration has come nearer to outline operational deterioration at run of the automobile 120…150 thousand in km at usual dust-laden air on an admission [5, 18].

Research of an intense condition of piston pin bosses and the bottom of the piston was loaded by hydraulic way with pressure from zero up to 11 MPa. At bench tests increased a corner of an advancing of injection that led to to increase in pressure in 1,5 times.

With the help of such technique were carried out(spent) test of pistons for the engine during 250 … 300 hours.

At a choice of necessary height of the crosspiece value combustion pressure p_z increased up to 12..14 MPa due to change of a corner of an advancing of injection (thus simultaneously in 2…3 times grow speed of increase of pressure of gases in the cylinder). As a result of a stresses in the piston grow up to 70 %; simultaneously on 50…60° C temperatures in a zone of the crosspiece raise, that on 25…30 % are reduced with durability of a piston aluminium alloy.

In case of need estimations of quality of a material or efficiency of heat treatment for reduction of time of tests height of the crosspiece reduce on 2…3 mm; thus K_1 = 25.

The estimation form crack the head of cylinders in a zone of crosspieces between valves was checked by speeding up of temperature stresses due to in-crease in a corner of an advancing of injection, and also periodic sharp change of temperature of a cooling liquid (heat strokes) and change of high-speed and force power setting. In Table 4 characteristics of cycles load the head of cylinders are resulted.

Table 4. Cycles load head of cylinders of Open Society "Autodiesel engine" (Yaroslavl)

Mode of tests	T cooling, °C	n, mines^{-1}	Ne, kw	Duration, mines
Start-up and warm up	20…40	1100	Idling	1
Nominal	40…90	2100	N_{enom}	25
Cooling	90…40	1100	Idling	10
The Stop/Washing of system of cooling	40… 20	-	-	15/60

Cracks in crosspieces have appeared already after 100…150 hours tests; and after 300 m-h took place in 50 % heads of of cylinders . We shall notice, that in operation of a crack appeared after 8000 engine hours. This way can be used at comparative tests (K_1= 8000/100=80).

Cavitation sleeves of cylinders arises in operational conditions only through 2…3 years. The method of its acceleration in 30…40 times is developed at work of a diesel engine idling due to increase in a backlash in pair «a piston - sleeve of cylinders » up to 0,6…0,7 mm and at reduction of temperature of a cooling liquid up to 55°C (the maximal intensity cavitation is observed just at such temperature of a cooling liquid). Cavitation destructions of sleeves and blocks of cylinders at such tests arise after 150 hours. In operation similar destructions occur through 150…200 thousand in km of run (6000…8000 hours), i.e. K_1 ≈ 40-50 [19,20].

The Central Research Automobile and Automobile Engine-Building Institute "NAMI" has developed a method of accelerated tests for the trouble-free operation of automobile engines, in which all additional units are involved. Duration of tests of petrol engines also makes 200…300 h for engines of cars and 300…400 h – lorries and buses, and for diesel engines – 350…800 h depending on their working volume. Tests for non-failure operation are carried out by the cycles consisting of modes, resulted in Table 5 [5,9].

The second type of tests of diesel engines is carried out with use of the forced modes. Duration of tests makes from 250 up to 600 h. The cycles resulted in Table 6.

Table 5. Cycles NAMI test for non-failure operation of petrol engines

	Frequency of rotation of a cranked shaft	The capacity, the twisting moment (gross)	Time, mines
1	$n_{ing.min} + 300$	0	5
2	n_{nom}	N_n	140
3	$n_{M_{e\max}}$	$M_{e\max}$	20
4	1,15 n_{nom}	0	10
5	$n_{ing.min} + 300$	0	5
TOTAL			180

Table 6. Cycles NAMI of test of diesel engines

	Frequencies of rotation of a cranked shaft	The capacity, the twisting moment (gross)		Time, minute
		Diesel engines without pressurization	Diesel engines with turbo-charging	
1	$n_{xx.min} + 300$	0	0	60
2	1,05 n_{nom}	1,05 N_H	1,1 N_H	300
3	$n_{M_{e\max}}$	$M_{e\max}$	$M_{e\max}$	210
4	1,15 n_{nom}	0	0	90
5	$n_{igle..min} + 300$	0	0	60
TOTAL				720

In research and design-technological Institute of engines (NICKID) was developed "Methodology of accelerated bench tests on the reliability" (OST 23.3.21-87) [11], and then with the aim of refining the developed diagram of loading shown in the figure, characterized in that the area CD, where the torque Me is a constant value, (in nominal conditions) imposed by its fluctuations from nominal to maximum values. Efficiency of a technique was checked at the accelerated tests of diesel engines of air and liquid cooling 8ЧН 15/16, 4ЧН 12/14 and 4ЧН 11/12.

4. The results of the study and their discussion

The developed modes load diesel engines have confirmed high efficiency at test for non-failure operation at thermocyclic loadings (K_1 =10 for the diesel engines established on universal tractors). At the same time, to receive the same factor of acceleration at definition of durability owing to wears by the given technique it was not possible. Wears after tests have made 40…60 % from limiting.

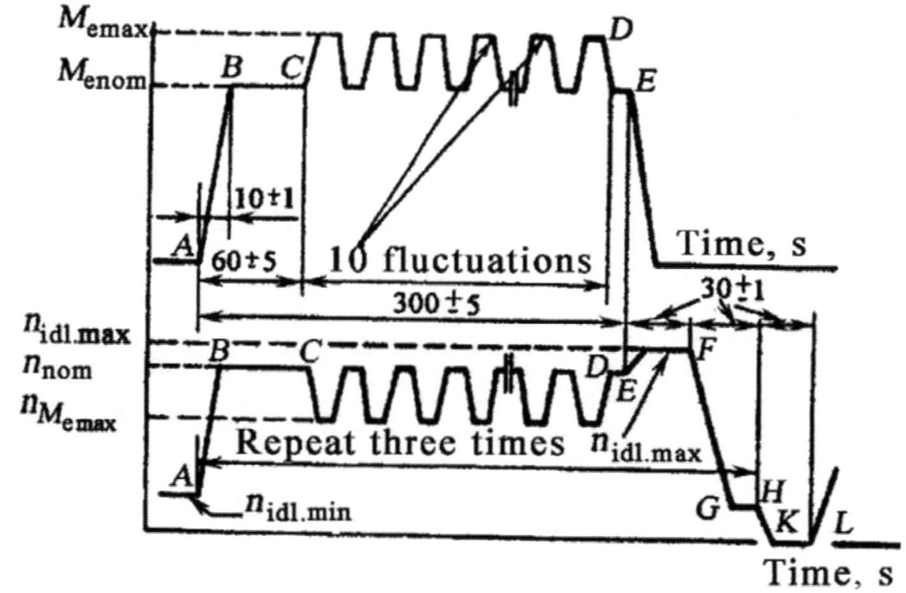

Figure 1. Sequence diagram of loading of a diesel engine when accelerated, which test method NICTID (area KL – stop diesel)

5. Conclusions

Despite the large number of methods used for accelerated bench testing, it is necessary to further study the physical phenomena that lead to failures. With all the proposed methods, one idea is laid down– with more strenuous work, products will receive failures in a shorter period of time. However, it is necessary to study the regularity of accelerating the process of loss of reliability, depending on the tightening of the regime. Usually, the hypothesis of the invariance of the laws of distribution of the product uptime under different modes, which is accepted in such studies, requires serious verification.

References

[1] Velichkin I N 1999 Accelerated testing in the General machine reliability testing system *Tractors and agricultural machines* **12** pp 40-41
[2] Velichkin I N 1999 Methodology and ways to reduce the duration of tests *Tractors and agricultural machines* **6** pp 30-33
[3] Velichkin I N 1999 Development of a complex of accelerated tests *Tractors and agricultural machines* **11** pp 34-36
[4] Velichkin I N 1999 On the need to improve the methods of accelerated testing *Tractors and agricultural machines* **4** pp 25-27
[5] Grigoryev M A 1996 Method of accelerated bench tests for reliability of gasoline engines of passenger cars *Dvigatelestroenyie* **1** pp 54-56
[6] Andropov B S and Mardaliev E J 1999 Research of parameters of the limit state of automobile engines YAMZ *Dvigatelestroenyie* **2** pp 32-33
[7] Abramchuk F I, Rjazantsev M K, Shechovtzov A F 2004 Internal combustion engine *6* p 324
[8] OST 23 3 21 87 1987 Diesel engines tractor and combine. Methods of the accelerated fail-safe tests p 19
[9] Puchkov V P and В.Ф. Rein V F 1992 Improving the methodology of accelerated testing for the reliability of tractor and combine diesels *Dvigatelestroenyie* **5** pp 44-46
[10] Kugel R V 1982 *Fail-safe test of machines and their elements* Moscow: Moscow Engineering

Press p 181
[11] Gotz A N 2004 Scientific of a basis of calculation and the accelerated tests of details crank-connecting-rod the mechanism of tractor diesel engines on a design stage *Abstract of the dissertation for the degree of doctor of technical Sciences* p 32
[12] Barzilovich E Yu and Gnedenko B V 1985 *Problems of aircraft reliability* Collection of articles. Moscow: Engineering Press p 280
13. Michurin A I 1966 *Accelerated tests for reliability of technical systems and publications* Moscow: Knowledge p 71
[14] Peshes L Ya and Stepanova M D 1972 Fund*amentals of the theory of accelerated reliability tests* Minsk: Science and techology p 176
[15] Iberla K 1980 *Factor analysis* Moscow: statistics p 397
[16] Solovyov A D 1975 *Fundamentals of mathematical theory of reliability* M. Knowledge p 62
[17] Pronikov A S 1978 *Reliability of machines* Moscow: Engineering Press p 600
[18] Weibull 1964 *Fatigue tests and analysis of their results* Moscow: Engineering Press p 279
[19] Bolotin V V 1984 *Forecasting the resource of machines and structures* Moscow: Engineering Press p 312
[20] Gotz A N, Prygunov M P, Frantzuzov I V 2013 Accelerated reliability tests of tractor diesel cylinder heads *J Modern problems of science and education* **3** pp 30-35

Analysis of the developed hinged joint assembly of the wheeled harvester

V E Klubnichkin[1,3], E E Klubnichkin[1], A B Kartashov[2] and K V Vasilyeva[1]

[1]Mytishi branch Bauman Moscow State Technical University, Department of Transport and technological means and equipment of the forest complex, 1st Institutskaya street, 1, 141005, Mytischi, Moscow region, Russian Federation
[2]Bauman Moscow State Technical University, Department of Wheeled Vehicles, 105005 Moscow, Russian Federation

e-mail: [3]vklubnichkin@gmail.com

Abstract. This work presents a kinematic analysis of the hinged joint assembly of the wheeled harvester which consists in determining limit folding angles of the joint assembly and travel of the hydraulic actuators against the developed electronic solid models. A resisting moment required for a vehicle spot folding on a hard bearing surface with a high friction factor with due consideration for the tires' interaction with the bearing surface has been found with use of the Pacejka semi-empirical model MF-Tyre 6.1. Parameters are for tire model 710/45-26.5. The harvester dynamic model has been created having the following features: all links of the dynamic system are absolutely rigid; all wheels are free and rotating; no friction in hinges; Modelling was carried out for fully loaded harvester; rear tandem axle is rigidly coupled through a drive shaft with the front tandem axle; axle differentials are not blocked. Presented are vertical loads in the front and rear axle wheels' patch area as well as lateral responses in the front and rear axle wheels' patch area.

1. Introduction

Review and analysis of designs of the control systems of the tractors equipped with articulated frame steering have demonstrated that they have certain advantages before steerable wheels that is strongly pronounced in the wheeled tractors with big wheels operating in forest off-road conditions [1 – 3].

The works [4 – 7] consider the issues of designing and researching configuration of special-purpose forest tractors with a hinge frame, they include researches aimed at improving fatigue endurance of the articulated framings used in multi-operational timber-harvesting vehicles.

The review and analysis of designs of the control systems of the wheeled articulated vehicles have demonstrated that hydraulically driven control systems are the most common.

In accordance with requirements of the project Statement of Work, the harvester's steering control shall be electromechanical or hydraulic. The wheeled harvester shall turn by folding the framing in a horizontal plane in the joint assembly.

The harvester steering control drive is hydrostatic. The maximum angles of the joint assembly turn are 44 deg. The steering control feedback is hydraulic.

Use of folding turn method allows for employment of wheels with large diameter, large cross-section and low inner pressure tires that in its turn provides better tractive properties, good clearance; lower soil pressure with high axle load [8 – 10].

IOP Conf. Series: Materials Science and Engineering 1086 (2021) 012007 doi:10.1088/1757-899X/1086/1/012007

2. Materials and methods

When designing a wheeled harvester hinge joint assembly, a traditional diagram of the hinged vehicle turn mechanism with steering control hydraulic cylinders tilted with reference to the vehicle longitudinal axis has been selected as a basic one. This diagram of the turn mechanism is employed by most foreign manufactures of harvesters.

Kinematic analysis of the hinged joint assembly of the harvester considers determining limit folding angles of the joint assembly and travel of the hydraulic actuators against the developed electronic solid models. The purpose of the kinematic analysis is to select hydraulic cylinders and to determine sizes of the joint mechanism. The main criteria for determining a kinematic diagram is a volume equality of the operating fluid which needs to be pumped to enter the corner and come out of the corner. This condition is provided under equal travel of both hydraulic cylinders during turning [11].

It is important to calculate a resisting folding moment of the harvester's half-frames. The purpose of calculation is to determine a moment required for spot folding of half-frames of the fully loaded vehicle on a hard bearing surface with a high friction factor [12 – 15].

The calculation has been made with use of a software package of automated analysis of the bodies system dynamics. This package describes a wheeled vehicle as a mechanical system made of solid bodies coupled by hinges and force links. This description is used to automatically generate a system of algebraic-differential equations governing a movement of the bodies system. Then a numerical method is employed to solve those equations to obtain time realizations of the kinematic and dynamic parameters of the mechanical system movement.

Figure 1 shows external view of the spatial dynamic model of the harvester.

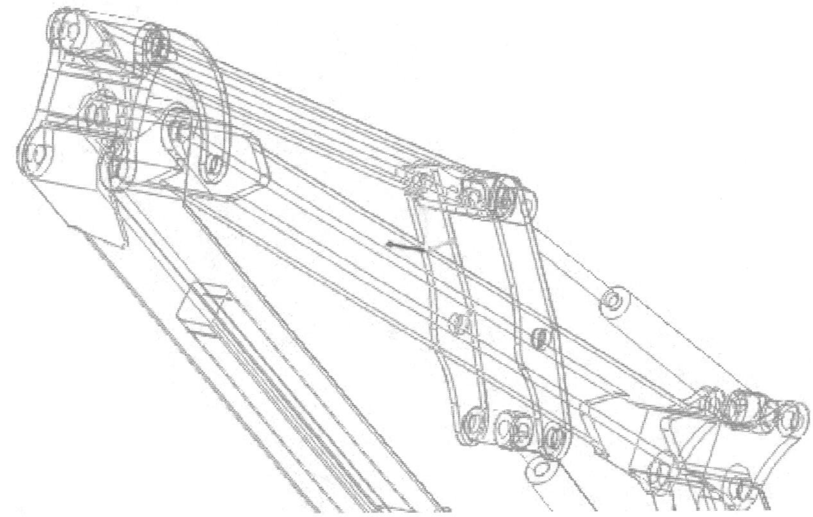

Figure 1. Harvester dynamic model

The model has the following features:

- all links of the dynamic system are absolutely rigid;
- all wheels are free and rotating;
- no friction in hinges;
- modelling was carried out for fully loaded harvester;
- rear tandem axle is rigidly coupled through a drive shaft with the front tandem axle;

- axle differentials are not blocked (free).

Tires interaction with the bearing surface was described with use of the Pacejka semi-empirical model MF-Tyre 6.1. Parameters of tire model 710/45-26.5 are given in Figure 2.

```
[DIMENSION]
UNLOADED_RADIUS      = 0.670        $ Свободный радиус, м
WIDTH                = 0.710        $ Ширина профиля, м
RIM_RADIUS           = 0.33655      $ Посадочный радиус, м
$--------------------------------------------------------------массово-инерционные параметры
[INERTIA]
MASS                 = 300          $ Масса колеса
IXX                  = 71           $ Моменты инерции колеса относительно поперечных осей, кг*м2
IYY                  = 104          $ Момент инерции колеса относително оси вращения, кг м2
GRAVITY              = -9.81        $ Ускорение свободного падения, м/с2
$--------------------------------------------------------------вертикальная сила
[VERTICAL]
VERTICAL_STIFFNESS   = 1301712      $ вертикальная жесткость, Н/м
VERTICAL_DAMPING     = 60000        $ вертикальное демпфирование Н с/м
FNOMIN               = 67689        $ Номинальная нагрузка на колесо, Н
$--------------------------------------------------------------продольная сила
[LONGITUDINAL_COEFFICIENTS]
PDX1                 =  1.1852      $ Коэф. продольного сцепления Mux при нагрузке Fznom
PDX2                 = -0.04        $ Изменение Mux при изменении нагрузки
PKX1                 = 20           $ Относительная жесткость х-ки продольного сцепления Kfx/Fz при Fznor
$--------------------------------------------------------------поперечная сила
[LATERAL_COEFFICIENTS]
PDY1                 =  0.8         $ Коэф. поперечного сцепления Muy при нагрузке Fznom
PDY2                 = -0.15        $ Изменение Muy при изменении нагрузки
PKY1                 = -15          $ Максимальный относительный коэф. сопротивления уводу Kfy/Fznom
PKY2                 =  2           $ Отношение нагрузки при максимуме коэф. сопр. бок. уводу Kfy к Fznor
$--------------------------------------------------------------сопротивление качению
[ROLLING_COEFFICIENTS]
QSY1                 = 0.015        $ Коэф. сопротивления каченю
$--------------------------------------------------------------стабилизирующий момент
[ALIGNING_COEFFICIENTS]
QDZ1                 = 0.12         $ Отношение плеча боковой силы к свободному радиусу при Fznom
```

Figure 2. Fragment of the file with parameters of model MF-Tyre 6.1 for tire 710/45-26.5

Figure 3 shows hinges and load-bearing components which describe kinematics of the joint assembly.

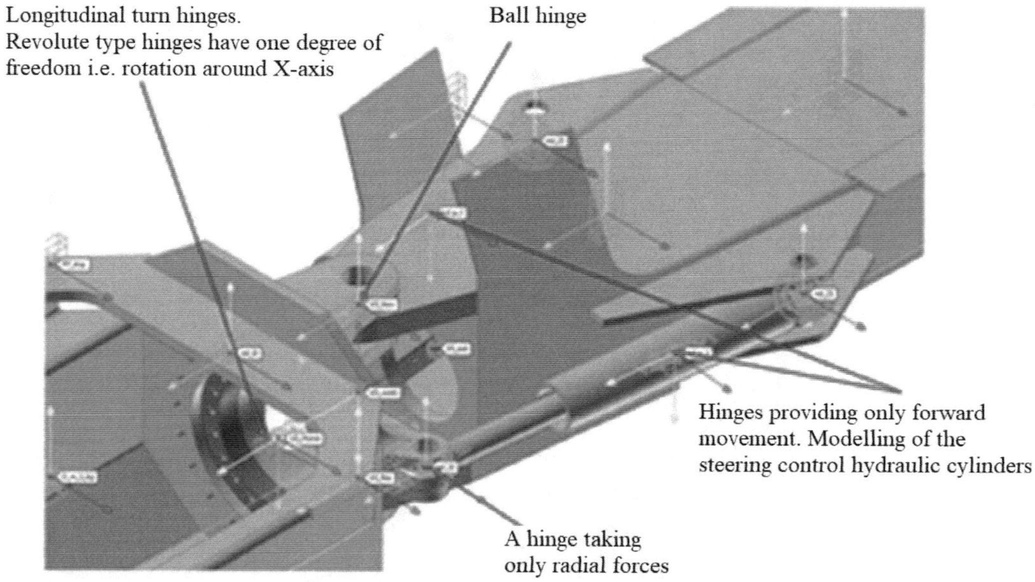

Figure 3. Hinges and load-bearing components of the joint assembly

3. Results and discussions

The results of the kinematic analysis of the joint assembly have shown that the main reason when an articulated wheeled vehicle which has entered the corner cannot come out of it at a sufficiently intensive rate is a variable quantity of fluid necessary for entry and coming out of corners which depends on the cylinders' position relative to a vertical hinge of the link joint assembly [16]. I.e. there is a necessity in a smooth travel of a piston required for entry and coming out of corner.

The second criterion is a minimum change in arms pushed by hydraulic cylinders. This is discussed in more detail in article [17].

The analysis has resulted in building a geometrical diagram of the joint assembly, obtaining the required parameters of the actuating cylinders and an amount of change of the cylinders' arms.

The kinematic analysis of the joint assembly has found that the maximum structurally achievable folding angle of the frame was 44 degrees.

When calculating a resisting folding moment of the vehicle half-frames, modelling was conducted during 50 seconds, provided that during the first 10 seconds the length of the steering control hydraulic cylinders remained unchanged and the vehicle was coming to static equilibrium during that time.

Figure 4 shows equilibrium position of the harvester on a horizontal surface. Arrows show wheel responses. Table 1 presents distribution of normal responses between wheels and axes as well as their total value.

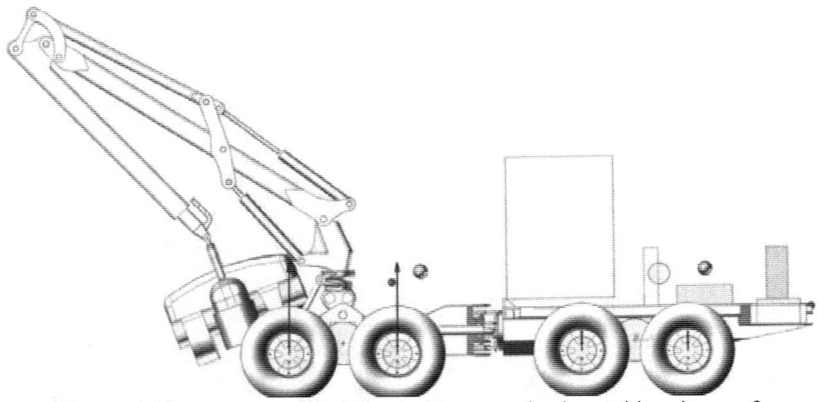

Figure 4. Harvester equilibrium position on a horizontal bearing surface

Table 1. Distribution of normal responses between the harvester wheels

	Axis 1	Axis 2	Axis 3	Axis 4
Portside	51,220	51,222	11,396	9,382
Starboard	51,774	51,776	10,944	10,943
Total for axes	102,994	102,998	22,340	20,325
Total	248,657 N (25,347 kg)			

Then the length of the left hydraulic cylinder changed in a predetermined manner, with that a moving force required for turning the vehicle's section automatically generated inside.

Results of modelling of the vehicle's sections spot folding are presented in figures 5 – 8.

The figure shows animation frames obtained at modelling the harvester's spot turn. The frames correspond to the specific positions of the vehicle. Positions at time T = 10 s correspond to the initial position, at time T = 24 s, to the maximum angle of right folding, T = 40 s, to the maximum angle of left folding, T = 50 s, to the final position.

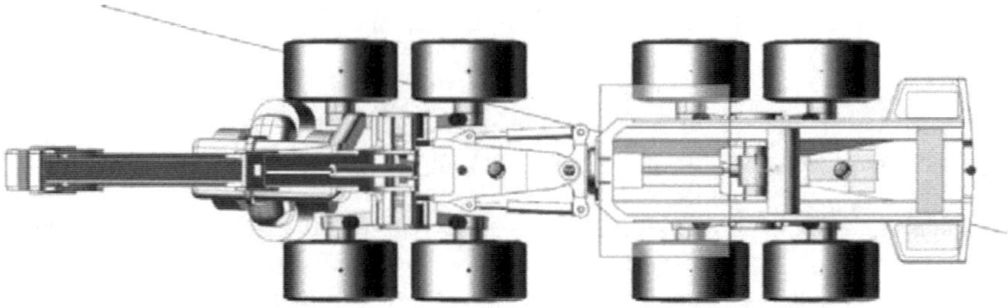

Figure 5. Vehicle positions at spot turn, T = 10 s

Figure 6. Vehicle positions at spot turn, T = 20 s

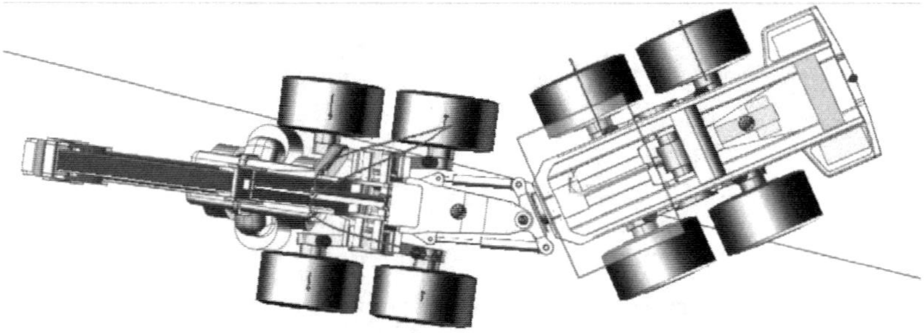

Figure 7. Vehicle positions at spot turn, T = 40 s

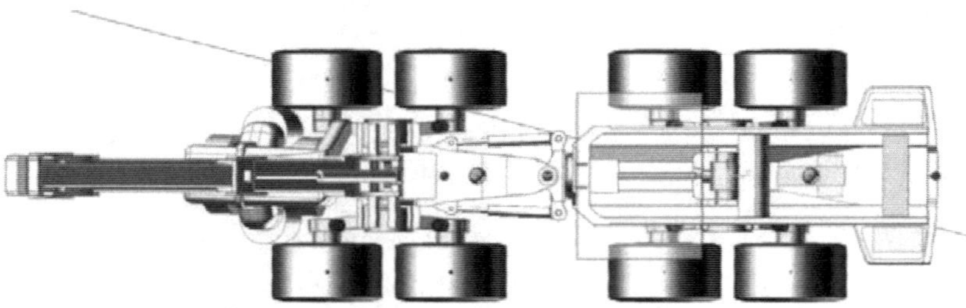

Figure 8. Vehicle positions at spot turn, T = 50 s

Figure 9 shows a resisting moment of the harvester's spot folding on a hard bearing surface with a high friction factor. Figure 10 shows the harvester frame sections' folding angle.

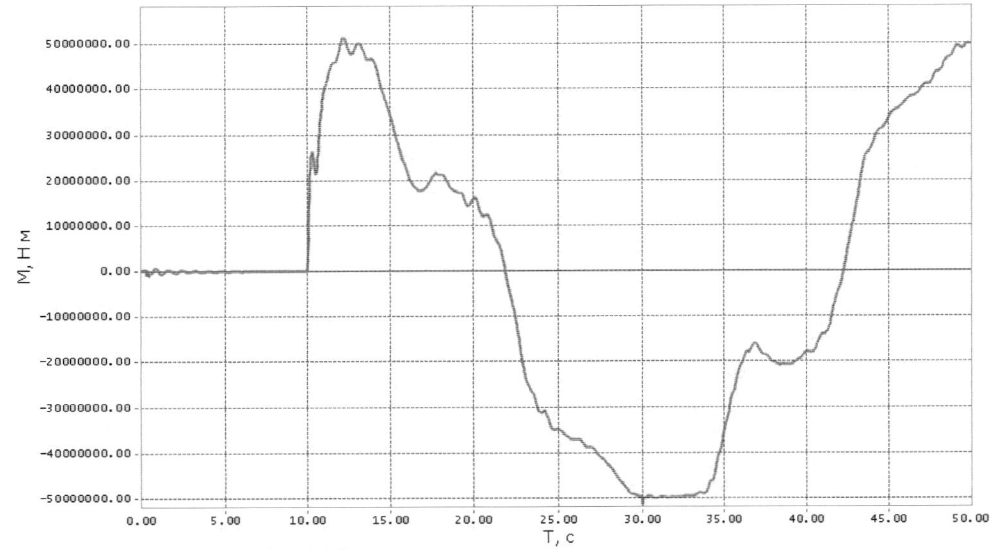

Figure 9. Harvester resisting folding moment

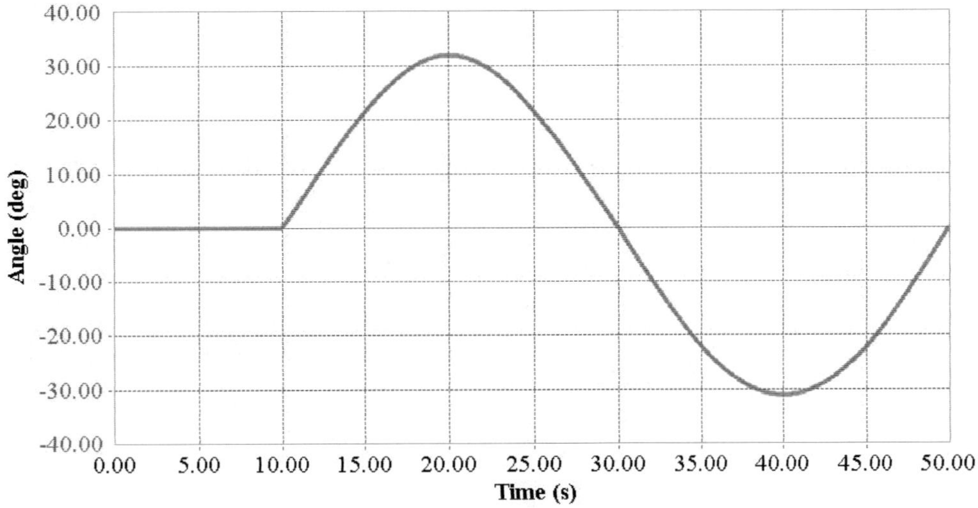

Figure 10. Harvester frame sections' folding angle, deg

Figures 11 and 13 show vertical loads vertical loads in the front and rear axle wheels' patch area, figures 12 and 14 show lateral responses in the harvester front and rear axle wheels' patch area

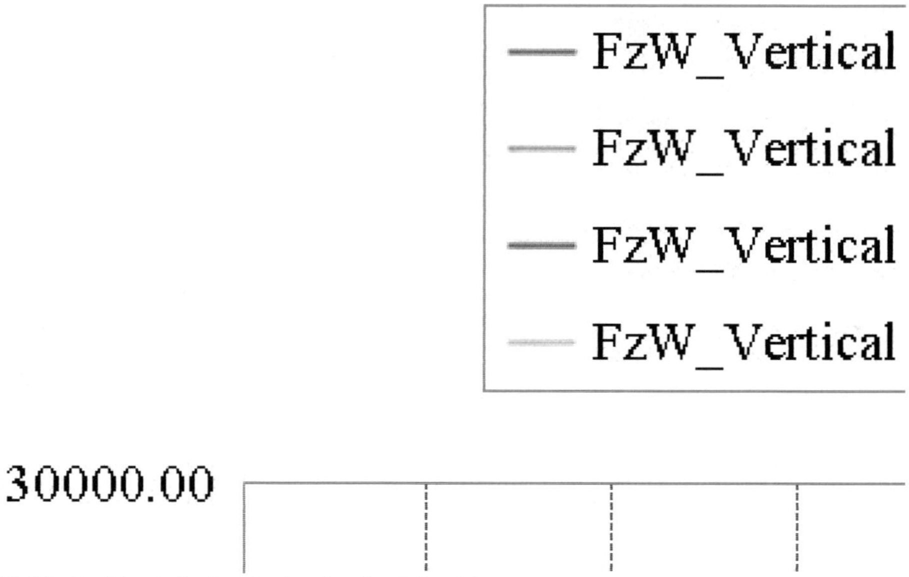

Figure 11. Vertical loads in the front axle wheels' patch area

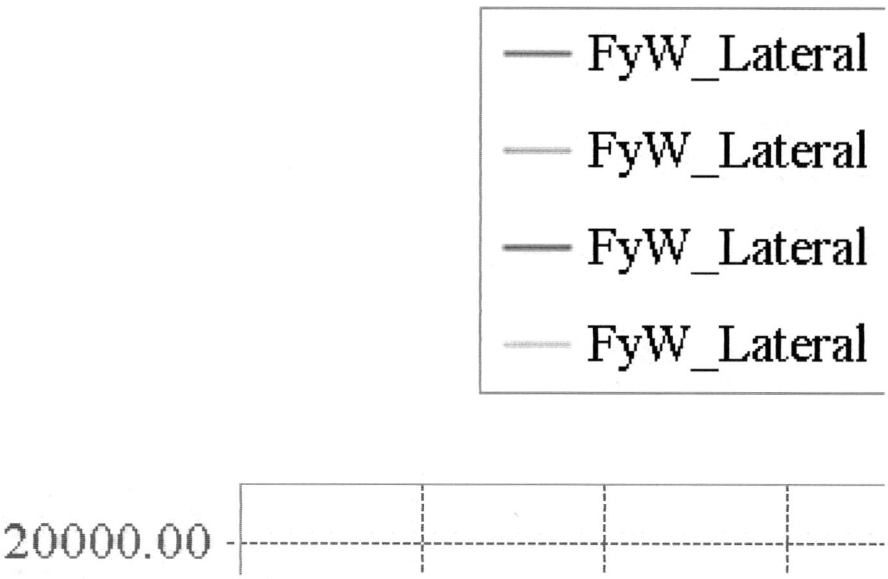

Figure 12. Lateral responses in the front axle wheels' patch area

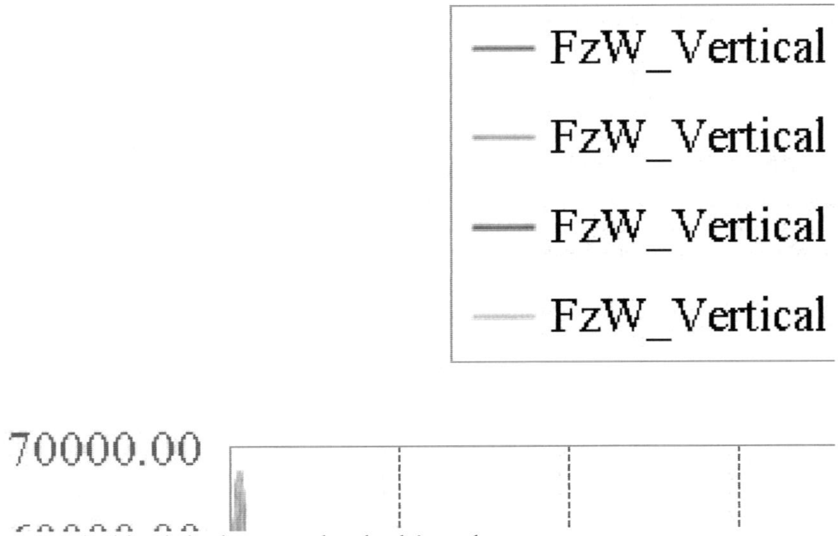

Figure 13. Vertical loads in the rear axle wheels' patch area

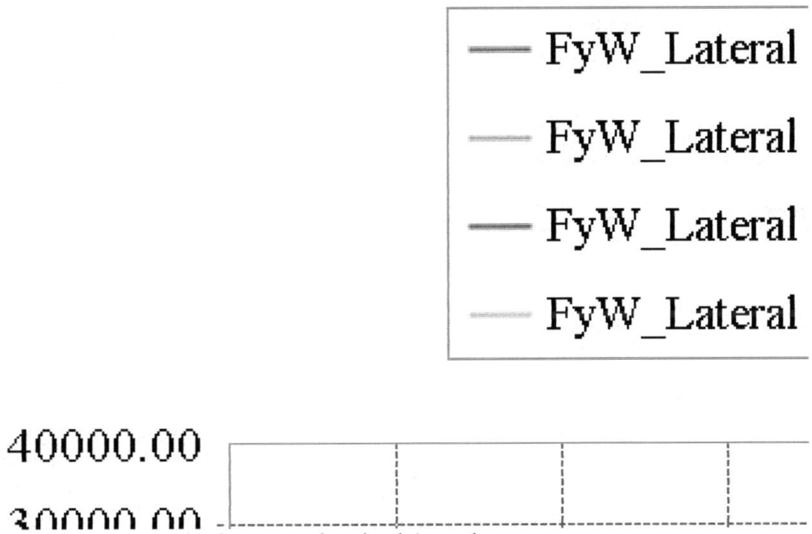

Figure 14. Lateral responses in the rear axle wheels' patch area

4. Conclusion

The kinematic analysis of the joint assembly has found that the maximum structurally achievable folding angle of the frame was 44 degrees. When a joint assembly folds by 44 deg, the stops bump into a frame. The difference in travel of the hydraulic cylinder pistons was 15.6 mm for the folding angle of 44 deg, or 6.8% of the travel. The arms changed by 99 mm for the external cylinder and 80 mm for the internal one.

Found is a resisting moment of the harvester spot folding on the hard bearing surface with a high friction factor. Presented are vertical loads in the front and rear axle wheels' patch area as well as lateral responses in the front and rear axle wheels' patch area.

Acknowledgments
The work has been accomplished at the Bauman Moscow State Technical University with financial support from the Ministry of Science and Higher Education of the Russian Federation under the agreement No. 075-11-2019-030 as of 22 November 2019.

References
[1] Klubnichkin V E, Klubnichkin E E and Kartashov A B 2020 A Brief Analysis of Trends in the Development of Logging Machines *Proceedings of NSTU im. R.E. Alekseeva* **3 (130)** pp 93-102
[2] Golyakevich S A and Goronovsky A R 2012 Experimental studies of the load-bearing structures of articulated logging machines *Proceedings of BSTU. Timber and woodworking industry* **2** pp 15-17
[3] Ringdahl O, Hellström T, Lindroos O 2012 Potentials of Possible Machine Systems for Directly Loading Logs in Cut-to-Length Harvesting *Canadian Journal of Forest Research* **42 - 5** pp 970-985
[4] Kozmin S F and others 2011 Study of the layout of forest wheeled tractors with an articulated frame of the LTA design. Forest tractors MTZ-82L and T-150KL with a 6x6 wheel arrangement and their modifications *Textbook GOU VPO "SPbGLTA". St. Petersburg* p 96
[5] Golyakevich S A and Goronovskii A R 2017 Evaluation of Loading Dynamics and Fatigue Life for a Forwarder Half-Frame Articulation *Journal of Machinery Manufacture and Reliability* **46 - 5** pp 463-471
[6] Golyakevich S A, Korobkin V A and Asmolovsky M K 2011 Operational tests of the hinges of the articulation of the semi-frames of the logging machines PO "MTZ" *Proceedings of BSTU. Timber and woodworking industry* **2** pp 122–126
[7] Proto A R, Macrì G, Visser R, Harrill H, Russo D and Zimbalatti G 2018 Factors Affecting Forwarder Productivity *European Journal of Forest Research* **137 - 2** pp 143-151
[8] Belyakov V V et al. 2014 The canvas of the path of transport and technological machines (reference materials to the theory "machine-locality") *Textbook "NSTU im. R.E. Alekseeva ". Nizhny Novgorod* p 447
[9] Allman M, Jankovský M, Messingerová V, Allmanová Z and Ferenčík M 2015 Soil compaction of various central european forest soils caused by traffic of forestry machines with various *Forest Systems* **24(3)**
[10] Klubnichkin E E, Klubnichkin V E and Kotiev G O 2018 Theoretical research of soil packing by timber harvester running gear *IOP Conference Series: Materials Science and Engineering* **286** 012025 doi: 10.1088 / 1757-899X / 386/1/012025
[11] Polungyan A A, Afanasyev B A, Bocharov N F and others 1999 Design of all-wheel drive vehicles *Textbook "MSTU im. N. E. Bauman" - Moscow* **1** p 488
[12] Makarov V S et al. 2019 Study of uneven surfaces distribution on forestry roads *Journal of Physics: Conference Series* **1177(1)** 012041 doi: 10.1088/1742-6596/1177/1/012041
[13] Laughery S, Gerhart G and Goetz R 1990 Bekker's Terramechanics Model for Off-Road Vehicle Research
[14] Wong, J.Y. 2010 Terramechanics and Off-Road Vehicle Engineering *Oxford Elsevier* p 488
[15] Klubnichkin V, Klubnichkin E, Podrubalov M, and Makarov V 2019 Researching rolling resistance of the wheeled forestry tractor at skidding *IOP Conference Series: Materials Science and Engineering* **695** 012003 doi: 10.1088 / 1757-899X / 695/1/012003
[16] Anisimov G M and Kochnev A M 2008 Testing of cutting machines *Textbook "SPbGLTA". St. Petersburg* p 488
[17] Kartashov A B and Gazizulin R L 2016 Development of all-terrain vehicles with an articulated support system based on a serial truck *Engineering bulletin* **12** p 10

Assessing load on support rollers of the tracked load-transport timber harvesting vehicle through simulated mathematical modeling

V E Klubnichkin[1,3], G O Kotiev[1] and V S Makarov[1,2]

[1]Bauman Moscow State Technical University, 2th Baumanskaya str., 5/1, 105005, Moscow, Russia
[2]Nizhny Novgorod State Technical University n.a. R.E. Alekseev, Minin str., 24, 603950, Nizhny Novgorod, Russia

E-mail: [3]vklubnichkin@gmail.com

Abstract. The article describes a process of simulated mathematical modeling of movement of a tracked load-transport vehicle in the applied software package followed by output of the results of the loads acting on the portside and starboard support rollers during movement at different speeds on flat surfaces, over single 100 to 400 mm obstacles located at a certain distance from each other, on sinusoidal unevenness and along a preset microprofile of the bearing surface. It also describes conditions of the virtual tests conducted.

1. Introduction

At the designing stage it is crucial for the manufacturers and customers to determine the service life of the structural elements of the tracked load-transport timber harvesting vehicles, including the support rollers. With that it is required to maintain the optimum ratio between the initial cost of the support rollers and maintenance and repair expenditures during the entire service life. Timber harvesting vehicles operate in hard climate-production conditions. During their operation they regularly meet variable obstacles like stumps, stones, fallen trees, pits etc. which in their turn negatively affect the life of the support rollers [1].

The movement dynamic of a tracked load-transport vehicle (LTV) has been modelled in the "Universal mechanism" package in accordance with the program and research method [2] and is based on the system of tests including equilibrium test, initial speed calculation, linear movement at different speeds on a flat surface, over single obstacles from 100 to 400 mm high, on sinusoidal unevenness and along a present microprofile of the skid roads [3 – 9].

To start modeling of the LTV movement it is necessary to conduct auxiliary tests including equilibrium tests and initial speed calculation. The aim of the auxiliary tests is to determine initial state of the LTV with due consideration for all forces acting on it in a static state. Then the main tests (rectilinear movement) are carried out which allow for calculation of the dynamic indicators of the LTV undercarriage during movement on the bearing surface unevenness [10, 11].

To carry out the tests an LTV simulated model was generated comprising a vehicle frame with installed process equipment, undercarriage, cabin, power plant, etc. [12, 13]. All model components were manufactured in accordance with the real geometrical and physical parameters of timber harvesting vehicle LZ-11.

2. Conditions of Virtual Tests

An equilibrium test was conducted to determine position of a static equilibrium to verify a model against the experimental data. Figure 1 shows a simulated model of LTV before a test and after a test.

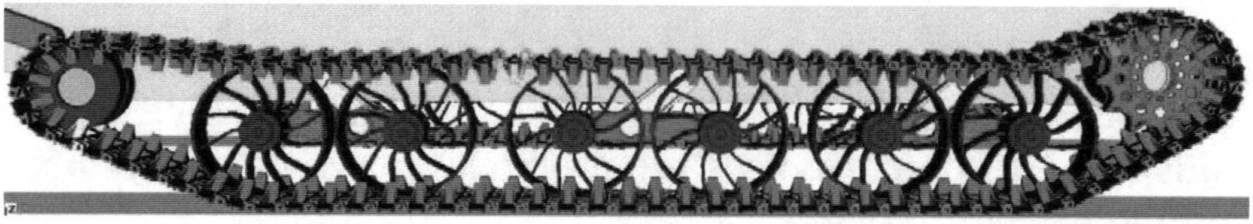

a)

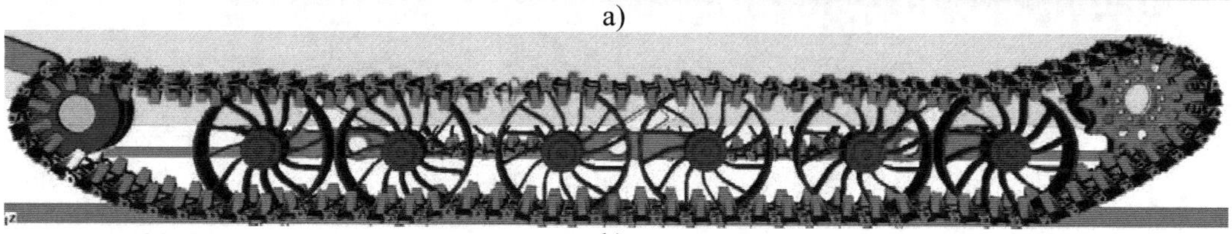

b)

Figure 1. Position of track links before a) and after b) an equilibrium test

During the test loads were determined which are applied to the starboard and portside support rollers of the LTV, from the side of the bearing surface with due consideration for a mass of the loaded vehicle, as well as the tracks which interact with the upper side of the support rollers at the beginning of modelling that leads to the track strike against the rollers. The analysis of the modelling processes demonstrates that it takes 3 s for the intensive transition processes in the LTV simulated model to pass and a model takes a state of equilibrium. The tests are interrupted when values become stable. During this test the highest load was noted at the beginning of the modeling process and was applied to the third support roller on the starboard and portside as the vehicle faces forward. The maximum peak load on a roller stood at 61 KN. By the end of the test when an LTV reached a state of complete equilibrium the total load applied to all portside support rollers was 138 KN, and 144 KN to all starboard support rollers.

The second test was a rectilinear movement on hard bearing surface. This test is among the main ones and allows for assessment of the LTV running gear dynamic. The LTV rectilinear movement is simulated in accordance with the given profile of the driving sprocket speed which is a relation between speed and time or passed way with due consideration for the preset profiles (unevenness) of the left and right track.

The tests to determine loads applied to the LTV support rollers were carried out in the following conditions (cases):

Case 1. LTV driving sprocket speed: V = 3 m/s. Surface profile characteristics: flat surface. Mass of transported assortment - 12 t.

Cases 2 and 3. LTV driving sprocket speed: V = 1 m/s; V = 2 m/s. Surface profile characteristics: single 100, 200, 300 and 400 mm obstacles located at an interval of 8 m. Assortment mass of 12 t.

Cases 4 5. LTV driving sprocket speed: V = 1 m/s; V = 2 m/s. Surface profile characteristics: sinusoidal uneveneses with the following parameters: A = 200 mm, L = 8 m, x0 = 2 m. Assortment mass - 12 т.

Cases 6 and 7. LTV driving sprocket speed: V = 1 m/s; V = 2 m/s. Surface profile characteristics: path profile unevenness for the starboard and portside tracks obtained during measurement of microprofile of the skid road in Murashinsly forestry of Kirovsk oblast [14]. Assortment mass of 12 t.

3. Results of Virtual Tests

The first case modeling results found that maximum loads on the support rollers were up to 40 KN during the LTV movement on a flat surface and were applied to the sixth starboard and portside rollers as the vehicle faces forward. Minor sudden variations of the support rollers' load were caused by interaction of the roller with a track which design includes an open hinge [15, 16].

In the second case the speed of the LTV driving sprockets was V = 1 m/s. The surface profile featured single 100, 200, 300 and 400 mm obstacles located at a distance of 8 meters. Assortment mass was 12 t.

During its operation the timber harvesting vehicle meets variable single obstacles like stumps, fallen trees, pits etc. that is why it is important to research load on the support rollers during movement over single obstacles. Conducted were experimental researches of the LTV movement over single obstacles to generate a respective loading mode [17 – 20].

The second case modeling results have found that during the LTV movement at a driving sprocket speed of V = 1 m/s, over single obstacles of variable height and located every 8 meters, the loads applied to individual support rollers were up to 48 KN with a vehicle moving over a 100-mm obstacle, up to 60 KN with a vehicle moving over a 200-mm obstacle, up to 123 KN with a vehicle moving over a 300-mm obstacle and up to 132 KN with a vehicle moving over a 400-mm obstacle. The maximum loads occurred in the second carriage and were applied both to the starboard and portside 3rd support rollers as the transport-load vehicle faces forward. Figure 2 shows a fragment case when a load-transport vehicle is overcoming single obstacles simulating a fallen tree.

Figure 2. A fragment case of an LTV simulation model overcoming single obstacles

In the third case during an LTV movement at a driving sprocket speed of V = 2 m/s, the loads applied to individual rollers were up 52 KN with a vehicle moving over a single 100-mm obstacle, up to 94 KN with a vehicle moving over a single 200-mm obstacle, up to 132 KN with a vehicle moving over a 300-mm single obstacle and up to 160 KN with a vehicle moving over a single 400-mm obstacle, however peak loads up to 232 KN were applied to the front support rollers of the first carriage which briefly occurred during the rollers' interaction with a 400-mm obstacle which is unlikely during movement at such a high speed. In a real-life operation an operator slows down a vehicle during movement over such a high obstacle.

Fourth case. The LTV driving sprocket speed was V = 1 m/s. Surface profile characteristics: sinusoidal unevenness with the following parameters: amplitude A = 200 mm, wave length L = 8 m,

IOP Conf. Series: Materials Science and Engineering 1086 (2021) 012008 doi:10.1088/1757-899X/1086/1/012008

phase displacement for the portside track x0 = 2 m. Assortment mass of 12 t. In this case the maximum loads applied to the support rollers reaches maximum values of 110 KN.

Fifth case. LTV driving sprocket speed V = 2 m/s. Surface profile characteristics: sinusoidal unevenness with the following parameters: A=200 mm, L=8 m, x0=2 m. Assortment mass of 12 t. The loads applied to the support rollers under such conditions reached 122 KN.

Sixth case. LTV driving sprocket speed V = 1 m/s. Surface profile characteristics: unevenness of the path profile for the portside and starboard tracks obtained during measurement of the microprofile [21 – 24]. Assortment mass of 12 t.

The virtual tests used data obtained during measurement of the skid road microprofile presented in the work [14]. Figure 3 shows fragment characteristics of the path unevenness for the portside and starboard tracks.

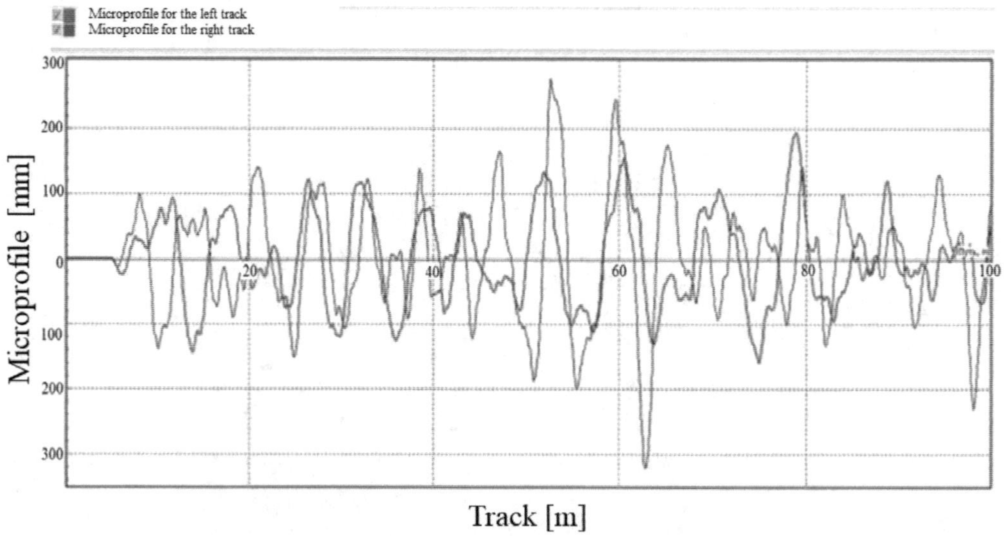

Figure 3. A fragment of the path unevenness for the portside and starboard track. Red for the portside track; blue for the starboard track

The sixth case modeling results have demonstrated that when an LTV moves along a microprofile, the maximum loads applied to the support rollers reach 106 KN.

The seventh case modeling results have found that the maximum loads applied to the support rollers during movement along a preset microprofile are single and reach 152 KN.

4. Conclusion

Two directions were employed in this research to justify experimental estimated data. The first one was based on that a real profile of a cutting area was composed of single unevennesses that is why the LTV undercarriage was designed by examining an LTV movement over single unevennesses. The second direction was based on experimental data and results of theoretical research demonstrating usefulness of using a path microprofile and harmonic shape path to design the LTV undercarriage.

It has been found that during the LTV steady movement along a periodic, close to harmonic path profile, the body angular oscillations are practically harmonic by their shape even in case of the undercarriage "breakthrough" causing acceleration of up to (4-5) g at the operator's workplace.

It has been found that the resonance characteristics of the LTV body oscillations are by their shape not corresponding to the nature of nonlinearity of characteristics of the elastic element in the undercarriage (track tension device).

It has been found that the LTV undercarriage may be considered as close to linear in practical estimates even during movement in the modes accompanied by hard strokes of the undercarriage beams against the restraints (frame).

The fundamental advantage of the conducted research of the LTV support rollers' loading is that it allows for a number of important practical general recommendations relevant not just for this LTV but the entire class of the vehicles featuring certain common properties.

It has been found that the 3rd and 6th support rollers of the vehicle undercarriage experience loads 1.5 times the loads applied to the rest rollers of the LTV undercarriage when the LTV is moving over the path microprofile unevenness and overcoming single 200, 300 and 400 mm obstacles.

References

[1] Klubnichkin E E, Klubnichkin V E 2015 Service life of track rollers of tracked forestry machines *Polythematic network electronic scientific journal of the Kuban State Agrarian University* **112** 1016-1026
[2] Klubnichkin VE, Klubnichkin EE, Kondratyuk DV, Beketov SA 2017 Program and methodology for experimental studies of a loading and transport machine in the applied software package *Forestry journal* **T 7 No. 4** (28) 175-182
[3] Klubnichkin E, Klubnichkin V 2017 Theoretical foundations of preliminary assessment of the impact of the timber harvesting vehicles on forest soils in the course of logging. *Proceedings of the 19th International & 14th European-African Regional Conference of the ISTVS*
[4] Klubnichkin V, Klubnichkin E, Kotiev G 2017 Developing a method of theoretical evaluation of the tracked timber harvesting machine undercarriage element loading. *Proceedings of the 19th International & 14th European-African Regional Conference of the ISTVS*
[5] Cambi M, Certini G, Neri F, Marchi E 2015 The impact of heavy traffic on forest soils. *Forest Ecology and Management* **338** 124-138
[6] Horn R, Vossbrink J, Peth S, Becker S 2007 Impact of modern forest vehicles on soil physical properties. *Forest Ecology and Management* **248** (1–2) 56-63
[7] Kotiev G, Padalkin B, Kartashov A, Dyakov A 2017 Designs and development of Russian scientific schools in the field of cross-country ground vehicles building. *ARPN Journal of Engineering and Applied Sciences* **12** (4) 1064-1071
[8] Rakheja S, Afonso M, Sankar S 1992 Dynamic analysis of tracked vehicles with trailing arm suspension and assessment of ride vibrations *International J. of Vehicle Design* **13** (1)
[9] Bekker M G 1969 *Introduction to Terrain-Vehicle Systems* (Ann Arbor: The University of Michigan Press)
[10] Pogorelov D J 2005 Computer simulation of the dynamics of technical systems using the software package "Universal mechanism". *J. of Computer and Information Technologies* **4** 27-34
[11] Bodin A 1999 Development of a tracked vehicle to study the influence of vehicle parameters on tractive performance in soft terrain *J. Terramechanics* **36** (3) 167-181
[12] Dhir A, Sankar S 1994 Analytical track models for ride dynamic simulation of tracked vehicles *J. Terramechanics* **31** (2) 107-138
[13] Dmitriev A A, Savochkin V A 1993 *Statistical dynamics of transport and traction tracked vehicles* (Moscow Mashinostroenie)
[14] Klubnichkin V E 2012 Improvement of design models of loading of transmissions of tracked logging machines depending on external traffic conditions *diss. Cand. tech. Sci.* Moscow State Forest University
[15] Goberman L A, Goberman V A, Ksenevich I P 2003 *Ground trailer transport systems Technical and economic bases of designing machines and processes Methodological aspects of project activities and decision making* 3 Moscow Encyclopedia
[16] Kochnev A M, Anisimov G M 2008 *Tests of harvesting machines* St Petersburg State Forest Technical University

[17] Said Al-Milli, Lakmal D, Kaspar Althoefer 2010 Track–terrain modelling and traversability prediction for tracked vehicles on soft terrain *J. Terramechanics* **47** (3) 151-160
[18] Wong J Y 2001 *Theory of Ground Vehicles* 3 (NY: John Wiley & Sons) ISBN 0-471-35461-9
[19] Wong J Y 2009 *Terramechanics and Off-Road Vehicle Engineering* 2 488
[20] Klubnichkin V E, Klubnichkin E E, Kotiev G O, Beketov S A, Makarov V S 2018 Interaction between elements of the track ground contacting area with the soil at curvilinear motion of the timber harvesting machine *IOP Conference Series: Materials Science and Engineering* **386** 012016
[21] Kotiev G, Sarach E, Beketov S 2018 Methods for road microprofile statistical data transformation *MATEC Web of Conferences* **224** 02096
[22] Klubnichkin E E, Klubnichkin V E, Kotiev G O 2018 Theoretical research of soil packing by timber harvester running gear *IOP Conference Series: Materials Science and Engineering* **386** 012025
[23] Alyabiev A F, Kotov A A, Klubnichkin V E, Klubnichkin E E, Bykovskiy M A 2019 Experimental determination of soil shear force by track model depending on track surface to soil friction coefficient *IOP Conf. Series: Journal of Physics: Conf. Series* **1177** 012033
[24] Klubnichkin V et al. 2019 Researching rolling resistance of the wheeled forestry tractor at skidding *IOP Conference Series: Materials Science and Engineering* **695(1)** 012003 doi: 10.1088/1757-899X/695/1/012003
[25] Makarov V S et al. 2019 Study of uneven surfaces distribution on forestry roads *Journal of Physics: Conference Series* **1177(1)** 012041 doi: 10.1088/1742-6596/1177/1/012041

IOP Conf. Series: Materials Science and Engineering 1086 (2021) 012009 doi:10.1088/1757-899X/1086/1/012009

Comparative study of impression and cutting into ice by helical blade

U Sh Vakhidov[1], I A Yerasov[1], Y I Molev[1,3] and D S Mokerov[2]

[1]Nizhny Novgorod State Technical University n.a. R.E. Alekseev, Minin st., 24, 603950, Nizhny Novgorod, Russia.
[2]Nizhny Novgrod State Pedagogical University n.a. K. Minin, Uyanova str., 1, 603950, Nizhny Novgorod, Russia

E-mail: [3] moleff@yandex.ru

Abstract The paper presents the results of energy efficiency studies of helical blade penetration types into ice. There are two principally different ways of helical groove generation in ice – by cutting and impression. It was found out that at a longer duration of the penetration process of the helical blade with a smaller lead angle the energy demand of the propulsion gets lower due to impression. At blade penetration by cutting, there is an extremum at the angle of cutting of 40-45^0, however, the energy demand in such condition is still higher than that of the penetration by impression. Considerable attention is paid in the paper to the measurement of the noise generation during the interaction with the penetrated surface which allows for prompt assessment of the energy efficiency of the considered parameter. The obtained results allow for efficiency improvement of the operation of vehicles with rotary screw propulsion units in real environments, as well as for their increased reliability due to new diagnostics methods.

1. Introduction

One of the most promising development ways of the Russian Federation is currently associated with the exploration of the Arctics which is scarcely possible without construction of appropriate infrastructure, transportation, communications. Thereby, ensuring safety of passengers and cargo transportation at low temperatures, as well as of ensuring the environmental safety of interventions into a vulnerable ecological system is a vital scientific mission. The simplest way to solve the described problem were to use water basin surfaces (also frozen), since this is the only method with the smallest possible environmental impacts produced by transportation vehicles. However, the development of the transportation system in this direction is impeded by problems with ensuring the safety of this process associated with machinery operation at low temperatures, dynamically changing ice conditions (also, occurrence of rain channels where vehicles could fall into), wind loads, icing, prolonged dark periods, remote location of the closest settlements. It means that consequences of emergencies in the Arctics can rapidly take a catastrophic scale. The present paper considers operation of vehicles with rotary screw propulsion units, the distinctive feature of which is the traction generation on ice exceeding the weight of the vehicle. This feature allows both to reduce the tractor weight and to drive it onto the ice out of an ice opening. Thereby, it should be noted that a wide use of such vehicles is impeded due to the high energy demand of their operation. To solve this problem is the objective hereof.

2. Overview of previous studies

The beginning of mobility studies of vehicles with rotary screw propulsion units in severe conditions dates back to the studies of such scientists as: B. Cole [1], A. Soltynski [2], M. Becker [3]. As the last studies of this matter we could consider papers by J. Wong [4]. It should be noted that the principal matter of interest of the above-listed was the locomotion of the subject vehicles on water basins, swamps and other highly humidified terrain. The development of the transportation infrastructure in the Arctic zone has lead to a growth of the number of studies dedicated to the interaction of the rotary screw propulsion unit with the ice and the snow. The development of this scientific research direction has been directly contributed by the research work of Nizhny Novgorod school of all-terrain vehicles [5-17]. The scientists of our school have defined and substantiated areas of us of rotary screw propulsion vehicles; a theory of steady locomotion was developed and systematized; equations were found for all forces acting on a rotary screw propulsion vehicle, also during driving on snow, etc. A thorough analysis of scientific studies associated with rotary screw propulsion units has demonstrated that recommendation on selection of rational geometry parameters of a rotary screw propulsion unit was made based on results of calculations with 4 to 6 considered propulsion unit parameters.

3. Methods

For generation of traction during rotary screw vehicle ride on ice, it is important that the helical blade penetrates into it. Thereby, the penetration depth of the helical blade is selected based on the necessity of generation of the required traction determined by the contact sensor area of the blade with the ice and the ice strength [18]. The required cross-section area of the penetration part of the blade is determined by such geometry parameters as OD of the propulsion element, contact angle of the propulsion element with ice and the lead angle of the helical blade (see figure 1).

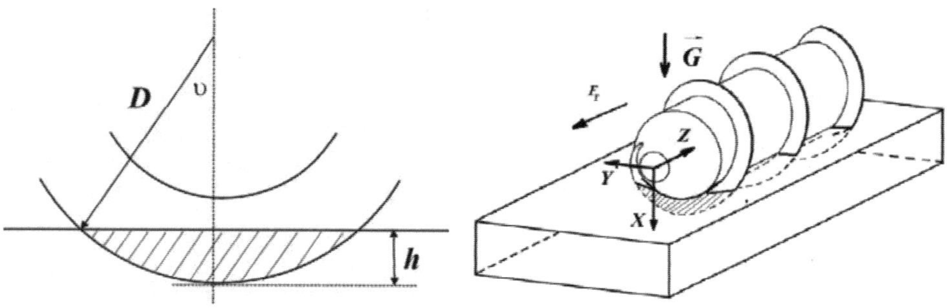

Figure 1. Diagram of geometry parameters of the helical blade penetration into ice

Setting the minimum required traction for vehicle propulsion F_T, as well as considering that this force should be directed longitudinally the following dependencies are obtained:

$$\frac{F_T}{\cos\alpha} \geq kG \text{ and } F_T \leq HB\frac{S}{\cos\alpha}$$

Whence it appears:

$$S \geq \frac{F_T}{HB}$$

on the other hand, the contact area of the helical blade is determined as double difference of the sector area with dia. D and υ angle and the area of a right triangle with sides D and D-h:

$$S_1 = (0{,}5D)^2\upsilon,\ S_2 = (0{,}5D - h_B)^2\sin\upsilon,\ \upsilon = \arccos\frac{0{,}5D - h_B}{0{,}5D}$$

$$S = S_1 - S_2 = (0{,}5D)^2\upsilon - (0{,}5D - h_B)^2\sin\upsilon$$

Considering that for small angles the equation $\upsilon = \sin\upsilon$ is valid, we obtain:

$$F_{\text{Т}} \leq h_{\text{В}}\{\text{D-} \text{h}_{\text{В}}\} \arccos\left(\frac{0{,}5D - h_{\text{В}}}{0{,}5D}\right) HB$$

presenting the height of the helical blade as part of the basic cylinder diameter - $h_{\text{В}} = kD$ we obtain:

$$F_{\text{Т}} \leq kD^2\{1 - k\} \arccos(1 - 2k) HB$$

The solution of the obtained dependency is shown in figure 2.

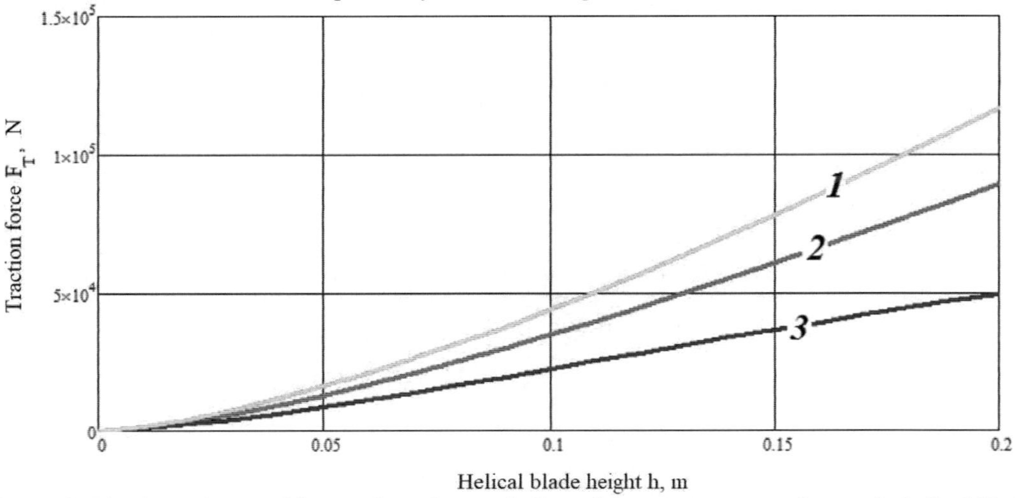

Figure 2. The dependence of the traction of one winding of the rotor screw on ice on the helical blade height; 1 - for basic cylinder dia. 0.5 m; 2 - for basic cylinder dia. 1.0 m; 3 - for basic cylinder dia. 1.5 m

For the time being, the profile of the helical blade is usually welded of two inclined flank plates. Fir such structure, the maximum bending moment can be calculated based on the following equation:

$$F_{\text{Т}} h_{\text{В}} = \sigma_{\text{И}} W_{\text{X}},$$

where W_{X} is the secant modulus relative to the central axis. For helical blade with triangular profile with height $h_{\text{В}}$ and height b:

$$W_{\text{X}} = \frac{bh^2}{24} \text{ and } F_{\text{Т}} = \frac{S\sigma_{\text{И}}}{48} \quad \text{b} = \frac{96F_{\text{Т}}}{\sigma_{\text{И}} h}$$

Substituting $F_{\text{Т}}$ values from the equation, we obtain:

$$b \geq \frac{96\{D\text{-}h_{\text{В}}\} \arccos\left(\frac{0{,}5D - h_{\text{В}}}{0{,}5D}\right) HB}{\sigma_{\text{И}}}$$

The obtained results unambiguously determine the geometry parameters of the helical blade allowing for unambiguous determination of the energy parameters of its penetration into the ice. If a flat helical blade is introduced as shown in fig. 3, the energy demand of the penetration will be determined as product of ice strength by single-axis compression of the cross-section of the helical blade and its penetration velocity.

The deformation velocity of the ice is one of the most important specifications determining both the pattern of its deformation and its strength. At low deformation velocities, the ice behaves as plastic medium with low yield strength. The higher the velocity, the higher the yield strength, finally, at a relatively high deformation velocity, the ice begins to behave as brittle-elastic medium. It is considered that the border between the zones of plastic and brittle destruction lies around $\zeta = 10^{-3}$ s^{-1}. This border corresponds to the maximum strength [18]. At 0.2 m blade width in the widest area, the maximum ice strength will be observed at a penetration rate of 2*10^{-4} m/s, which is much less than real penetration rates of the helical blade. At a penetration rate above 2*10^{-2} m/s, the influence of the penetration rate on the compression resistance of the ice is absent [17]. Then, the penetration power of the helical blade into the ice can be determined as follows:

$$S = \frac{48F_{\text{T}}}{\sigma_{И}} \rightarrow N_{\text{B}} = S \cdot HB \cdot V_{\text{B}} = S \cdot HB \cdot V_{\text{X}} / \sin\alpha = S \cdot HB \cdot V_{\text{X}} \cdot (1+\cos\lambda)/2\sin\alpha,$$

where V_{B} is the penetration rate of the helical blade, correlating with the travel speed of the vehicle in accordance with the dependence $V_{\text{B}}=V_{\text{X}}/\sin\alpha$, where $\alpha$is the lead angle of the helical blade. Angle λ is the penetration angle of the helical blade into the ice, as shown in figure 3. At a penetration angle λ above 90^0 the penetration of the helical blade into the ice occurs by impression, at smaller penetration angles cutting prevails.

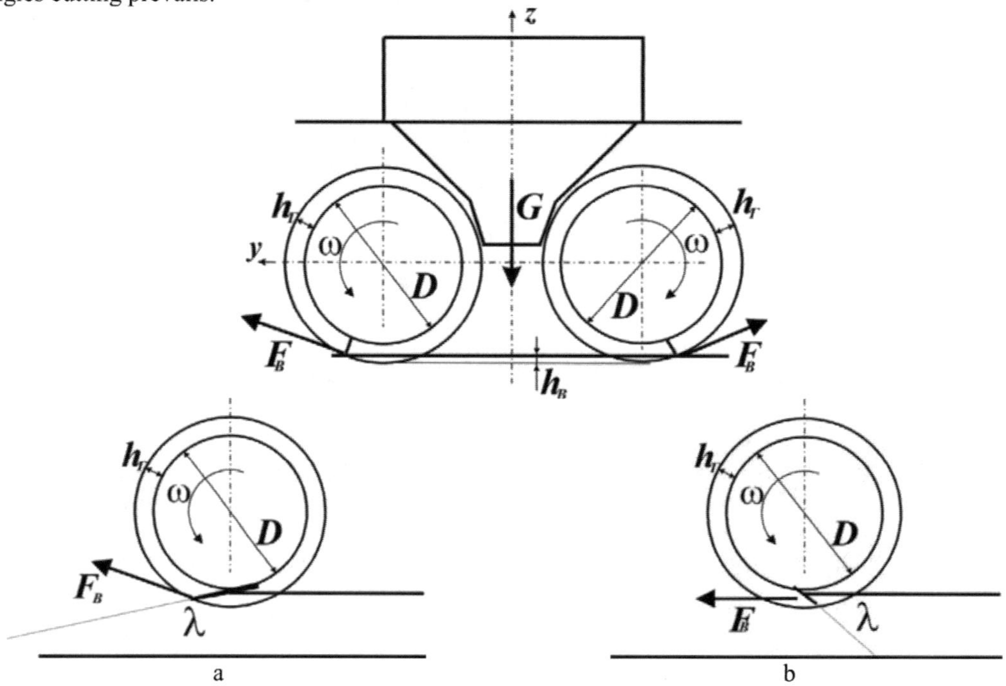

Figure 3. Ice penetration diagram of the helical blade; a – by impression; b – by cutting

The cutting force are used to be found by formulas suggested by V.F. Kulepov [19-21]:

$$F_{\text{B}} = \frac{P_{Э}}{1+kt_0} * h^{\Omega}[1+\Gamma(b-1)](1+kt)(1+\xi V)[1+\varsigma(u-40^o)^2]\gamma),$$

where $P_Э$ is the reference cutting force obtained during cutting of an ice master of 0.01 m thickness, 0.01 m width, at 0^{0}C ice temperature, 90 deg. cutting angle, and 0 m/s cutting velocity; k is the coefficient accounting for changing of strength properties of the ice versus temperature, equal to 0.04 N/deg.; $\xi$is the coefficient accounting for changing of strength properties of the ice at different load accommodation rate equal to 9.87 Ns/m; Ω is the coefficient accounting for changing of the cutting force vs. cutting thickness, equal to 0.47; Γ is the coefficient accounting for changing of the cutting force vs. cutter width, equal to 0.865; ζ is the coefficient accounting for changing of the cutting force vs. cutting angle, equal to 3.2*10^{-3} (deg.$^{-1}$); γ is the coefficient of blocked cutting assumed equal to 1.5 for "angular in slot" cutting condition; t_{B} is the width of the helical blade; V is the cutting velocity determined by the distance of the cutting point from the rotation axis and rotation speed of the rotor: $V=(h_{\Gamma}+D/2)\omega$, where D is the basic cylinder dia.; ω is the angular rotation speed of the rotor.

Then, the power absorbed for the helical furrow cutting can be calculated from the following equation:

$$N_{\text{B}}=\frac{VP_0}{1+kt_0}*h^{\Omega}\left[1+\Gamma(t_{\text{B}}-1)\right](1+kt)(1+\xi V)\left[1+\varsigma(u-40^o)^2\right]\gamma),$$

The influence of the shape of the helical blade can be determined by substitution of h_B and V_P values in the obtained dependencies as mathematical functions of traction. Then, the energy absorption by the penetration of the helical blade into ice versus the inclination angle of the helical blade edge penetration into ice will appear like the dependency illustrated in figure 4.

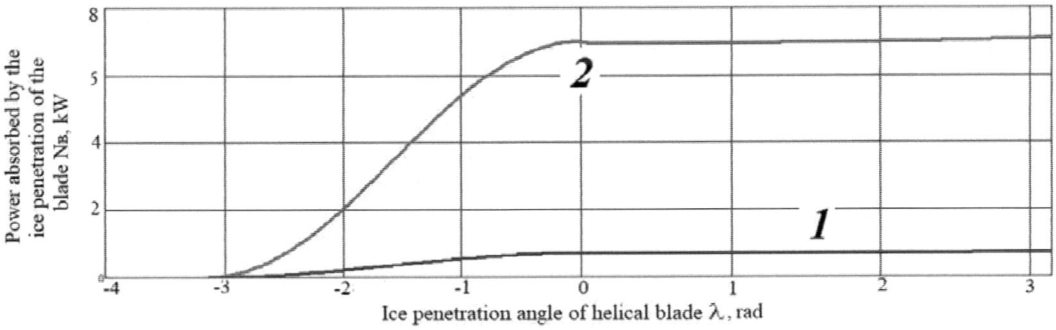

Figure 4. Energy absorption of the ice penetration of a helical blade of 0.1 m height at 0^0C ice temperature (with 0.3 MPa one-axis compression strength) versus front edge inclination angle; 1 – for 1 m/s penetration velocity; 2 – for 10 m/s penetration velocity

4. Conclusions

The obtained results unambiguously demonstrate that the minimum power absorption during the locomotion of a rotary-screw propelled vehicle on ice is provided by the ice penetration of the helical blade by impression. Thereby, the less is the inclination angle of the front edge of the helical blade the less is the power absorption of the penetration, however, the time required for the blade penetration is thereby longer. That is, the energy demand of the helical blade penetration by impression with kW*h dimension remains the same. On the other hand, since the energy demand of the ice penetration of the helical blade by cutting remains almost unchanged, the growth of the inclination angle of the front edge of the helical blade will be followed by a longer cutting process and its increased energy absorption.

References

[1] Cole B N 1961 Inquiry into amphibious screw traction *University of Birmingham* **175** p 19
[2] Soltynski A 1963 Ocena "Pizektadni glebowej" modelowego pojaz du tere-nowego *Technika motoryzacyjna* **13 - 10** pp 321-329
[3] Bekker M G 1986 Parametric analyses of tracks and tracklayers update-sample of engineering problems and their solutions in off road locomotion *SAE Technical Papers*
[4] Wong J Y, Jayakumar P, Toma E, Preston-Thomas J 2020 A review of mobility metrics for next generation vehicle mobility models *Journal of Terramechanics* **87** pp 11-20
[5] Kulyashov A P and Kuklina I G 2011 Estimation of the Rotor Screw Machine Vibration Depending on Suspension Parameters and the Mover Line Contact with the Ground *Journal Road and Construction Vehicles* **7** p 45
[6] Krasheninnikov M et al. 2013 The concept and methodology of creating the universal life-saver with rotary-screw mover *Lecture Notes in Electrical Engineering* **195 - 7**

[7] Molev Y et al. 2019 Sound power spectra modeling of the vehicle in motion equipped with rotary-screw propulsion unit *IOP Conf. Series: Journal of Physics: Conf. Series* **1177** 012034
[8] Belyakov V et al. 2017 Ground vehicle for ice conditions monitoring *Thirteenth International MEDCOAST Congress on Coastal and Marine Sciences, Engineering, Management and Conservation (MEDCOAST 2017)* pp 775-785
[9] Strizhak A et al. 2019 Modelling of vehicles with rotary-screw propulsion unit along water-flooded substructure *Journal of Physics: Conference Series* **1177** 012039
[10] Abramova E et al. 2018 The simulations of helical blade interaction with ice *MATEC Web of Conferences* **245**
[11] Vahidov U et al. 2020 Case study: Regulation of noise produced by a rotary-screw propulsion unit in an all-terrain vehicle *6th International Conference on Vehicle Technology and Intelligent Transport Systems (VEHITS 2020)* pp 548-551
[12] Kurkin A et al. 2015 Autonomous Robotic System for Coastal Monitoring *Twelfth international conference on the Mediterranean coastal environment (MEDCOAST 15)* **1 and 2** pp 933-943
[13] Kurkin A et al. 2017 Unmanned Ground Vehicles for Coastal Monitoring *International Journal of Imaging and Robotics* **17-1** pp 64-75
[14] Tyugin, D Yu 2018 Development of a mobile robot group for coastal monitoring *IOP Conference Series: Materials Science and Engineering* **386** 012009 doi:10.1088/1757-899X/386/1/012009
[15] Zaytsev A et al. 2017 Coastal monitoring of the Okhotsk sea using an autonomous mobile robot *Science of Tsunami Hazards* **36-1** pp 1-12
[16] Zeziulin D et al. 2017 Development of a ground mobile robot for motion in conditions of coastal zones *19th International and 14th European-African Regional Conference of the ISTVS*
[17] Erasov I A et al. 2018 Determining friction coefficient between the base cylinder of rotary screw propulsion unit and various ice types in different daylight surface *IOP Conference Series: Materials Science and Engineering* **386** 012006 doi:10.1088/1757-899X/386/1/012006
[18] Stepanyuk I A 2001 *Testing and modeling technique of sea ice* St. Petersburg: Hydrometeoizdat p 78
[19] Kulepov V 2002 Development of a parametric series of disk-mill ice cutters *Journal Road and Construction Vehicles* **11** pp 16-24
[20] Kulepov V 2002 Determining the loads imposed on the attachments of ice-cutting machines to cut slots in the ice cover *Journal Road and Construction Vehicles* **10** pp 42-48
[21] Kulepov V F 2002 Development and creation of ice-cutting machine of new generation *Mekhanizatsiya Stroitel'stva* **9** pp 20-22

Conceptual model of the automated pedestrian simulation system for unmanned vehicles and ADAS systems testing

A A Tumanov, A V Zolotov, L O Fedosova[1] and A Y Sizov

Nizhny Novgorod State Technical University n.a. R.E. Alekseev, Nizhny Novgorod, Russian Federation

E-mail: [1] fedosovaludmila@list.ru

Abstract. This article is devoted to the issue of creating the concept of an automated system for simulating pedestrians on the road surface, intended for testing unmanned vehicles and active driver assistance systems. The article examines in detail the prerequisites for the creation of such a system and its relevance for Russia. The existing analogues of this system are demonstrated, their strengths and weaknesses are analyzed. A conceptual model of the system is proposed, its functional capabilities are described, the design of the main units and components of the system is presented. Two possible options for the operation of this system during testing are considered: at regulated and unregulated pedestrian crossings. Variants of test methods in accordance with the named options are proposed.

1. Introduction
The modern development of unmanned vehicles is carried out in two main areas:

- introduction and expansion of the functionality of various driver assistance systems (which are currently standardly equipped with cars of all classes);
- creation of methods and control systems for fully BPS (which are currently at the stage of testing prototypes, including operational ones).

A modern car contains many electronic driver assistance systems with varying degrees of automation of the driving process. They include systems for cruise control, directional stability, collision avoidance, parking distance control, preventive preparation for an accident, etc. These systems take over part of the car control functions, for example, the functions of automatic speed control, acceleration, braking, taxiing, engine and transmission operating modes. Due to this, in some conditions (for example, when driving on a motorway), the can move autonomously, and the driver can not interfere with the driving process. Further improvement and expansion of the functions of driver assistance systems provides an increase in the autonomy of the car, bringing it closer to a completely unmanned vehicle [1].

In 2014, the Society of Automotive Engineers (SAE) decided to standardize the requirements for these systems and dividing them into 6 levels according to the degree of their automation [2]. Currently, on public roads in Russia, you can find cars not higher than 3 levels of automation, which are not yet autopilot, but the system is already equipped with many different software and hardware that control the car under the driver's control. At the moment, Russia is already testing level 4 unmanned vehicles on public roads with a driver inside. So, in 2019, 35 auto-pilot cars of the joint work of Mobis and Yandex Hyundai Sonata engineers have already appeared on the roads of Moscow, and in March 2020, their number has already increased to 100 cars.

IOP Conf. Series: Materials Science and Engineering 1086 (2021) 012010 doi:10.1088/1757-899X/1086/1/012010

For the active implementation of driver assistance systems, especially in the context of using them as control subsystems for unmanned vehicles, it is advisable to develop methods for testing and testing them to obtain reliable information about their technical characteristics. One of the active driver assistance systems is the pedestrian detection system, testing of which is an urgent task, since it is directly related to the safety of all road users [3].

2. System for detecting pedestrians on the road surface

The people safety on the roads is one of the most important aspects of road traffic. Car manufacturers are increasingly looking to protect everyone involved. An important aspect here is the prevention of collisions between cars and pedestrians and cyclists.

For the first time such systems were used on Volvo cars in 2010. The system currently has various variations such as Subaru's EyeSight, Volvo's Pedestrian Detection System (PDS), TRW's Advanced Pedestrian Detection System (APDS). Let's take a closer look at the PDS system presented on Volvo cars. The PDS Pedestrian Detection System is used to reduce the risk of a collision between a vehicle and a pedestrian. Pedestrian detection near the vehicle is its main function. In this case, the system automatically reduces the speed, which reduces the impact force in the event of an imminent collision. The system performs such functions as: detection of people in the direction of the vehicle; warning the driver about the risk of collision; automatic reduction of vehicle speed. After detecting a pedestrian on the radars of a car, the system automatically calculates what the position of a person will be at the moment of close proximity to the vehicle. Then a signal is sent to the driver on the car display. If the system determines that there is a risk of collision, an audible alarm sounds. When braking, the PDS system enhances its action by applying emergency braking. If the driver does not respond to the signal, the system is triggered automatically and stops the car by itself. At the moment, the PDS system guarantees complete traffic safety at a speed not exceeding 35 kilometers per hour. At higher speeds, the system reduces the force of the impact by reducing the vehicle speed. But there is a significant disadvantage, it lies in the fact that recognition of people at night or in conditions of poor visibility is difficult [4].

3. EuroNCAP tests for detection of pedestrians on the road surface

The EuroNCAP (The European New Car Assessment Program) is a European committee for conducting independent car crash tests with an assessment of active safety and passive safety. NCAP is paying close attention to tests involving pedestrians. This is closely related to developments in the field of unmanned vehicles, because most pedestrian tests are done to test a smart braking system that is unattended. To detect pedestrians, Euro NCAP tests different types of scenarios:

- a pedestrian or cyclist crosses the path of the test vehicle (figure 1);
- a pedestrian or cyclist walks / drives in the same direction as the vehicle (figure 2);
- a pedestrian or cyclist crosses the road onto which the test car is turning (figure 3);
- a pedestrian or cyclist is behind and the car is going in reverse (figure 4);

The first two scenarios are selectively repeated in low light conditions as well. For trials in which a pedestrian or cyclist walks or drives in the same direction as the vehicle, there are also several scenarios: a pedestrian walks in the center of the road or is displaced to one side.

There are 3 kinds of scenarios with a pedestrian crossing the path tested by a vehicle (figure 5,6):

• an adult running from the driver's side of the vehicle (a pedestrian runs at a speed of 8 km / h, the speed of an approaching car is 20-60 kmph);

• an adult walking from the passenger side (a pedestrian runs at a speed of 5 kmph, and the speed of an approaching car is 20-60 kmph);

• a child runs in front of cars on the passenger side of the car (a pedestrian child runs at a speed of kmph, the speed of an approaching car is 20-60 kmph);

For tests in which a pedestrian or a cyclist crosses the road onto which the car is turning, two types of tests are used: the car turns to the right or to the left. These two versions of the same scenario present

IOP Conf. Series: Materials Science and Engineering 1086 (2021) 012010 doi:10.1088/1757-899X/1086/1/012010

different problems due to the relative angles and distances between the test vehicle and the pedestrian [5].

The tests use a specially designed pedestrian target on a platform that has articulated limbs to simulate a person's walking. According to the specification, male and 6-year-old mannequins are used for the tests. Pedestrian mannequins meet such requirements as: the presence of an installed platform; directional following the path; the ability to reach speeds up to 15 kmph; impact resistance up to 60 kmph with minimal vehicle damage; reconstructability for retesting.

Figure 1. A pedestrian crosses the path of a test vehicle

Figure 2. A pedestrian walks / drives in the same direction as the vehicle

Figure 3. Pedestrian crossing a road into which a car is turning

Figure 4. Pedestrian crossing a road into which a car is turning

Figure 5. Test with an adult walking on the passenger side

Figure 6. Test with a child running in front of cars

IOP Conf. Series: Materials Science and Engineering 1086 (2021) 012010 doi:10.1088/1757-899X/1086/1/012010

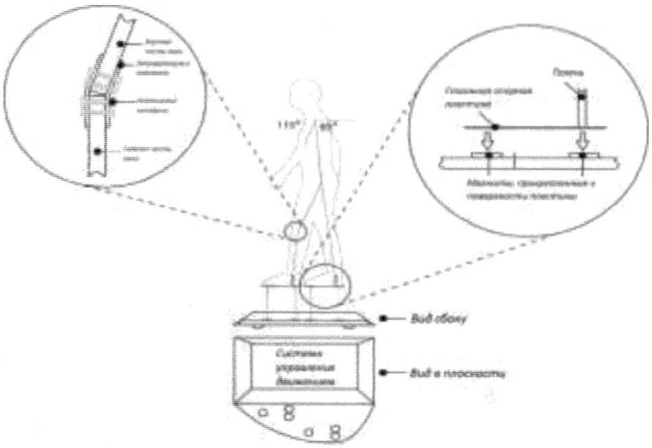

Figure 7. Pedestrian model used by EuroNCAP

The crash test dummy (figure 7) has a controlled platform with programmable speed and trajectory following capability. The mannequin has a modular assembly system, so it has a fast recovery time. The dummy rail is detached from the platform on impact, and the knee joints are designed to adjust its position. They flex on impact to minimize damage. The pedestrian platform holds an impact-resistant, lightweight pedestrian dummy that attaches with magnets. The dummy has high acceleration typical of real pedestrians and can reach speeds of up to 15 kmph. The mobile platform (figure 8,9) has dimensions of 840x1115x90 mm, a reinforced drive structure, and minimal radar and visual signals.

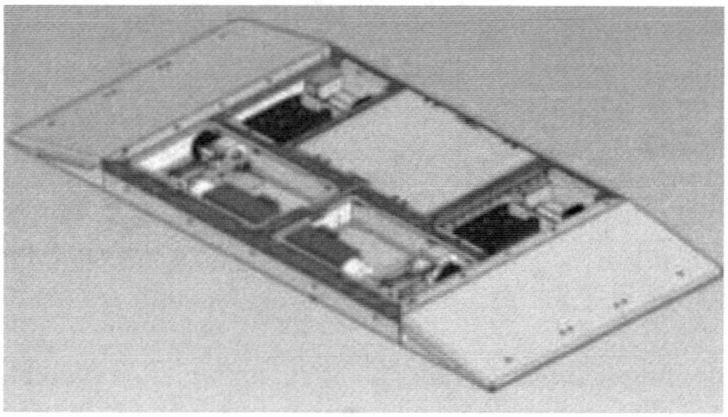

Figure 8. 3D model of a movable platform

Figure 9. Movable platform Euro NCAP

At the moment in Russia there is no certified domestic equipment for conducting such tests, and there are only prototypes.

4. Development of the concept of an automated system for imitating pedestrians on the road surface

When developing the concept of an automated pedestrian imitation system (the APIS), it is necessary to consider the requirements:

- compliance of the system with the requirements of the ISO 19206-2 standard [6,7];
- the height of the movable platform should be no more than 70 mm, so that in an emergency the platform would pass freely under the vehicle;
- the mass of the movable platform should not exceed 30 kg;
- in a frontal collision with a car, the dummy should be disconnected from the cart in order to reduce damage to the dummy and the vehicle upon impact;
- a pedestrian dummy must be equipped with heating elements to simulate infrared radiation of the human body;

- movement of the dummy is carried out in both directions;
- pedestrian speed must be up to 10 km / h with the possibility of regulation;
- movement of the mannequin should be carried out from left to right and from right to left;
- for the system to work in autonomous mode, it is necessary to provide for the detection of a car in the area of a pedestrian crossing;
- the system should exclude a collision of a pedestrian with a vehicle moving or stopped at a pedestrian crossing.

At the initial stage of the development of APIS, the two most typical scenarios of the behavior of a pedestrian on the carriageway are considered: a pedestrian crossing at a regulated pedestrian crossing (figure 10) and a pedestrian crossing at an unregulated pedestrian crossing (figure 11).

Based on the formulated requirements, the APIS concept was developed, which implies the presence of the following components: a movable platform with a dummy; dummy rotation system; system for detaching the mannequin from the cart; heating system for dummy parts; UMV detection unit. Each component of the system is discussed in more detail below.

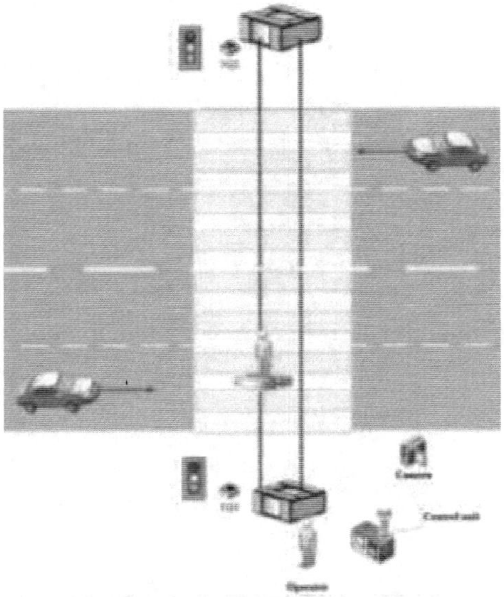

Figure 10. Scenario of a pedestrian crossing a regulated pedestrian crossing

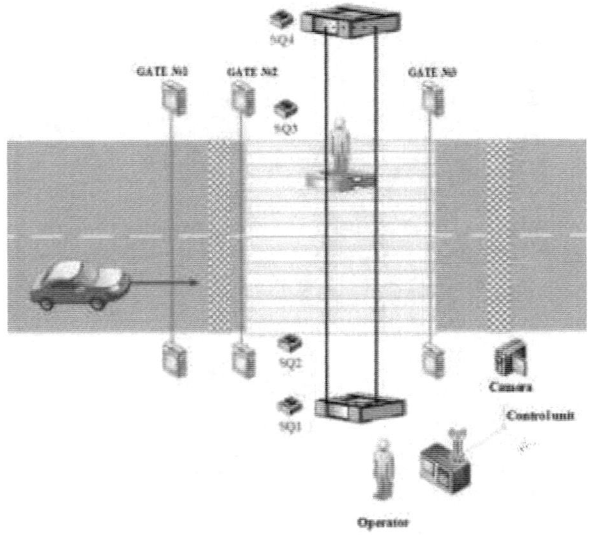

Figure 11. Scenario of a pedestrian crossing an unregulated pedestrian crossing

5. Movable platform with a mannequin

A movable platform with a dummy (figure 12) includes: a movable cart (1), a drive unit (2), a cable tension unit (3), a dummy (4), a dummy rotation system (5), a heating system for dummy parts (6) , a system for detaching the mannequin from the trolley (7).

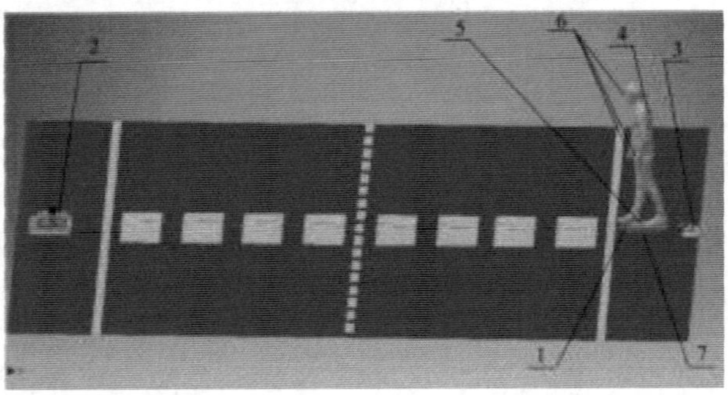

Figure 12. 3D model of a platform with a mannequin

Movable trolley (figures 13,14,15) is a mobile platform on casters designed to install mannequins on it and simulate the movement of a mannequin on a pedestrian crossing. The bogie frame is flat, it consists of 4 spars and 2 cross beams. The connection between the frame elements is made by means of fasteners for the structural profile. Trolley suspension is not provided. The bogie frame material was a 40x20 mm structural aluminum profile. Polyurethane casters were chosen as wheels for the trolley. The wheel is mounted on a smooth shaft on two bearings; positioning on the shaft is carried out by means of the presence of bushings. The shaft itself is fixed in supports, which are installed on the side members, and the supports are installed on a threaded connection. Also, a dummy rotation mechanism is mounted on the trolley, thanks to which, when it approaches the end of the transition, the dummy has the ability to turn.

The tension of the cable (figure 18, 19) is carried out by means of the tension knot. It is a spatial frame mounted on a metal sheet, inside which there are two rolling rollers fixed in bearing assemblies. The rollers are stacked discs mounted on shafts. The top of the frame is covered with a 3 mm sheet steel cover. The tensioning unit is mounted on a support sheet, which is reinforced with concrete blocks for immobility. From above, the tension unit is covered with a PVC plastic cap covered with a decorative film. All sheet metal parts are covered with moisture resistant enamel.

Figure 13. 3D-model of the trolley (bottom view)

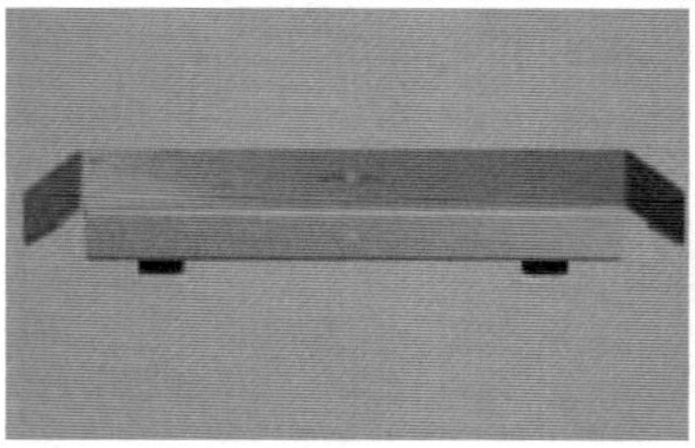

Figure 14. 3D-model of the trolley (side view)

Figure 15. Movable trolley

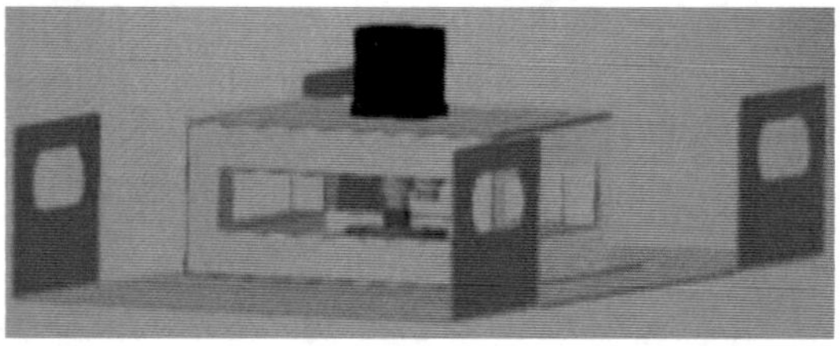

Figure 16. 3D model of the drive unit

Figure 17. Drive unit

The tension of the cable (figure 18, 19) is carried out by means of the tension knot. It is a spatial frame mounted on a metal sheet, inside which there are two rolling rollers fixed in bearing assemblies. The rollers are stacked discs mounted on shafts. The top of the frame is covered with a 3 mm sheet steel cover. The tensioning unit is mounted on a support sheet, which is reinforced with concrete blocks for immobility. From above, the tension unit is covered with a PVC plastic cap covered with a decorative film. All sheet metal parts are covered with moisture resistant enamel.

Figure 18. 3D- model of cable tension unit

Figure 19. cable tension unit

The adult mannequin (figure 20) is a foam rubber manikin on a steel wire frame. The child mannequin (figure 21) is a fabric-frame mannequin with the ability to dismantle the arms and legs.

Figure 20. Adult pedestrian mannequin

Figure 21. Baby pedestrian mannequin

6. Manikin turning system

The turning system (figure 22, 23) is one of the main components of the system, thanks to it, the manikin turns at the end points of the transition when starting the reverse movement "facing" in the direction of travel. To solve this problem, a dummy rotation system was developed. The development was complicated by the fact that the trolley itself is not driven, as a result of which no power sources and batteries are installed on the trolley, which would help equip the swing system with an individual drive. Therefore, the energy of the main traction drive installed in the drive unit is used to rotate the bogie. The principle of operation is that the cable is attached not to the mobile cart itself, but to the turning mechanism, and when the movement starts at one of the end points of the route, the mechanism is activated due to the high inertia of the cart with a dummy, due to which the turn is carried out. The dummy turning system includes: a toothed rack on a slider sliding along the groove of the bogie spar; two bearing assemblies to ensure low friction when turning the platform; bottom disc with toothed wheel; the upper disc, on which the swivel plate is mounted, on which the mannequin is mounted.

Figure 22. 3D-model of the turning system

Figure 23. Manikin turning system

IOP Conf. Series: Materials Science and Engineering 1086 (2021) 012010 doi:10.1088/1757-899X/1086/1/012010

7. System for detaching the mannequin from the trolley

In the process of testing the BPTS, situations may occur with the knocking down of the dummy by a vehicle. In this case, if the dummy and the mobile trolley are a single rigid structure and do not have the ability to disconnect, in the event of a frontal collision of the UMV with the dummy, an emergency situation may arise with damage to all the components of the APIS, as well as the UMV itself. To avoid such consequences, a system was developed that provides for the detachment of the dummy from the cart in case of a frontal collision with the UMV.

The system includes a reinforced dummy frame and a magnetic holder for mounting the dummy on the turntable. The reinforced frame is a structure of a profiled aluminum corner, designed to provide additional support for the mannequin in an upright position during movement. The magnetic holder consists of permanent neodymium magnets installed in the sole of the mannequin's shoe to mount it on the swivel plate. The frame was installed only on adult mannequins, since they have greater cantilever and windage. Children's mannequins remained stable without a frame. In the event of a frontal collision, the dummy is detached from the platform and falls on the asphalt in front of the vehicle that hit it, without damaging the APIS and UMV units.

8. Heating system for dummy parts

The manikin heating system is designed to simulate the infrared radiation of the human body, so that it will be visible in the infrared range. The components of the system are installed in the face of the mannequin, as well as on the arms, to simulate the thermal radiation of open areas of the human body in winter. The heating system for one dummy should include: a power supply for the heating elements; flexible heating element for gloves "palm" (figure 24); flexible heating element on the "face" (figure 25).

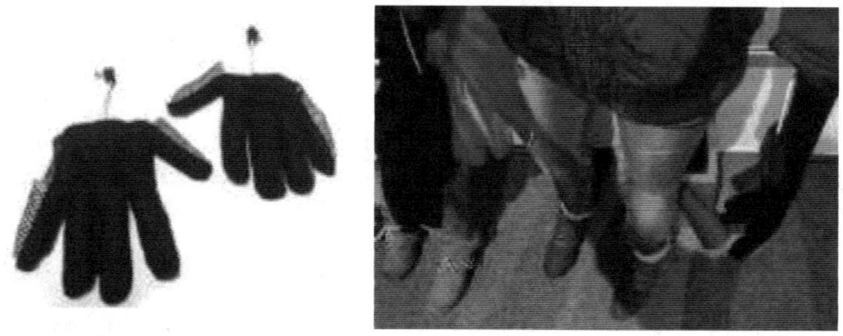

Figure 24. Flexible heating element for palm gloves on the mannequin

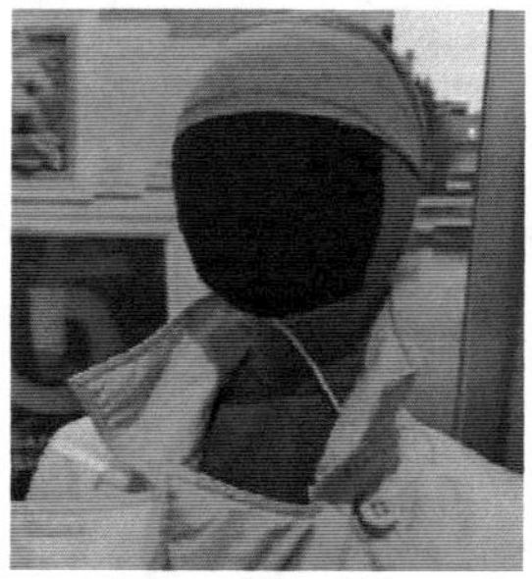

Figure 25. Flexible heating element on the "face"

9. UMV detection unit

Designed to determine the location of the UMV relative to the APIS. The block is equipped with optical barriers (photo-gates), allowing to determine the entry of the UMV into the APIS zone.

10. Conclusions and Future work

The implementation of the concept embodied in the described system will allow testing unmanned vehicles and active driver assistance systems in the conditions of test sites. The purpose of the system is to work out the algorithms for automatic registration and development of the response of the control system of an unmanned vehicle to the appearance of a pedestrian on the roadway along the route of the vehicle. Given the high interest of the scientific and technical community of the country in the development of unmanned vehicles, the development and implementation of this system looks like an important and demanded task.

Currently, the automated pedestrian simulation system on the road surface has been implemented in the form of a conceptual prototype, and test tests of the concept have been carried out. At the moment, work is underway to modernize the drive unit, and a universal methodology for testing unmanned vehicles and active driver assistance systems is being developed in order to introduce ASIP into the test ranges for unmanned vehicles. In the more distant future, this system may be certified for testing active driver assistance systems in order to certify vehicles.

References

[1] Saykin A M et al. 2017 Development of unmanned ground vehicles, driver assistance systems and components according to patent publications *International Automobile Scientific Forum MANF-2017 " Intelligent transport systems "*
[2] SAE J3016 2018 *Taxonomy and Definitions for Terms Related to Driving Automation Systems for On-Road Motor Vehicles* SAE p 35
[3] Pandey P and Aghav J V 2020 Pedestrian–Autonomous Vehicles Interaction Challenges: A Survey and a Solution to Pedestrian Intent Identification *Lecture Notes in Networks and Systems* **94**
[4] Fu H, Zhang Z, Zhang Y and Lin Y 2019 Design and Implementation of Pedestrian Detection System *Advanced Hybrid Information Processing. ADHIP 2018. Lecture Notes of the Institute for Computer Sciences, Social Informatics and Telecommunications Engineering* **279**
[5] Seeck A, Shi E and Wiggerich A 2019 *Automated Driving – Misunderstandings About Level 3 and Euro NCAP's Activities in this Field.* Fahrerassistenzsysteme
[6] ISO 19206-2:2018 Road vehicles – Test devices for target vehicles, vulnerable road users and other objects, for assessment of active safety functions – Part 2: Requirements for pedestrian targets.
[7] Gaevsky V V et al. 2019 Creation of a pedestrian model for use within the framework of the national rating of an independent assessment of safety RUNCAP *Materials of the 106th International Scientific and Technical Conference* Irkutsk. Irkutsk National Research Technical University

IOP Conf. Series: Materials Science and Engineering 1086 (2021) 012011 doi:10.1088/1757-899X/1086/1/012011

Damping of pressure pulsations in the fuel system of the motor

A A Lepeshkin, L A Berdnikov, M G Korzhachkin[1] and V A Panov

Nizhny Novgorod State Technical University n.a. R.E. Alekseev, Minin st., 24, 603950, Nizhny Novgorod, Russia.

E-mail: [1]kormg@list.ru

Abstract. The paper presents results of works carried out to damp pressure pulsations in the fuel line of a gasoline engine to provide for the required environmental friendliness level. The paper describes testing of a modernized damped injector rail. Based on the obtained test results, conclusions were made on the efficiency of use of a fuel line damper reducing fuel pressure pulsations.

1. Introduction
During the operation of motor vehicle engines, their environmental performance is influenced by quite a big number of factors. Partially, these factors are operational and can be traced during the operation of motors in different conditions of road, traffic and climate. Another group of factors is related to the design features of both the vehicle and the engine proper. Design options of the engine feed system also influence a number of the engine operation parameters. One of the phenomena is the presence of of a short-time fuel pressure change in the fuel line. [1] The elimination of pressure pulsations in the fuel system of UMZ-42164 engine is the main objective of the works summarized herein.

2. Description of the experimental setup
The pressure pulsation problem has occurred after introduction of a metal injector rail supply pipe. Due to high pulsations in the fuel system, the engine ceased to be compliant with the rules of UNECE/ECE No. 83-05B. [2]

In order to troubleshoot this problem, technicians of PEKAR Company have developed an injector rail with a damper Part No. 4216.1104010-20. [3] A necessity occurred for efficiency testing of the suggested measures to reduce the pulsation effects.

To assess the efficiency of the injector rail under test, comparative tests were conducted of the new injector rail with a damper and the conventional existing injector rail.

Fuel systems were tested in the following configurations:
- standard configuration (PEKAR 4216.1104010-14 injector rail);
- test configuration with a damper (PEKAR 4216. 1104010-20 injector rail).

The injector rail was tested on UMZ engine Serial No. 421640E0802055 on a GAZ-322132 lorry. [4-6] The general arrangement of GAZ-322132 fuel system is shown in Figure 1.

GAZ-322132 fuel system under test consisted of a fuel module, a fine fuel filter, an injector rail and connection tubes. The fuel system under test is shown in a configuration without a return fuel line. The fuel return to the tank is provided for only by the fuel line section from the fuel tank outlet to the

IOP Conf. Series: Materials Science and Engineering 1086 (2021) 012011 doi:10.1088/1757-899X/1086/1/012011

fuel module. The injector ramp is installed in straight design, without a fuel pressure regulator. The fuel pressure regulation in the system is provided for by the pressure regulator in the fuel pump module. The fuel pressure built-up in the system shall be 373 to 392 kPa. During fuel system operation, the rail operates as a liquid pressure accumulator maintaining the pressure on a constant level. Thereby, a phenomenon of short-time pressure changes (pulsations) can be observed which affects the carburation process.

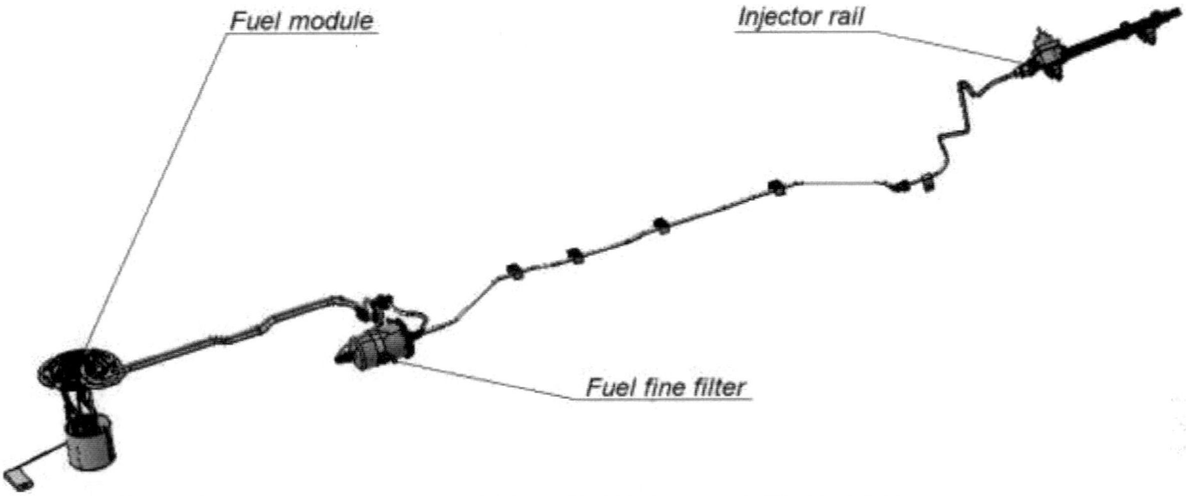

Figure 1. General view of a fuel system of GAZ-322132 lorry with UMZ-42164 engine

For the appearance of PEKAR 4216.1104010-20 injector rail with a damper, see figure 2. During testing, the injector rail consisted of a damper, fuel nozzles connected to the rail with rubber rings, a fuel pressure relief valve, fuel supply nipple and two holders for rail support on the cylinder head.

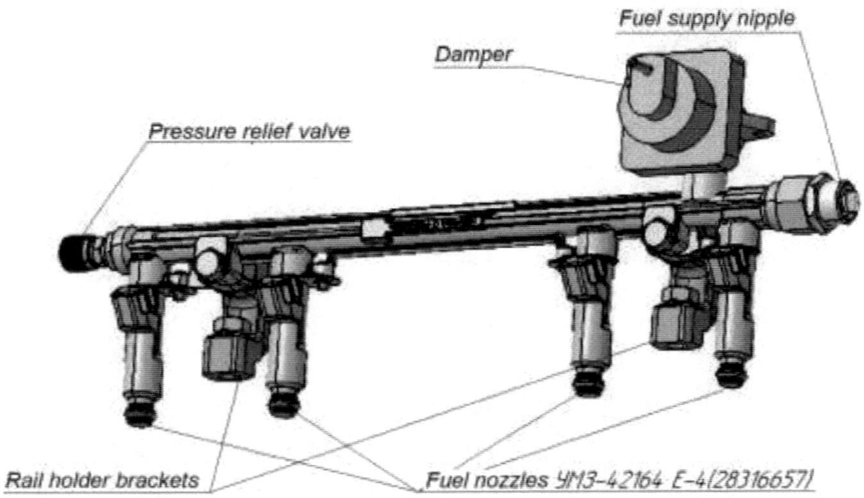

Figure 2. PEKAR 4216.1104010-20 injector rail with a damper

During testing, a fuel module by SOATE was used, see Figure 3. The fuel module consists of a fuel pump inside an accumulator tank, a fuel pressure regulator, nipples for connection of fuel lines with quick couplings, an electrical connector and a fuel gauge sending unit.

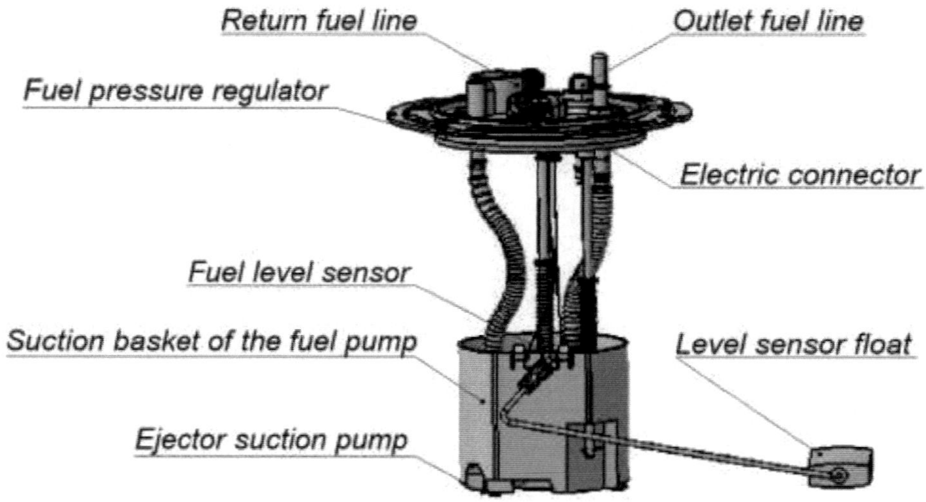

Figure 3. SOATE Fuel module

Appearance of the fine fuel filter, see figure 4. The fine fuel filter features 2 nipples for fuel pipes with quick couplings, the filter contains a paper filter element trapping particles of dirt and rust suspended in the fuel.

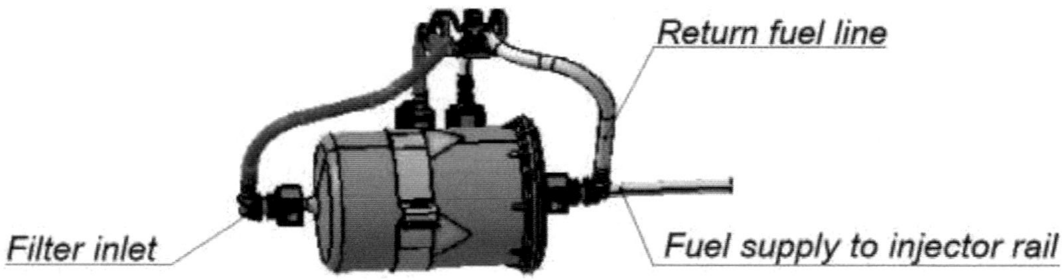

Figure 4. Fuel fine filter

3. Experimental research

The fuel system tests were conducted in the following modes:

- measurements and evaluation of high-frequency fuel pressure pulsations in the injector rail from the minimum idle rpm up to 4000 rpm;
- measurements and evaluation of high-frequency fuel pressure pulsations in the injector rail during testing for compliance with the requirements of Rules 83-05B.

As a fuel for testing and evaluation of the reduction of the pulsations in the fuel line, AI-95 gasoline was used.

For measurement of pressure pulsations, an absolute pressure sensor with 0...10 bar (a) measurement range was installed in the rail via an adapter nozzle and connected to an amplifier. The pressure pulsations in the rail was evaluated versus the rotation speed of the engine. For this purpose, one of the amplifier channels was a tachometer signal tapped from the electronic controller module. The measurement data were recorded by a laptop PC.

During testing, for all versions pressure pulsation vs. time and engine rpm diagrams were plotted.

Figure 5 shows a rail pressure pulsation diagram for the initial configuration of the fuel system in the driving cycle (exhaust toxicity).

IOP Conf. Series: Materials Science and Engineering 1086 (2021) 012011 doi:10.1088/1757-899X/1086/1/012011

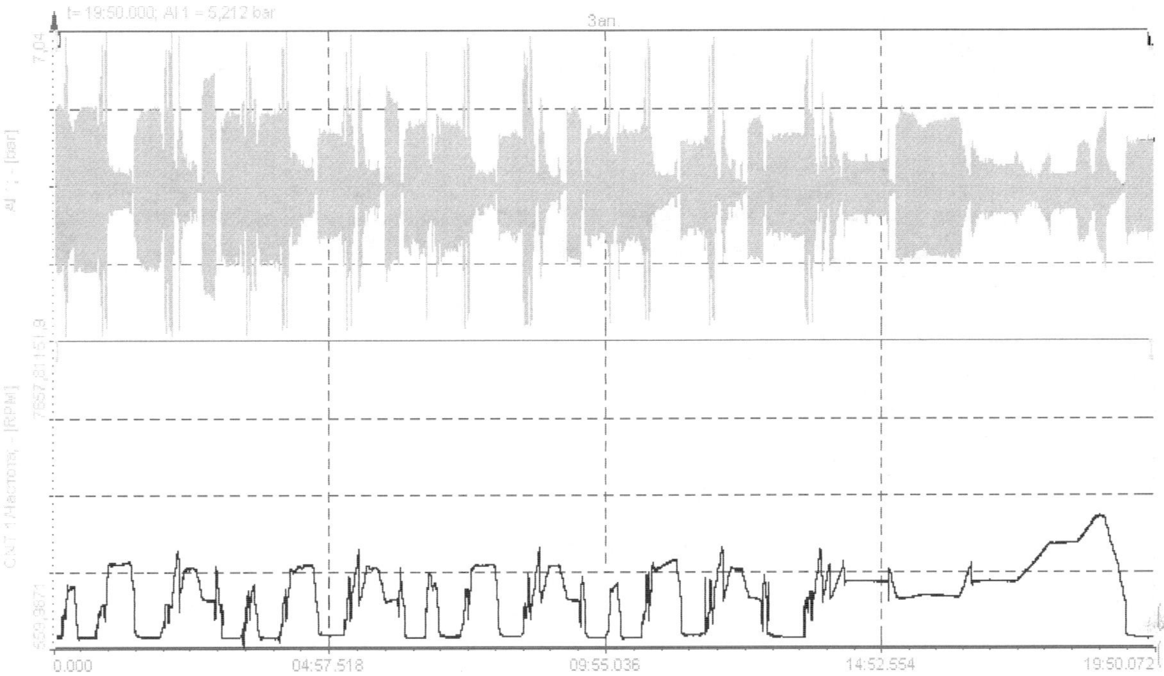

Figure 5. Rail pressure pulsation (initial configuration, in driving cycle (exhaust toxicity), second curve showing engine rpm).

Figure 6 shows a rail pressure pulsation diagram for the modified configuration with a damped rail, in the driving cycle (exhaust toxicity).

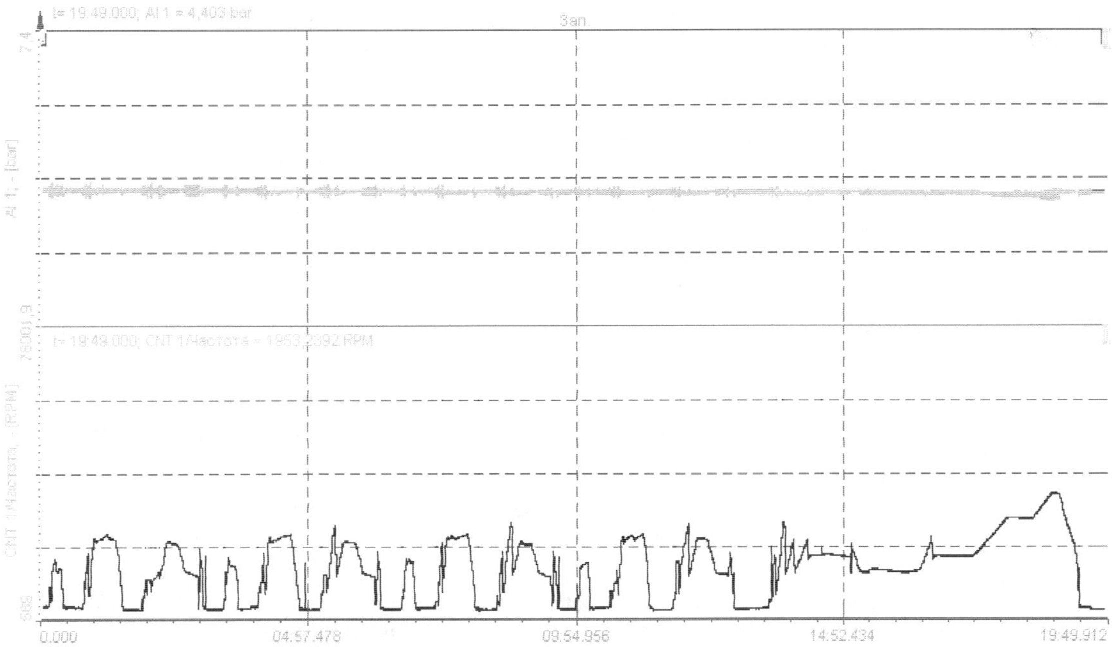

Figure 6. Rail pressure pulsation (with damper, in driving cycle (exhaust toxicity), second curve showing engine rpm).

Figure 7 shows a comparative diagram of the rail pressure pulsations vs. engine rpm (black curve: initial version, gray curve: with a damper).

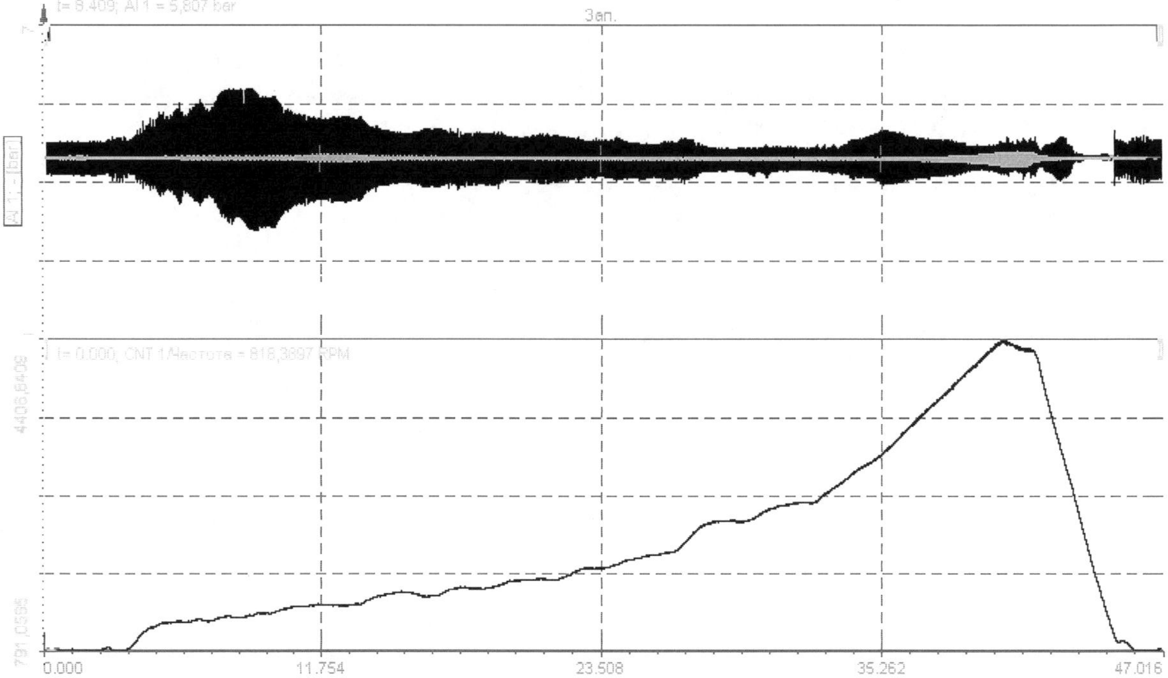

Figure 7. Injector rail pressure pulsations vs. engine rpm (black curve: initial version, gray curve: with a damper diagram).

4. Results and conclusions

Values of maximum pressure pulsations for all fuel system configurations are given in Table 1. The average rail pressure at working fuel pump during testing was 4.4 bar.

Table 1. Maximum fuel pressure pulsation values

Fuel system configuration	Maximum pressure amplitude, bar	Frequency of pulsations, Hz	Engine rpm
Initial configuration	3.13	15.2	1,200
Experimental injector rail with a damper	0.24	20	4,000

The table and the diagrams demonstrate, that:

- in the initial configuration, the pressure is pulsing from 2.83 to 5.96 bar with 15.2 Hz frequency at 1,200 rpm;
- in the experimental configuration, the pressure is pulsing from 4.22 to 4.46 bar with 20 Hz frequency at 4,000 rpm.

Based on the analysis of the obtained test data, it is obvious that in comparison with the initial configuration, the damper on the injector rail reduces the pulsations by 13 times.

Thus, the results of the comparative testing of UMZ-42164 engine fuel system demonstrate that the use of PEKAR 4216.1104010-20 injector rail with a damper reduces the pressure pulsations by 13 times and provides for compliance of the engine to UNECE / ECE Rules No. 83-05B in terms of toxic exhaust emissions.

References

[1] Korchazhkin M G and Khoryayev E A 2016 Studies of fuel system reliability of automotive diesel [engines] *94th International Scientific and Technical Conference of Automotive Engineers' Association "Unmanned vehicles: problems and prospects"* pp 240-245
[2] Berdnikov L A et al. 2019 Problems of sediment formation in internal combustion engines *Journal of Physics: Conference Series* **1177** 012009 doi:10.1088/1742-6596/1177/1/012009
[3] Berdnikov L A et al. 2019 Improved methods of hydrodynamic characteristics for calculating and profiling input, transition and outlet pipes of the GTE transport *Journal of Physics: Conference Series* **1177** 012008 doi:10.1088/1742-6596/1177/1/012008
[4] Beresnev P et al. 2017 Simulation of road impacts for forecasting of durability of vehicle's suspension elements and supporting system *19th International and 14th European-African Regional Conference of the ISTVS*
[5] Leliovsky K Y et al. 2019 Vibration load of transmission units at vehicle's motion over different roads. *Journal of Physics: Conference Series* **1177** 012053 doi:10.1088/1742-6596/1177/1/012053
[6] Zezyulin D et al. 2014 Modeling of roads impacts for life prediction of light commercial vehicles parts *FISITA 2014 World Automotive Congress*

IOP Conf. Series: Materials Science and Engineering 1086 (2021) 012012 doi:10.1088/1757-899X/1086/1/012012

Determination of the moment of resistance to turning of a forestry articulated machine

D Rogachev[1,3], A Alyabiev[1] and V Latysheva[2]

[1]Bauman Moscow State Technical University, Department of Wheeled Vehicles, 105005 Moscow, Russian Federation
[2]Moscow State University of Civil Engineering, MSUCE or MGSU, 105005 Moscow, Russian Federation

E-mail: [3]d2345@live.ru

Abstract. This article considers the calculation of the moment of resistance to turn off a wood articulated machine. The total moment is determined from the sum of the moment of resistance to turning the tire in place, the moment that occurs when the ground is cut off by the side surface of the tire, the moment of resistance to rolling the wheel.

1. Introduction

The initial parameter for calculating the articulated frame is the moment of resistance to mutual turn of the links. For the calculation, it is necessary to consider the turn from a static position, since in this case resistance to turn is maximum. The calculation scheme is shown in figure 1.

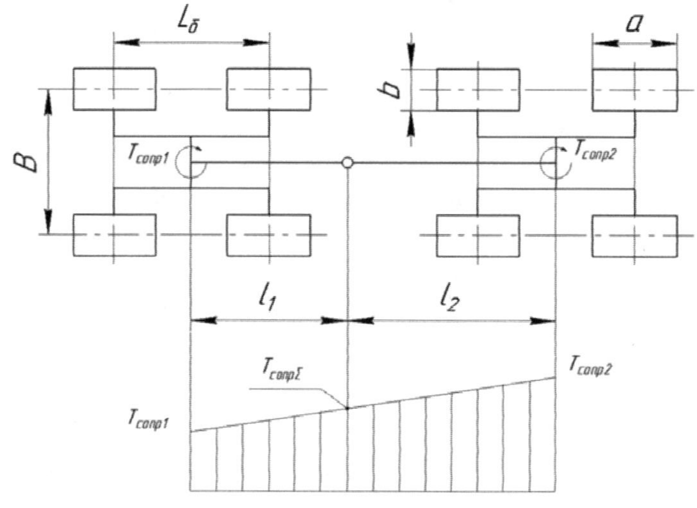

Figure 1. Calculation scheme for determining the moment of resistance to rotation.

In Figure 1: B – track of the designed machine, m; $L_б$ – distance between the wheels of the balancing cart, m; a, b – respectively, the longitudinal and transverse axis of the tire contact patch, reduced to an equal rectangle, m; l_1, l_2 – distances, respectively, of the first and second links from the axis of the vertical hinge of the articulated module assembly to the axis of the tandem bridge.

Moment for the implementation of the relative turn of the links, $N \cdot m$ [1-4]:

$$T_{\text{сопр}\Sigma} = T_{\text{сопр}1} + \frac{\left(T_{\text{сопр}2} - T_{\text{сопр}1}\right) \cdot l_1}{l_1 + l_2}$$

where $T_{\text{сопр}1}, T_{\text{сопр}2}$ – moments of resistance to turn of the first and second modules, respectively, $N \cdot m$.

When the module is turned, the wheel performs a complex movement, which can be represented as the sum of two movements: turning the wheel in place and rolling the wheel. Thus, the moment of resistance to turn of the n-module is equal to, $N \cdot m$:

$$\mathrm{T}_{\text{сопр}_n} = \mathrm{M}_{\text{тр}\,n} + \mathrm{M}_{\text{ср}\,n} + \mathrm{M}_{\text{к}\,n}$$

where $M_{\text{тр}}$ – is the moment of resistance to turn the tire in place, $N \cdot m$; $M_{\text{ср}}$ – the moment arising when cutting off the soil by the side surface of the tire, if the module is in the track, $N \cdot m$; $M_{\text{к}}$ – is the moment of resistance to wheel rolling, $N \cdot m$.

To determine the moment of resistance to turning the tire on the spot, let us assume that the coefficient of adhesion φ and specific pressures at each point of the tire print with the supporting surface are the same. We will take into account that the wheel axis is located parallel to the reference surface perpendicular to the longitudinal axis of the module.

2. The moment of resistance to turn the tire in place

The car is equipped with diagonal low-profile tires, so the contact spot of the tire with the support surface will have the shape of a rectangle. To determine the size of the contact spot, we use the calculation scheme (Figure 2).

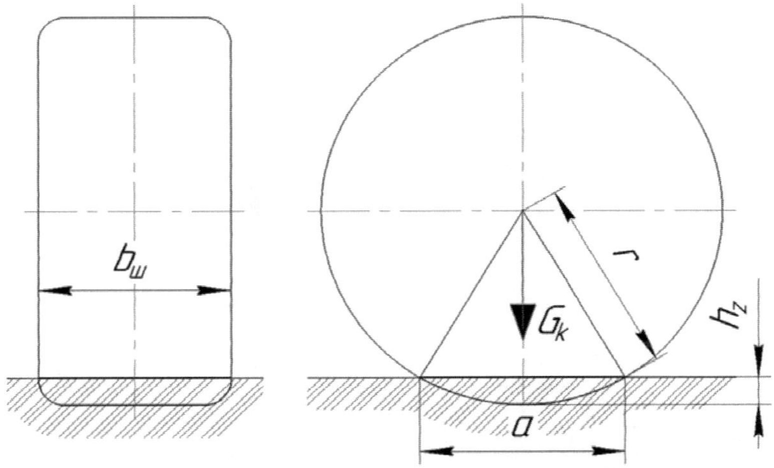

Figure 2. Calculation scheme for determining the tire contact spot

Figure 2 shows: $b_ш = b$ – width of the tire profile; r – radius of the tire; $G_к$ – weight per wheel of the module; h_z – deflection of the tire. The deflection of the n-module tire is determined by the formula, m:

$$h_{z\,n} = \frac{0{,}01 \cdot K \cdot G_{\text{к}\,n}^{3/4}}{1 + p_w},$$

where K– is the empirical coefficient; p_w – is the tire pressure, kPa.

We determine the value of the coefficient K, knowing the load carrying capacity $G_ш$ and the radius under static load r_G:

$$K = \frac{(r_{\text{СТ}} - r_G) \cdot (1 + p_w)}{G_{\text{Ш}}^{3/4}}$$

Calculation of the length of the longitudinal axis of the contact patch of the n-module tire, m:

$$a_n = 2 \cdot \sqrt{r^2 - (r - h_{z\,n})^2}.$$

Let's determine the pole of the tire turn. Let's select an elementary area with sides dx and dy on the contact spot ab (Fig. 3).

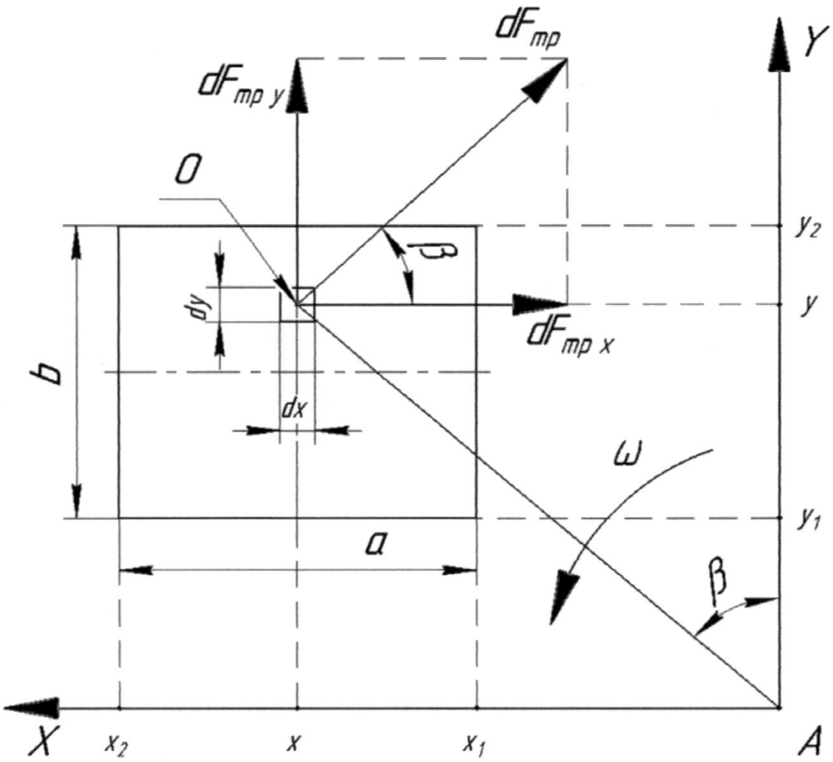

Figure 3. Calculation scheme

Figure 3 shows: A is the axis of turn of the module, and AX is the longitudinal axis of the module. The elementary frictional force acting in the contact patch when turning the tandem bridge is equal to:

$$dF_{\text{тр}} = dG_{\text{К}\,n} \cdot \varphi = \frac{G_{\text{К}\,n} \cdot \varphi}{a_n \cdot b_n}\, dxdy.$$

The projection of the elementary friction force on the AX axis is equal to:

$$dF_{\text{тр}x} = \frac{G_{\text{К}\,n} \cdot \cos\beta}{a_n \cdot b_n}\, dxdy;\ \cos\beta = \frac{y}{\sqrt{y^2 + x^2}}.$$

The projection of the elementary friction force on the axis will take the form AX:

$$dF_{\text{тр}x} = \frac{G_{\text{к}\,n} \cdot \varphi}{a_n \cdot b_n} \cdot \frac{y}{\sqrt{y^2 + x^2}}\, dxdy.$$

dF_{mpx} will tend to rotate the wheel about its axis. Thus, when turning the wheel in place, the turning pole will be at the intersection of the projection of the wheel axis on the support surface and the direction of action of the projection of the resultant friction force in the contact spot when turning the tandem bridge. Formula for determining the Y turn pole coordinate:

$$Y = \frac{\int_S y\, dF_{\text{тр}x}}{\int_S dF_{\text{тр}x}} = \frac{\int_{y_1}^{y_2}\int_{x_1}^{x_2} \frac{y^2}{\sqrt{y^2 + x^2}}\, dxdy}{\int_{y_1}^{y_2}\int_{x_1}^{x_2} \frac{y}{\sqrt{y^2 + x^2}}\, dxdy}.$$

The center of the tire contact spot is determined by the coordinates $(L_6/2; B/2)$. As a consequence, the limits of integration of y_1, y_2, x_1, x_2 can be represented as:

$$x_1 = \frac{L_6 - a_n}{2};\ x_2 = \frac{L_6 + a_n}{2};\ y_1 = \frac{B - b_n}{2};\ y_2 = \frac{B - b_n}{2}$$

Let's determine the moment of resistance to turning the tire in place. The calculation scheme is shown in (Fig. 4), where O is the pole of turning the tire in place.

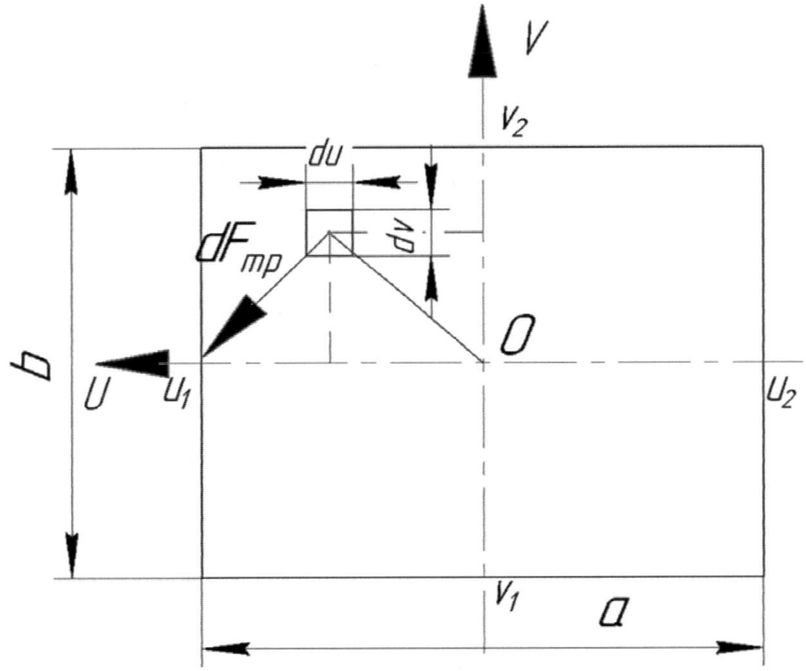

Figure 4. Calculation scheme for determining the moment of resistance to turning the tire in place

Contact tire boundary coordinates u_1, u_2, v_1, v_2 are defined as:

$$u_1 = -\frac{a_n}{2};\ u_2 = \frac{a_n}{2};\ v_1 = \frac{B + b}{2} - Y;\ v_2 = \frac{B - b}{2} - Y.$$

Resistance to turning the tire in place: $dM_{\text{тр}\,n} = dF_{\text{тр}\,n} \cdot w$;

$$dF_{\text{тр}\,n} = dG_{\text{к}\,n} \cdot \varphi = \frac{G_{\text{к}\,n} \cdot \varphi}{a_n \cdot b_n} dudv.$$

Arm of the moment: $w = \sqrt{u^2 + v^2}$. Substitution:

$$dM_{\text{тр}\,n} = \frac{G_{\text{К}\,n} \cdot \varphi}{a_n \cdot b_n} \cdot \sqrt{u^2 + v^2} du dv.$$

Module frictional moment:

$$M_{\text{тр}\,n} = 4 \cdot \int_{v_1}^{v_2} \int_{u_2}^{u_1} \frac{G_{\text{К}\,n} \cdot \varphi}{a_n \cdot b_n} \cdot \sqrt{u^2 + v^2} du dv.$$

Let's determine the moment that occurs when the soil is cut by the side surface of the tire (Fig. 5).

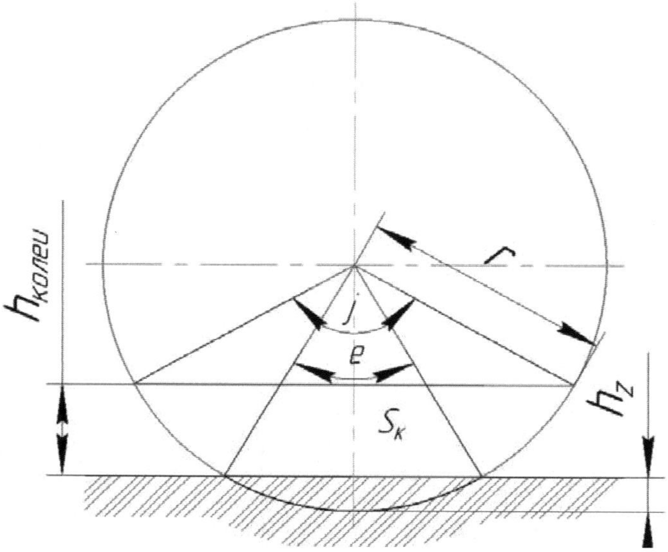

Figure 5. Calculation scheme for determining the track

The depth on the drag can be equal to the clearance. However, on the drag, the turn is made in motion. Turning in place occurs when loading assortments, where the track size is determined by the deformation of the soil under the tire of the machine during a single pass and depends on the tire pressure on the soil and the strength properties of the soil. On drained soils with a single pass, the depth of the track does not exceed 5 cm [4-8].

The depth of the formed track, m: $h_{\text{колеи}\,n} = 2{,}2 \cdot \frac{1-\mu_{\text{п.р}}^2}{\text{E}} \cdot \frac{G_{\text{К}\,n}}{b_{\text{ш}}}$, where $\mu_{\text{п.р.}}$ – is the modulus of longitudinal expansion of the soil; E – is the modulus of soil deformation, Pa.

Contact area of the tire with the track, m²: $S_{\text{К}\,n} = \frac{r^2}{2} \cdot \left(\left(\frac{\pi \cdot j}{180} - \sin j\right) - \left(\frac{\pi \cdot e}{180} - \sin e\right)\right)$, where $j = 2 \cdot \arccos\frac{r - h_{\text{колеи}\,n} - h_{z\,n}}{r}$; $e = 2 \cdot \arccos\frac{r - h_{z\,n}}{r}$.

The moment that occurs when the ground is cut off by the side surface of the tire, $N \cdot m$:

$$M_{\text{ср}\,n} = \text{T}_{\text{сопр}_n} \cdot \mu_{\text{б}} \cdot \tan\lambda + C_0 \cdot l_n \cdot \mu_{\text{б}} \cdot 4 \cdot S_{\text{К}\,n},$$

where $\mu_{\text{б}} = 0{,}4 \ldots 0{,}9$ – depends on the type of surface [8-10]; C_0, λ – strength parameters of the deformable surface, depending on the soil consistency index.

Wheel rolling resistance moment, $N \cdot m$:

$$M_{\text{К}\,n} = 2 \cdot G_{\text{К}\,n} \cdot f \cdot B$$

where f – is the rolling resistance coefficient.

As a result, we obtain a system of two equations:

$$\begin{cases} M_{\text{ср}\,n} = \text{Т}_{\text{сопр}_n} \cdot \mu_{\text{б}} \cdot \tan\lambda + C_0 \cdot l_n \cdot \mu_{\text{б}} \cdot 4 \cdot S_{\text{к}\,n}; \\ \text{Т}_{\text{сопр}_n} = M_{\text{тр}\,n} + M_{cp\,n} + M_{\kappa\,n}. \end{cases}$$

Let's determine the end formula:

$$\text{Т}_{\text{сопр}_n} = \frac{M_{\text{тр}\,n} + \text{М}_{\text{к}\,n} + 4 \cdot C_0 \cdot l_n \cdot \mu_{\text{б}} \cdot S_{\text{к}\,n}}{1 - \mu_{\text{б}} \cdot \tan\lambda}.$$

3. Conclusion

A formula for determining the moment of resistance to turn of the module is derived and calculation schemes and formulas for determining the depth of the track depending on the type of surface are proposed.

Acknowledgments
Work has been performed in Bauman Moscow State Technical University with a financial support of the Russian Ministry of Education under Agreement No. 075-11-2019-030 dd 22 November 2019.

References

[1] Afanasiev B A, Belousov B N, Zheglov L F et al. 2008 *Textbook for universities* MSTU publishing house named after N.E. Bauman p 441
[2] SmirnovA G 1990 *Theory of movement of wheeled vehicles* M.:Mechanical engineering p 351
[3] Alyabiev A F, Kotov A A, Klubnichkin V E, Klubnichkin E E and Bykovskiy M A 2019 Experimental determination of soil shear force by track model depending on track surface to soil friction coefficient *IOP Conf. Series: Journal of Physics: Conf. Series* **1177** 012033 doi: 10.1088 / 1742-6596 / 1177/1/012033
[4] Klubnichkin E E, Klubnichkin V E and Kotiev G O 2018 Theoretical research of soil packing by timber harvester running gear *IOP Conference Series: Materials Science and Engineering* **386** 012025 doi: 10.1088/1757-899X/386/1/012025
[5] Klubnichkin V E, Klubnichkin E E, Kotiev G O, Beketov S A and Makarov V S 2018 Interaction between elements of the track ground contacting area with the soil at curvilinear motion of the timber harvesting machine *IOP Conference Series: Materials Science and Engineering* **386** 012016 doi: 10.1088/1757-899X/386/1/012016
[6] Belyakov V, Makarov V, Klubnichkin V and Klubnichkin E 2020 Analysis of the diagrams and main parameters of the existing structures of the articulated tracked vehicles *IOP Conference Series: Materials Science and Engineering* **820** 012001 doi: 10.1088/1757-899X/820/1/012001
[7] Klubnichkin E, Klubnichkin V, Erkhov A and Alyabiev A 2019 Researching impact of pressure re-distribution along the length of the track ground contacting area on resistance of the timber-hauling vehicles' travel over snow *IOP Conference Series: Materials Science and Engineering* **695** 012004 doi:10.1088/1757-899X/695/1/012004
[8] Makarov V S, Markovnina A I, Papunin A V, Klubnichkin E E and Belyakov V V 2019 Study of uneven surfaces distribution on forestry roads *Journal of Physics Conference Series* **1177** 012041 doi: 10.1088/1742-6596/1177/1/012041
[9] Klubnichkin V, Klubnichkin E, Makarov V and Podrubalov M 2019 Researching rolling resistance of the wheeled forestry tractor at skidding *IOP Conference Series: Materials Science and Engineering* **695** 012003 doi: 10.1088/1757-899X/695/1/012003
[10] Bazhenov E E 2000 *Theory of a Vehicle and a Tractor* Yekaterinburg: UGTUUPI p 125

IOP Conf. Series: Materials Science and Engineering 1086 (2021) 012013 doi:10.1088/1757-899X/1086/1/012013

Determining of optimum composition of water-fuel mixture for a spark-ignition internal combustion engine with electronic control

A P Kartoshkin[1], A L Birykov[2], Sh V Busikov[3], D G Sergeev[3] and S.A.Plotnikov[3]

[1] St.-Petersburg State Agricultural University, St. Petersburg, Russia
[2] Vologda State Dairy Farming Academy n.a.N.V. Vereschagin
[3] Vyatka State University, Kirov City, Russia

E-mail: [1] akartoshkin@yandex.ru

Abstract. The relevance of the study is stipulated by the necessity for determining of optimum composition of water-fuel mixture (WFM) for a spark-ignition internal combustion engine with electronic control. The objective of the study is the determining of the optimum composition of water-fuel mixture (WFM) during the operation of a VAZ-21114 engine which allows for the best quantitative performance specifications. For this purpose, an experimental test fixture was developed, consisting of a KI-2139 electrical brake bench with a pendulum arm machine and an "AVTOTEST-02.03" 5-component gas analyzer, with equipment for bench testing in accordance with the requirements of GOST 14846-81 "Automotive engines. Bench test methods". As a result, for the first time, optimum quantitative WFM composition was obtained for operation of VAZ-21114 spark-ignition engine with electronic control providing for the best quantitative performance specifications. The analysis of the obtained data demonstrates lowering of the concentration of toxic exhaust gas (EG) components of the engine, specific effective (g_e) and hourly (G_m) fuel consumption rate, as well as the EG temperature (T_{oz}) during its operation by the external rpm vs. load curve within the utilization range of WFM with 0 to 12% water content. At water content growth in the WFM from 12 to 20%, a growth of carbon monoxide (CO) and unburnt hydrocarbons (*CH*) in the EG is observed, as well as a growth of g_e and G_m. The minimum quantitative values of the parameters *CO*, *CH*, NO_x, T_{oz}, g_e and G_m were: 6.0 and 1.65% of *CO* at water content in WFM of 10 and 13%, respectively; 173 and 142.5 ppm of *CH* at 10% water content in WFM; 113 and 215 ppm of NO_x and T_{oz} 650 and 730 K at up to 20% water content in WFM; g_e 330 and 303 g/kWh at 12.5 and 11.5% water content in WFM, respectively, G_m 6.5 kg/h at 11.5% water content in WFM. The maximum quantitative value of the effective engine power N_e of 44.2 kW was obtained with WTM with 20% water content. The obtained results demonstrate that, for the operation of a spark-ignition type VAZ-21114 engine with electronic control, the WTM should contain 9…12% water as an optimum providing for achievement of the best quantitative performance values.

1. Introduction

For the time being, the world practice offers no single method allowing for simultaneous improvement of both the performance and the environmental friendliness of internal combustion engines. The improvement of these specifications is advisable to get started along with the solving of the exhaustion problem of conventional fuel sources. Using of non-conventional fuels in their natural forms leads to a number of drawbacks, such as: lowering of environmental friendliness, energetic and economic performance, sophistication engine design concepts, variance of physical and chemical fuel properties and reluctance of the fuel and energy branch to produce new fuel types. In this connection, the use of fuel mixtures consisting of conventional fuels and different non-petroleum additives is a matter of vital importance.

Fuel mixtures are used both in diesel and in spark-ignition internal combustion engines [1]. The studies of water-fuel mixtures (WFM) (emulsions) are carried out in two directions: 1) individual water supply to the combustion chamber prior to the fuel ignition; 2) timely preparation of the water-fuel emulsion [1-24]. Emulsions are thermodynamically instable of two or more mutually immiscible liquids, which are stabilized kinetically with the absorption film [2]. As emulsion stabilizers surfactants can be used (SAS) [3]. Mechanical agitation is the most frequently used method of emulsion preparation [4].

Use of WFM in emulsion form in internal combustion engines leads to the emission of the total nitrogen oxides (NO_x). This fact is stipulated by high specific heat capacity and specific heat of evaporation of about 2256 kJ/kg [5]. The cooling effect of a thermodynamic system occurring as a result of water injection into the combustion chamber improves the mixture build-up and combustion on the fuel injection step, minimizing the ignition delay and influencing the chemical reactions generating soot [6-10]. The addition of water to the fuel also reduces the maximum combustion temperature which, in its turn, decreases the build-up of thermal NO_x [5, 11]. Using of emulsions to reduce NO_x emission has been studied for quite a prolonged time. The conclusions based on the result of studies [12, 13] in which three different methods of NO_x emission reductions were evaluated, such as: recirculation of exhaust gases, water injection and WFM use by means of a multi-zone imitation models demonstrate that the use of emulsions is optimal from the point of view of decreasing the toxicity of the exhaust gases and and the fuel consumption rate.

As a result of the conducted experimental studies of three different water contents in the water-gasoline emulsions 5%, 10% and 15%, respectively, during test runs of the vehicle at 90 km/h, a reduced concentration of NO_x by 35,0% was measured, compared with pure gasoline [14]. Paper [15] also notes the mutual dependency of the thermal build-up efficiency of NO_x while using emulsified fuel.

The basic reduction method of NO_x emissions is using water-based fuel additives [16, 17]. This method allows for a reduction of the maximum temperature in the combustion chamber [17]. The results of [18] demonstrate that water-gasoline emulsions feature antiknock properties. The results of studies [19] demonstrate that using of a water-gasoline emulsion scores no direct fuel savings. At optimum water content of 20%, the efficiency in most operation modes is almost the same the same as with pure gasoline. Emulsions with water contents of up to 40% cannot be used in contemporary engines without their re-designing. At equal values of the air excess coefficient, the concentration of CO in the exhaust gases does not depend on the water content, the concentration if NO_x goes back by appr. 1.3% per each additional percent of water; the CH contents rises in certain operation modes several-fold.

The results of studies [20, 21] of water-gasoline mixture supply to the intake manifold demonstrate a reduced rate of combustion, a shorter ignition delay, as well as a shortening of the combustion process phases [22-24].

Studies of WFM with water contents 5, 10 and 15% in a 4-stroke engine with 125 cm^3 displacement, with injector, spark ignition and electronic control at crankshaft rpm rate n = 2000 – 7000 rpm, at completely open gasoline throttle, in idle run demonstrate lower carbon monoxide (CO) and nitrogen oxides (NO_x) concentrations on water content growth in the emulsion [13, 22]. The content values of the unburnt hydrocarbons (CH) were growing with the growth of the water content, stipulated by the combustion chamber cooling [17, 22]. At 5 and 10% water content in WFM, higher torques and improved fuel consumption rates were observed. However, at 15% water content in the emulsion, reduced torques, uncreased fuel consumption rates and EG temperatures were registered, due to a slower combustion process rate [5].

The results of the experimental studies of engine fueling with water -gasoline emulsion with 5, 10 and 15% water content, in a one-cylinder 4-stroke spark-ignition engine with 376 cm^3 displacement show that *CO*, NO_x and *CH* contents in the EG went back with the growth of the water content in the fuel, whereas the power with 5 to 15% water content grew from 3.8 to14% [23, 24].

The analysis of the results of the studies makes it evident that the search for optimum WFM composition at which optimum quantitative performance specifications of spark-ignition engines with electronic control can be provided for is, for the time being, a promising initiative.

2. Research objectives

The objective of the present study is to determine an optimum WFM composition for VAZ-21114 spark-ignition engine with electronic control system ensuring the best quantitative specifications of the environmental friendliness.

3. Methods and materials

The device under test is a 4-stroke spark-ignition VAZ-21114 engine with sequential multi-point fuel injection, modified by integration of a WFM feed system. The bench testing method corresponds to GOST 17.2.2.02–86, GOST 17.2.2.05–86.

The experimental text fixture consisted of a KI-2139 electrical braking bench with a pendulum arm machine, an engine feed system adapted for WFM supply and measurement instruments. The equipment included into the experimental test fixture was compliant with GOST 14846–81 "Automotive engines. Bench test methods".

The coolant temperature was maintained within the range from 93 to 95°C by means of a closed-circuit cooling system with forced circulation. The lubrication system of the engine was standard. The motor oil pressure measurement in the lubrication system of the engine was carried out by means of an MT type pressure gauge with 1.0 MPa scale range and A accuracy class.

The diagram and the general appearance of the experimental test fixture, see Figure 1. The test fixture consists of a gasoline fuel tank (1), a water cylinder (2), a gasoline pump (3), a measurement cylinder (4) for water flow rate measurement, filters (5), (6) for gasoline and water filtering, a fuel pressure regulator (7), a mixer (8) for WFM mixing, a scale (9) for gasoline quantity measurement, a gasoline tank (10) for fuel feed system flushing, a load simulation device (11) for brake torque generation, an electronic engine controller (EEC) (12), a laptop PC (13) with software for real-time parameter monitoring of the electronic engine control system (EECS), an "AVTOTEST-02.03" 5-channel gas analyzer (14), an exhaust gas temperature sensor (EGTS) (15), a 16 UKT-38SchCh EGTS transmitter.

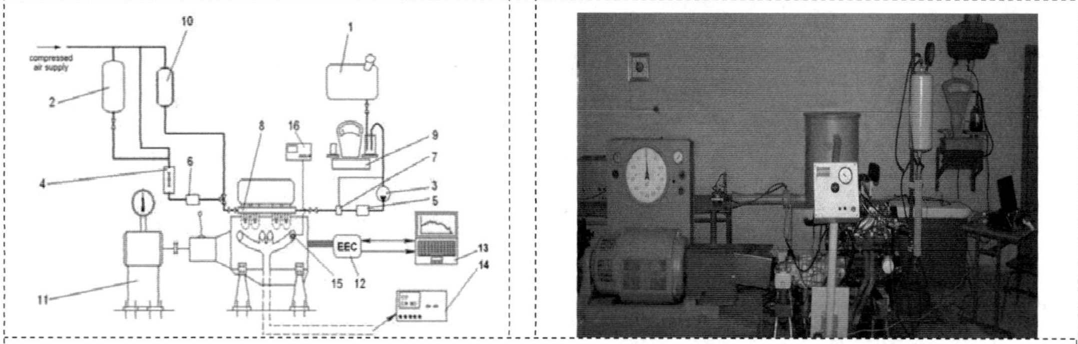

Figure 1. Experimental test fixture: a) diagram; b) general appearance: 1 – fuel tank; 2 – water cylinder; 3 – gasoline pump; 4 – measurement cylinder; 5,6 – filters; 7 – fuel pressure regulator; 8 – WFM mixer; 9 – scale; 10 – flush gasoline tank; 11 – brake device for load simulation; 12 – electronic engine controller (EEC); 13 – laptop PC; 14 – "AVTOTEST-02.03" 5-channel gas analyzer; 15 – exhaust gas temperature sensor; 16 – EGTS transmitter

The experimental tests were carried out in terms of the external speed (at n = 4000 rpm) and load (at n= 2000 rpm and N_e = 20 kW) curves. During the tests, the influence of different WFM compositions on the environmental friendliness specifications of VAZ-21114 engine.

During the tests, the rotation frequency of the crankshaft (CS) of the engines, the torque, the effective power, the hourly and the specific water and gasoline consumption, as well as the hourly air consumption. To determine the contents of carbon monoxide (CO), unburnt hydrocarbons (CH) and the total nitrogen oxides (NO_x), samples of EG were taken.

J5 On-line Tuner Software installed in the laptop PC connected to the diagnostic interface of the EECS (OBD-II) allowed for on-line measurement of the engine's CS rpm, of the air mass flow, of the coolant temperature and throttle valve position. Based on standard methods set forth in GOST 17.2.2.02–86, GOST 17.2.2.05–86, a calculation of the basic parameters, such as torque, effective engine power, as well as specific fuel and water consumption rates was carried out. The measurement of the consumed water quantity was volumetric. The concentration of toxic EG components was measured with an "AVTOTEST-02.03" 5-channel gas analyzer. The total nitrogen oxides concentration was determined by a 3NF/F Nitric Oxide CITIcel electrochemical cell.

During the test, EG temperature was measured in the exhaust manifold. For this purpose, a DTPK 135-0314.250 thermocouple cpl. with a UKT 38-Sch4-TP transmitter. After conducted experimental tests, content dependencies of CO (%), CH (ppm), NO_x (ppm) in EG and EG temperature on the fuel mixture composition fed to the engine.

During the experiment, simultaneous recording of all monitored parameters was carried out, which, in its turn, could ensure the trustworthiness of the obtained data. In accordance with the methods set forth in GOST 17.2.2.02–86, GOST 17.2.2.05–86, the limit error value of single measurements and the maximum possible statistical error of the mean absolute error in multiple measurements were determined.

4. Result and discussion

As a result of the studies, dependencies of the quantitative concentrations of toxic components of EG of the spark-ignition engine with a control system were obtained with WFM with 0 to 20% water content during operation by the external rpm (n = 4000 rpm) and in the load mode (n = 3000 rpm and N_e = 20 kW).

Figure 2 shows the dependencies of the influence of WFM composition on the performance parameters of VAZ-21114 engine at operation by the external rpm curve (at n = 4000 rpm).

The analysis of the obtained dependencies has shown that along with the growth of the water content in WFM from 0 to 20%, the effective engine power (N_e) also grows from 42 to 44.2 kW.

The influence of the use of different WFM compositions on CO and content CH in the EG of the engine, as well as on the specific effective fuel consumption g_e is unambiguous. The increase of the water content in WFM from 0 to 13% was leading to the reduction of CO concentration in ОГ from 8.5 to 6.0%, whereas the subsequent growth of the water content in the WFM from 13 to 20%, was leading, on the contrary, to increased concentrations of CO in the EG from 6.0 to 6.9%. A reason for that could be improved combustion efficiency due to the turbulization of the mixture during water vaporization in the WFM and to the action of the active radicals H and OH [5-11]. In general, at all studied WFM compositions, the concentration of CO in the EG was lowered, however, at the growth of the water content in the WFM beyond 13%, the growth of the CO concentration was observed, due to the lower combustion efficiency [5].

The minimum CH concentration in the EG of 173 ppm was observed at water content in the WFM equal to 10%. The reduction of the water content from 10 to 0% increased the CH content up to 214 ppm, whereas the increase of the water content from 10 to 20% lead to an increase of CH concentration from 173 to 203 ppm. It can be explained by completer combustion due to better mixing during water evaporation [17, 22]. However, if the water в content in WFM is increased beyond 10%, the water considerably cools down the wall layers of the working mixture inside the ICE cylinder making the combustion less effective [22].

The increasing of the water content in WFM from 0 to 12.5% has lead to a reduction of the specific effective fuel consumption g_e from 363 to 330 g/kWh. The increasing of the water content in WFM s 12.5 to 20% has lead to a growth of the specific effective fuel consumption g_e from 330 to 348 g/kWh. This fact can be explained by a slower combustion process [5, 23, 24].

The reduction of NO_x concentration in the EG, as well as of the EG temperature T_{oz} occurred within the entire range of the water content in the WFM.

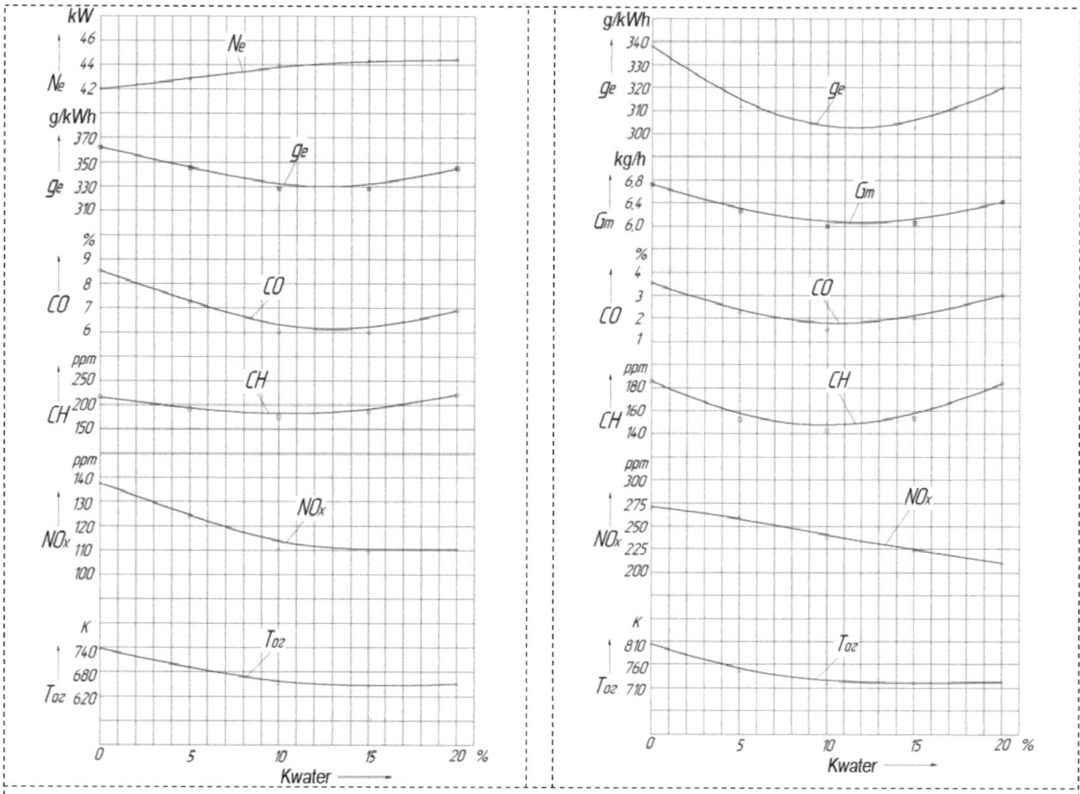

Figure 2. Influence of fuel composition on VAZ-21114 engine performance: a) by external rpm curve (n = 4000 rpm); b) by load curve (n = 2000 rpm and N_e = 20 kW)

The biggest reduction of NO_x emission from 138 to 113 ppm was observed at the increase of the water content in the WFM from 0 to 20% of the total weight of the fuel. The EG temperature was also going down from 740 to 650 K at the increase of the water content in the WFM from 0 to 20%. It occurred due to the lower temperature of the operation cycle stipulated by the cooling of the work charge by the water contained in the WFM [5, 6].

The influence of the variance of WFM composition on CO and CH contents in the engine's EG, as well as on the specific effective g_e and the hourly G_m fuel consumption rates is unambiguous. The diagram makes it obvious that, along with the growth of the water content in the WFM from 0 to 10%, the CO concentration went down from 3.46 to 1.65%, and the CH concentration wend down from 186.7 to 142.5 ppm. Along with the further growth of the the water content from 10 to 20%, the CO concentration grew from 1.65 to 3.02%, and the CH concentration grew from 142.5 to 184 ppm.

The minimum specific effective g_e = 303 g/kWh an the hourly G_m = 6 kg/h fuel consumption rates are achieved using WFM with 11.5% water content. The reduction of the water content in the WFM from 11.5 down to 0% and the increase of the water content from 11.5 up to 20% were leading to the increase of g_e values up to 338 and 320 g/kWh, and to the increase of G_m values up to 6.75 and 6.41 kg/h, respectively.

Observed was also the reduction of NO_x concentration in the EG, as well as of the EG temperature T_{oz} within the entire range of the water content in the WFM.

Nox_x concentration reduction in the EG from 270 down to 215 ppm and EG temperature reduction from 810 down to 730 K were observed after the increase of the water content in the WFM from 0 to 20%.

Based on the analysis of the obtained data, it is getting obvious that *CO* and *CH* contents in the EG get better with WFM water content of 9…12%, whereas at further increase of the water percentage they begin to worsen, again. It can be explained, that up to the optimum point of the function, each of the parameters under test is influenced positively by the addition of water, due to the intensification of the turbulization of the mixture in the ICE cylinder, the intensification of the chemical reaction rates and the more effective fuel combustion [5-11, 17, 22-24]. However, having passed the optimum peak, that is, the minimum of the function for *CO* and *CH* values, the further increasing of the water content in the mixture begins to produce a negative impact mainly due to the unwanted heat dissipation and the decreased combustion rate and efficiency [25-28]. The NO_x content, on the contrary, goes down rather intensively within the whole water content range 0…20%, since the build-up of the nitrogen oxides is mostly stimulated by high cycle temperatures which get lower with the growth of the water content in WFM [28-30].

5. Conclusions

1. As a result, for the first time, optimum quantitative WFM composition was obtained for operation of VAZ-21114 spark-ignition engine with electronic control providing for the best quantitative performance specifications.

2. The lowest quantitative values of *CO*, *CH* concentrations in the EG, of the specific effective g_e and hourly G_m fuel consumption during the engine operation by the speed and the load curve were: 6.0 and 1.65% of *CO*, at 10 and 13% of water in the WFM, respectively; 173 and 142.5 ppm of *CH* at 10% of water in the WFM, 330 and 303 g/kWh g_e at 12.5 and 11.5% of water in the WFM, respectively; 6.5 kg/h G_m at 11.5% of water in the WFM, respectively.

3. The minimum quantitative values of NO_x concentration and T_{oz} during the engine operation by the speed vs. load curve were: 113 and 215 ppm of NO_x, and 650 and 730 K T_{oz}, respectively, at BTC до 20% of water in the WFM.

4. The maximum quantitative value N_e = 44.2 kW was obtained during the motor operation by the speed curve at 20% of water in the WFM.

5. The obtained results demonstrate that, for the operation of a spark-ignition type VAZ-21114 engine with electronic control, the WTM should contain 9…12% water as an optimum providing for achievement of the best quantitative performance values.

References

[1] Dłuska E et al. 2007 The influence of helical flow on water fuel emulsion preparation *Chemical Engineering Communications* **194(10)** pp1271-1286 doi: 10.1080/00986440701293959

[2] Park J W, Huh K Y and Lee J H 2001 Reduction of NOx, smoke and brake specific fuel consumption with optimal injection timing and emulsion ratio of water-emulsified diesel *Proc. Inst. Mech. Eng. D* **215** pp 83–93

[3] Piaseczny L and Zadrag R 2005 Researches of influence of water delivery to cylinder on parameters of combustion process and toxicity of CI engine *Proceeding of Polish Congress-PTNSS, Bielsko Biala=Szczyrk* Poland

[4] Piaseczny L and Zadrag R 2006 Water fuel emulsions properties for naval diesel engines *Proceeding of Polish Conference on Gas Engines* Czestochowa, Poland

[5] Wu Y-Y, Chen B-C, Hwang J-J and Chen C-Y 2009 Performance and emissions of motorcycle engines using water-fuel emulsions *International Journal of Vehicle Design* **49(1-3)** pp 91-110

[6] Farfaletti A, Astorga C and Martini G 2005 Effect of water/fuel emulsions and a cerium-based combustion improver additive on HD and LD diesel exhaust emissions *Environmental Science & Technology* **39(17)** pp 6792-6799

[7] Marwan A A, Rajakaruna N H and Wagstaff S A 2001 The use of emulsion, water induction and EGR for controlling diesel engine emissions *SAE Technical Papers* 2001-01-1941

[8] Janakiraman G, Johnson J H, James ST, Cuong P W, Khan T H A and Leddy D G 2002 Oxidation catalytic converter and emulsified fuel effectson heavy-duty diesel engine emissions *SAE Technical Papers* 2002-01-1277

[9] Hall D, Thorne C and Goodier S 2003 An investigation into the effect of a diesel /water emulsion composition of the particulate emissions from a heavy-duty diesel engine *SAE Technical Paper* 2003-01-3168

[10] Bertola A and Boulouchos K 2003 Influence of water-diesel fuel emulsions and EGR on combustion and exhaust emissions of heavy-duty, di-diesel engines equipped with common-rail injection system *SAE Technical Paper* 2003-01-3146

[11] Ozcan H and Soylemez M S 2005 Experimental investigation of the effects of water addition on the exhaust emissions of a naturally aspirated, liquefied-petroleum-gas-fueled engine *Energy and Fuels* **19** pp1468–1472

[12] Hountalas D T et al. 2006 Use of water emulsion and intake water injection as NOx reduction techniques for heavy duty diesel engines, *SAE Technical Papers - SAE World Congress* 90162

[13] Hountalas D T, Mavropoulos G C and Zannis T C 2007 Comparative evaluation of EGR, intake water injection and fuel/water emulsion as NOx reduction techniques for heavy duty diesel engines *SAE Paper* No 2007-01-0120.

[14] Nguyen Q A and Wu Y Y 2009 Experimental investigations of using water-gasoline emulsions as a NOx treatment and its effects on performance and emissions of lean-burn spark-ignition engine *Proceedings of the International Conference on Power Engineering-09 (ICOPE-09)*

[15] Yamamoto H 1999 Investigation on Relationship between Thermal Efficiency and NOx Formation in Ultra-Lean Combustion *SAE Small Engine Technology Conference and Exposition* 1999-01-3328

[16] Kim S S et al. 1996 A Study on Efficiency and Emission Enhancements in a 4-Stroke Natural Gas Lean Burn Engine *SAE International Congress & Exposition* 960849

[17] Tabata M, Yamamoto T and Fukube T 1995 Improving NOx and Fuel Economy for Mixture Injected SI Engine with EGR *SAE International Congress & Exposition* 950684

[18] Bruce P and Stebar R F 1976 Water Gasoline Fuels – Their effect on Spark Ignition Engine Emissions and Performance *Meeting of the SAE on combined fuels and power plant* 760547

[19] Zavodov V S, Luksho V A and Neudakhin V I 1982 Operation of carburetor engines on water gasoline emulsions Chemistry and Technology of Fuels and Oils **18(1-2)** pp 25-29

[20] Langer D A, Petek N K and Schiferl E A 2001 Maximizing the effectiveness of waterblended fuel in reducing emissions by varying injection timing or using after-treatment device *SAE* 2001-01-0513

[21] Harrington J A 1982 Water addition to gasoline–effect on combustion, emissions, performance, and knock *SAE Paper* 820314

[22] Peters B D and Stebar R F 1976 Water-gasoline fuels–their effect on spark ignition engine emissions and performance *SAE Paper* 760547

[23] Tsao K C and Li C 1984 Performance comparison of gasoline–water and gasoline-methanol emulsions as spark ignition engine fuels *SAE Paper* 840241

[24] Tsao K C, Wang C L and Miller E M 1984 Performance of gasoline–water fuel in a modified SI engine *SAE Paper* 841399

[25] Plotnikov S A, Buzikov Sh V, Kozlov I S 2020 Study of functionality of a diesel injection nozzle with mixed fuels with insufficient low-temperature properties *Tractors and Agricultural Machines* **1(1)** pp 10-16

[26] Plotnikov S A, Buzikov Sh V, Kozlov I S 2019 Determining of effective performance of a tractor diesel engine running with a fuel composition with rape seed oil additions *Advanced Science* **3 (14)** pp 15-18
[27] Plotnikov S A, Buzikov Sh V 2019 *Theoretical pre-requisites for determining of optimum compositions of* mixed fuels *Advanced Science* **4 (15)** pp 34-38
[28] Plotnikov S A, Buzikov Sh V, Motovilova M V 2019 Study of operation of a diesel-powered TPA with highly-activated fuel *Transportation systems* **1 (11)** pp 12-18
[29] Plotnikov S A, Buzikov Sh V, Cheremisinov P N 2019 Determining of the influence of mixed fuel on the combustion process mechanism for automotive and tractor ICE's *Transportation systems* **2 (12)** pp 4-8
[30] Buzikov Sh V, Plotnikov S A, Kozlov I S 2020 Optimizing of rape seed oil content in mixed fuel used in tractor diesel engines *Transport Bulletin of Volga Region* **5 (83)** pp 72-77

IOP Conf. Series: Materials Science and Engineering 1086 (2021) 012014 doi:10.1088/1757-899X/1086/1/012014

Determining of optimum operation modes of a diesel engine with a multi-component bio-fuel composition

S A Plotnikov[1], A N Kartashevich, M V Simonov and M N Glushkov

Vyatka State University, 36, Moskovskaya Str., Kirov, 610000, Russian Federation

E-mail: [1] plotnikovsa@bk.ru

Abstract. A problem of reduced number of tests was identified during the determining of optimal load and speed performance of a diesel engine with a multi-component bio-fuel composition. The planning method of the experiment for determining of effective ICE performance was improved. For factor space description, graph analyses were carried out using a semi-rotatable 2nd order Box-Behnken statistical design for three factors. The operation studies of a 4ChN 11,0/12,5 diesel engine were performed with the most stable multi-component bio-fuel composition, including the following ingredients, % wt.: RSO – 34.5; ethanol – 31.0; DF – 34.5. The analysis of regression models and 2D sections of response surfaces provided for determining of the optimum advance angle of fuel injection $\Theta_{впр.} = 23.5$ deg. to TDC. By method of superposition of the 2D sections of the response surfaces of the effective efficiency η_e and the specific effective fuel consumption g_e a search for a compromise solution on optimum combination of the levels of the factors to be studied. It was found out that the optimum area of the combination of factors is located in the range of the crankshaft rpm $n = 1400...1550$ min^{-1} of the diesel engine and in the effective load range $P_e = 0.68...0.85$ MPa. Under recognition of the variable nature of the load and speed modes of a diesel engine in a real operation environment, the obtained data are of practical interest.

1. Introduction

Among all known alternative energy sources, a special role is assigned to bio-fuels [1-4]. They are usually manufactured from renewable vegetable and animal origin [5], due to that they arouse keen interest of investigators. Due to substantially different physical and chemical properties of bio-fuels from those of the conventional diesel fuel (DT) [6] an adaptation of ICE design is required. The most acceptable way could be the extended applicability of alternative fuels due to the properties of bio-fuels getting closer to DF properties. Using multi-component bio-fuel compositions can also compensate the deviation of the motor property of one ingredient on account of the other one. As a result of prolonged tests, the following composition of the multi-component bio-fuel was recommended, % wt.: RSO – 34.5; ethanol – 31.0; DF – 34.5 [7]. The stability of the above-mentioned composition to coalescence is up to 35 hours. The use of the recommended composition makes it possible to preserve the viscosity/temperature properties of the bio-fuel composition at ambient temperature changes in the framework of the standard in effect [8].

For thorough analysis of the working processes in the diesel engine, modern methods of experiment planning look attractive. It allows for clear definition of admissible operation ranges of the engine at considerably reduced number of test points. Beside that, a possibility appears for assessment of the influence of certain operation mode factors on the other operation mode factors.

2. Literature Review

In paper [9], an algorithm is reported to have been developed for optimization of multi-factor test plans by branch-and-bound method with introduced limitation by an additional criterion. The algorithm allows for accounting of the limiting criterion, superimposed, e.g., on the temporal realization of the plotted plan of the multi-factor experiment, or on the plan value number. The realization of the reported algorithm makes it possible to find two alternative plans: with the minimum value of the first criterion; with the minimum value of the additional limiting criterion.

For analytical and graphical description of the processes under test, as well as for studies of the influence of the factors on the optimization criteria as well as for the search for an optimum combination of the factors under study, paper [10] reports on use of the response surfaces method. Based on the analysis of the obtained data, it appeared possible to model the parameter scattering of transistors during manufacturing.

The authors of paper [11] were studying the composition of Bentonite used in self-reinforced cement for improvement of its purposes. Using of Box-Behnken statistical design of the test allowed for determining of an optimum combination of ingredients in the self-reinforced cement to obtain the maximum compression strength (above 45 MPa). It was found out that the cement composition shall contain 20% of Bentonite, with 0.45 water-to-powder ratio.

The Department of Technologies of the University of Nis [12] reports on studies on optimization of sunflower oil methanolysis conditions in an ultrasonic reactor. Based on Box-Behnken statistical design of the experiment, response surfaces were obtained. The conducted qualitative analysis of the results has delivered a trustworthy substantiation of the temperature value of about 40°C, of 0.7% wt catalyst quantity, of 92.2% wt. of fat-acid methyl ethers (FAME) content and of 7.5:1 methanol-to-oil ratio.

The authors of paper [13] studied the operation mode of an inductor including the following parameters: frequency (f), gap (h) and flux density (J). The planning method of the experiment based on Box-Behnken statistical design, the following optimum combination of factors was identified: f = 1000 Hz, h = 5 mm, J = 45 mm, to provide for the required inductor performance under conditions defined by the process mode.

Paper [14] reports on a study of wheat gluten hydrolysis process (moisture content: 6.8%, protein content (N 5.70): 78.52% in terms of dry substance). Gluten is used for bread quality improvement, as well as as functional protein supplement in various non-bakery products for their structural improvement. The utilization of Box-Behnken statistical design of the experiment enabled the authors to find optimum working process parameters for ferment hydrolysis of wheat gluten. The data have demonstrated that optimum functional properties of gluten hydrolysates may be achieved at a reaction temperature equal to 40°C, pH = 9 and at a ferment-gluten ratio equal to 0.5 a.u./g of gluten.

The authors of paper [15] report on a study of lead removal process from coal (leaching) excavated in Enugu, using different acids as leaching agents in different conditions. The leaching of coal samples prior to combustion or use in a power plant is required to remove the most part of microelements in this coal to mitigate the risk of environment pollution. Based on orthogonal Box-Behnken statistical design of 2nd order, parameters were obtained allowing for maximum lead removal. It was found out that nitric acid HNO3 in 0.5 mole concentration should be used, thereby, the leaching time is 32 hours with 63 μm admissible particle size of the coal and the required volume of the leaching agent not exceeding 40 ml (solid/liquid ratio).

3. Research objectives

The objective of the study was to determine admissible mode performance specifications of a diesel engine with a multi-component bio-fuel composition.

The mission of the study was to determine the most efficient fuel injection advance angle, optimum load and speed operation ranges of the diesel engine.

4. Methods and materials

General appearance and list of equipment and instruments, see figures 1, 2 and Table 1.

Table 1. Instrumentation and equipment as part of experimental fixture.

Type of test	Equipment	Designation	Note
Time of physical stability	Stop-watch	SOP Pr-2a-2-010 "AGAT"	Accuracy ± 0.1 sec
Bench tests	Automotive/tractor diesel engine	D-245.S2	Power 70 kW
Fuel efficiency	Flowmeter	ACE-50	Accuracy ± 1 %
Frequency of rotation	Tachometer	AVL DIS-peed 492	Accuracy ± 5 rpm
Loading of diesel engine	Scale	RAPIDO	Error ± 0.2 kg

Figure 1. Control panel of diesel engine.

Figure 2. Diesel engine D-245 in the test bench.

The laboratory tests of the stability assessment of the bio-fuel composition were carried out in the chemical laboratory of FGBOU VO Vyatka State University. Experimental studies were carried out in the test laboratory of UO Byelorussian Sate Agricultural Academy.

For description of the factor space with regression models, revealing of regularities of the influence of the loading and the speed mode of the diesel engine on its performance (effective efficiency ηe, specific effective fuel consumption ge) and determining of optimum factor values, a semi-rotatable Box-Behnken statistical design of 2nd order was selected and implemented for three factors. Factors and their levels (Table 2) were selected by a priori ranking method based on the results of one-factor experiments.

Table 2. Factors, levels of fixation and intervals of their variation.

Coded designation of factors	Names of factors, their designations and units of measure	Levels of factors			Intervals of variation
		-1	0	+1	
x_1	Effective loading, P_e, MPa	0	0.47	0.94	0.47
x_2	Rotation frequency of the crankshaft, n, rpm	1,400	1,600	1,800	200
x_3	Advancing angle Θ of fuel injection, deg.	10	18	26	8

The statistical processing of the experimental data, the calculation of regression coefficients, the plotting of the response surfaces and their 2D sections were realized by means of "STATGRAPHICS+" software [16].

5. Result and discussion

As a result of the experimental series of the test design, trustworthy (p = 0.95 after Fisher's F-criterion) [17,18] models of a 2nd order regression analysis of effective efficiency η_e measurement and r=the specific effective fuel consumption g_e measurement were obtained:

$$\eta_e = 33{,}0 + 15{,}3x_1 - 1{,}3x_2 + 0{,}5x_3 - 12{,}3x_1^2 - 0{,}7x_1 \cdot x_2 + 0{,}6x_1 \cdot x_3 - 0{,}06x_2^2 - 0{,}1x_2 \cdot x_3 - 0{,}7x_3^2 \quad (1)$$

$$g_e = 258{,}7 - 230{,}6x_1 + 9{,}8x_2 - 3{,}8x_3 + 208{,}1x_1^2 + 4{,}3x_1 \cdot x_2 - 4{,}2x_1 \cdot x_3 + 0{,}3x_2^2 + 0{,}6x_2 \cdot x_3 + 4{,}8x_3^2 \quad (2)$$

After exemption of negligible coefficients from the regression models (1) and (2), and inclusion of the rest coefficients, get the following appearance:

$$\eta_e = 33{,}0 + 15{,}3x_1 - 1{,}3x_2 + 0{,}5x_3 - 12{,}3x_1^2 - 0{,}7x_1 \cdot x_2 + 0{,}6x_1 \cdot x_3 - 0{,}7x_3^2 \quad (3)$$

$$g_e = 258{,}7 - 230{,}6x_1 + 9{,}8x_2 - 3{,}8x_3 + 208{,}1x_1^2 + 4{,}3x_1 \cdot x_2 - 4{,}2x_1 \cdot x_3 + 4{,}8x_3^2 \quad (4)$$

The analysis of the regression models (3) and (4) demonstrates (fig. 3) that in the experimental are of studies, the maximum value of the effective efficiency η_e = 39.6%, whereas the minimum specific effective fuel consumption g_e = 180.2 g/kWh is observed at almost equal values of the factors being studied.

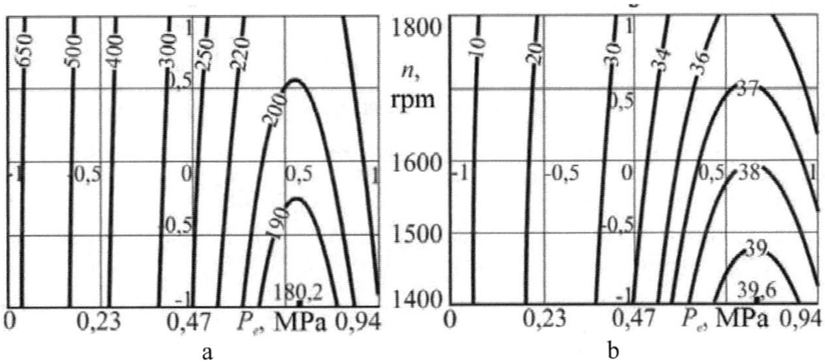

Figure 3. Bidimensional sections of response surfaces of effective efficiency η_e (a) and specific effective fuel consumption g_e (b) at advance angle of fuel injection Θ = 23.5 deg.

Thus, the maximum efficiency η_{emax} is achieved at x_1 = 0.7 (P_e = 0.66 MPa), x_2 = -1 (n = 1,400 rpm) and x_3 = 0.6 ($\Theta_{inj.}$ = 23.5 deg. to TDC). The minimum value of the specific effective fuel consumption g_{emin} is obtained at the following factor values x_1 = 0.6 (P_e = 0.56 MPa), x_2 = -1 (n = 1,400 rpm) and x_3 = 0.6 ($\Theta_{inj.}$ = 23.5 deg. to TDC).

The obtained results rest on real physical sense.

The presence of ethanol and raps seed oil in the multi-component bio-fuel composition causes a certain reduction of its total cetane number. During the operation of the diesel engine, a growth of the fuel ignition delay period will be observed, as well as the shift of the combustion beyond the TDC. To compensate the occurring processes, a certain upturning of the advance angle of fuel injection relative to its standard value $\Theta_{inj.}$ = 22 deg.

The operation of the automotive/tractor diesel engine is the most efficient in overload mode, that is, in the crankshaft rpm decrease range from the rated value n = 1,800 rpm down to the maximum torque rpm n = 1,400 rpm.

Naturally and significantly, both the effective efficiency η_e, and the value of the specific effective fuel consumption g_e, are influenced by the change of the effective load P_e. Thereby, the dependency curves of the parameters η_e and g_e on the parameter P_e, demonstrate a kink at P_e = 0.5…0.7 MPa and a variance of the two remaining factors across the entire experiment area.

This result can be explained by the circumstance that the most economical operation of a piston ICE is usually obtained in the loading range of 60…80% of the rated load.

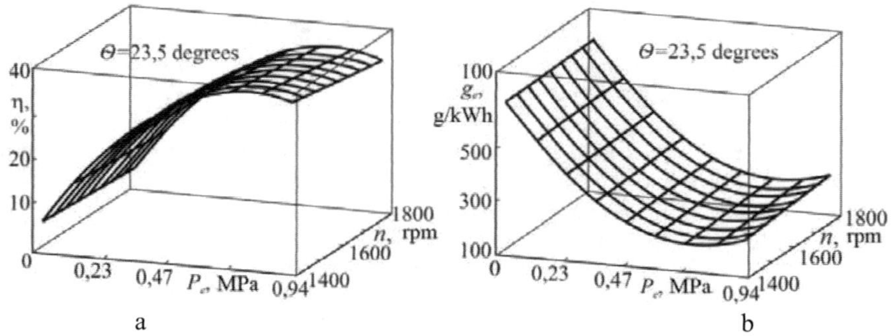

Figure 4. Dependences of effective efficiency η_e (a) and specific effective fuel consumption g_e (b) on effective load P_e and crankshaft rpm value n at advance angle of fuel injection $\Theta_{inj.}$ = 23.5 deg.

For example, at diesel engine crankshaft rpm n = 1,400 rpm and advance angle of fuel injection $\Theta_{inj.}$ = 23.5 deg to TDC, a change of the effective load from 0.47 MPa to 0.6 MPa leads to an increase of the effective efficiency η_e from 5.6% to 39.6% and to the decrease of the specific effective fuel consumption g_e from 694 g/kWh to 180 g/kWh (fig. 4). At further growth of the effective load P_e up to 0.94 MPa, the value of the effective efficiency η_e goes back to 38.2%, whereas the value of the specific effective fuel consumption g_e grows up to 220 g/kWh.

A search for a compromise solution of an optimum combination of the level of the studied factors was carried out by the method of superposition of 2D sections of the response surfaces of the effective efficiency η_e and the specific effective fuel consumption g_e (figure 5).

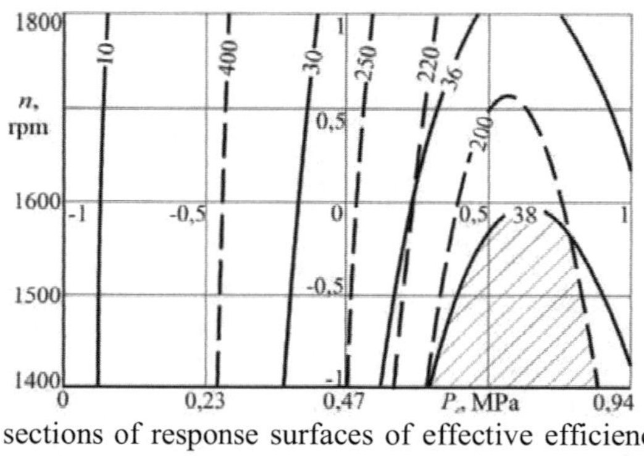

Figure 5. Bidimensional sections of response surfaces of effective efficiency η_e (——) and specific effective fuel consumption g_e (– – –) at advance angle of fuel injection $\Theta_{inj.}$ = 23.5 deg.

It is evident that the optimum area of the combination of the studied factors (shaded area in fig. 5) is located in the variable crankshaft rpm range of the diesel engine n = 1,400…1,550 rpm and in the variability range of the effective load P_e = 0.68…0.85 MPa at advance angle of fuel injection $\Theta_{inj.}$ = 20…25 deg. to TDC. At such combination of the factors, an almost practical effective efficiency η_e in the experimental range (at least, 38%) at a low specific effective fuel consumption g_e of max. 200 g/kWh is provided for.

Conclusions

For description of the factor space and reduction of the number of tests during the determining of optimal load and speed performance of a diesel engine with a multi-component bio-fuel composition, the method of experiment design was improved.

The optimum area of the combination of the factors is located in the variable crankshaft rpm range of the diesel engine n = 1,400…1,550 rpm and in the variability range of the effective load P_e = 0.68…0.85 MPa at advance angle of fuel injection $\Theta_{inj.}$ = 23.5 deg. to TDC.

Under recognition of the alternating nature of the load and the speed modes of the diesel engine in real operation environments, the data of diesel engine operation as part of a traction vehicle are of practical relevance.

References

[1] Plotnikov S A 2011 Improving the performance of diesel engines by creating new alternative fuels and improving fuel supply equipment (Nizhny Novgorod: Abstract. thesis Doct. Eng. science)

[2] Plotnikov S A, Smol'nikov M V and Cheremisinov P A 2019 Expanding the multi-fuel capacity of automotive diesel using alternative fuels *Izvestiya MSTU «MAMI»* **3 (41)** 66–72

[3] L'otko V Lukanin V N and Hachiyan A S 2000 *Use of alternative fuels in internal combustion engines* (Moscow: Publishing house of MADI) p 311

[4] Plotnikov S A, Kartashevich A N and CHeremisinov P N 2017 Advanced mixtures of diesel fuel with rapeseed oil for use in tractor diesels *Engine building* **4** 21–24

[5] Markov V A, Devyanin S N, Zykov S A and Gajdar S M 2016 *Biofuels for internal combustion engines* (Moscow: SIC «Engineer») p 292

[6] Terent'ev G A, Tyukov V M and Smal F V 1989 *Motor fuels from the alternative raw material resources* (Moscow: Chemistry) p 272

[7] Plotnikov S A, Glushkov M N, Kartashevich A N and Shaporev V A 2020 *Multi-component biofuel composition* (The application for a patent of the Russian Federation No. 2020123967(W20041379))

[8] Plotnikov S A, SHipin A I, Kartashevich AN and Malyshkin P YU 2020 *Method for obtaining a multicomponent biofuel composition* (The application for a patent of the Russian Federation No. 2020120544(W20035021))

[9] Koshevoj N D, Burleev O L and Kostenko E M 2010 Optimal planning of the experiment with the introduction of restrictions on an additional criterion *Bulletin of the Samara State Technical University. (Series «Technical Sciences»* vol 2*)* pp 63–67

[10] Denisenko V V 2004 Simulation of the spread of transistor parameters in CMOS SBI *Components and Technologies* **1** 28–34

[11] Arabi N S, Qadi A L, Mahmoud B A, Alhasanat Ahmad, Dahamsheh A L and Sleiman A L 2016 Using of Box-Benken method to predict the compressive strength of self-compacting concrete containing wadi musa bentonite *American Journal of Engineering and Applied Sciences* **9 (2)** 406–11

[12] Jelena M Avramović, Dragana B Radosavljević, Ana V Veličković and Ivan J. Stojković 2019 Statistical modeling and optimization of ultrasound-assisted biodiesel production using various experimental designs *Zastita Materijala* **60** 70–80

[13] Leuca T, Novac M, Stanciu B, Burca A and Codrean M 2014 Using minitab Box-Benken software to optimize the induction heating process *Journal of Electrical and Electronics Engineering* **7-1** 73-6

[14] Jovanovic J, Stefanovic A, Sekuljica N and Knezevic-Jugovic Z 2018 Production of wheat gluten hydrolysates with improved functional properties: optimization of operating parameters by statistical design *Journal of Hygienic Engineering and Design* **24** 90–100

[15] Ezenwa O S, Oluchukwu A C and Agaba A M 2019 Optimization and kinetic modeling of the removal of lead from enugu coal by acid leaching *Journal of Energy Research and Reviews* **3-1** 1-13

[16] State Standard Ru R 50.1.040–2002 Statistical technology. Design of experiments. Terms and definitions (Moscow) p 39

[17] Grachev YU P 1979 *Mathematical methods for planning experiments* (Moscow: Food industry) p 200

[18] Mel'nikov S V, Alyoshkin V R and Roshchin P M 1980 *Experiment planning in researches of agricultural processes* (Leningrad: Ear) p 168

Developing an unmanned vehicle local trajectory using a reinforcement learning algorithm

T S Evdokimova[1,3], A A Sinodkin[1], M I Tiurikov[1], L O Fedosova[1] and A V Kalyashina[2]

[1] Nizhny Novgorod State Technical University n.a. R.E. Alekseev, Nizhny Novgorod, Russian Federation
[2] Department of Laser Technologies of KNRTU KAI

E-mail: [3]evdokimovats97@gmail.com

Abstract. This article describes the algorithm development for constructing a local trajectory for an unmanned vehicle or for implementation in an ADAS system using the reinforcement learning method. A special part is dedicated to reinforcement learning. One of the methods that is best suitable for the task conditions will also be implemented. This method will allow bypassing obstacles and reaching the specified short target points.

1. Introduction

Currently, one of the promising areas in the robotics field is the unmanned robots development and in particular - vehicles. The purpose of developing unmanned transport systems is to prevent accidents. Road users' protection from unforeseen situations and the human factor minimization is of particular importance. One of the main functional purposes of active driver assistance systems is to prevent and avoid an emergency, both with static and dynamic obstacles.

At the moment, in addition to completely unmanned vehicles, driver assistance systems (ADAS) are being created. Such assistance systems are aimed at predicting possible unforeseen situations for the driver while driving a vehicle. Apart from notifying the driver of an existing danger, this system involves intervention in the vehicle control, applying engine control algorithms, braking system and steering.

One of the most important stages in the unmanned vehicle development is constructing a route. Path developing algorithms require up-to-date and timely environment data in which the unmanned vehicle will be located. The route is divided into global and local. A global route is a trajectory from a starting point to a target. The local planner is a separate fragment of the global planner path that allows achieving short goals, taking into account the presence of both static and dynamic obstacles.

2. Hardware equipment

To implement the local trajectory development the vehicle was equipped with a certain set of information measuring devices. Depending on the conditions in which the solution of the task will take place and taking into consideration the hardware equipment, a set of information measuring devices was selected. At the same time, it is important to take into account the systems expediency of the systems used. Since the tasks of developing local trajectories for unmanned vehicles are considered, the most accurate data on the mobile platform movement is needed. Therefore, reliable and accurate rangefinders are required to determine the map of obstacles, and to determine the movement in space, only those measuring devices that will determine the direction and location relative to themselves at different points in time should be considered, and thus it one should realise that odometric, inertial data are suitable.

The following equipment was used to develop the local trajectory:

• Ultrasonic sensor HC-SR04;
• NVIDIA Jetson Nano Developer Kit;
• Laser rangefinder URG-04LX-UG01;
• Controller NUCLEO-F031K6;
• Sensor with inertial system BNO055 Xplained Pro.

3. Implementation of a local trajectory development for an unmanned vehicle

To implement the local trajectories development it was necessary to configure the data exchange between the Unity simulation environment and the training environment and to configure the processing of various data coming from the simulation environment for the subsequent application of this data to learning through the Asynchronous Actor-Critical Agents algorithm (Figure 1).

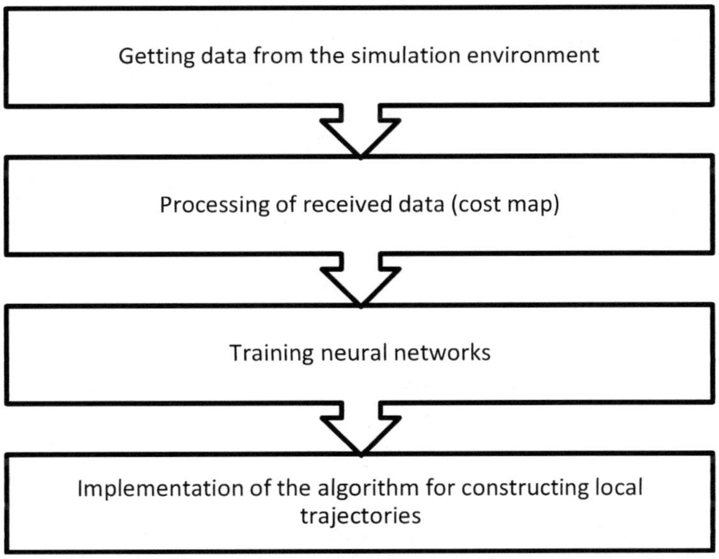

Figure 1. Algorithm for the development sequence for building a local trajectory

3.1 Receiving and exchanging data with the Unity simulation environment

The general data exchange algorithm between the simulation environment and the learning environment (Figure 2). The algorithm has two components - the simulation environment and the training environment. They interact in such a way that data about the world in which the model exists (obstacle map, data on the position, direction and the simulated robotic platform speed) comes from the simulation environment. The learning environment, in turn, conveys data about the actions that the platform should take in in the simulated environment. Interaction with the simulation environment is done through the ZeroMQ messaging library. ZeroMQ is a high-performance asynchronous messaging library designed for use in distributed or parallel applications.

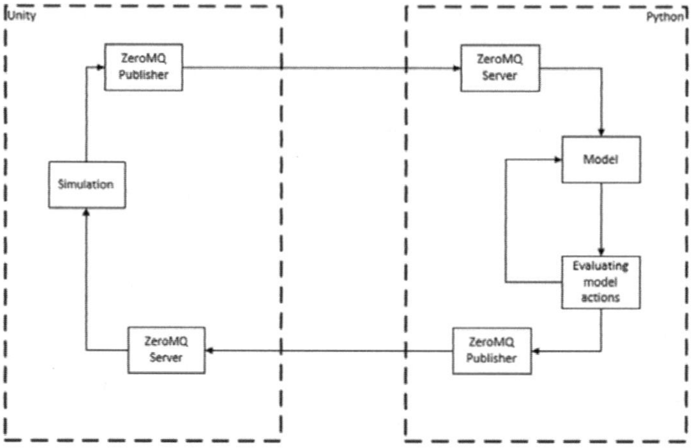

Figure 2. Data exchange between the simulation environment and the training environment

3.2 Data preparation required for training a neural network

To implement neural network training for developing an unmanned vehicle local trajectories, it is necessary to use more than one type of input data, but to take into account all the factors and parameters necessary for the neural network correct training.

Implementation requires the following data and parameters:

- Cost card;
- Position (x);
- Position (y);
- Movement speed (υ);
- Wheels rotation angle (φ);
- Current orientation relative to the global system;

To process the cost map, a convolutional neural network is used, the rest of the data represents the kinematic state of the vehicle (Figure 3) at the current moment and a multilayer perceptron is used for these parameters.

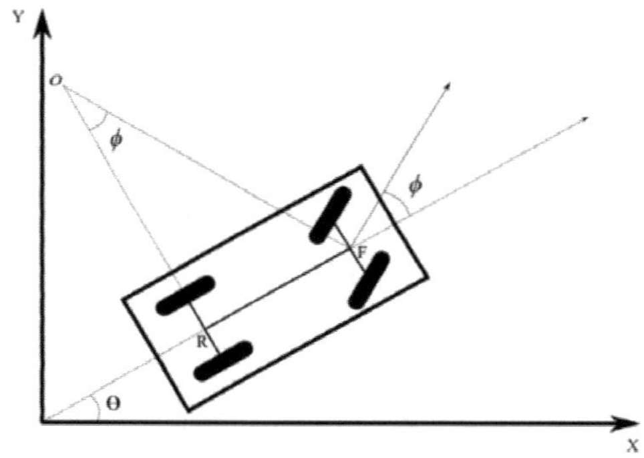

Figure 3. Kinematic diagram of a non-holonomic vehicle position (x), position (y), speed (y), wheel angle (φ), current orientation relative to the global coordinate system (θ).

One of the main stages is the cost map creation (Figure 4), since it is one of the leading parameters designed to solve the local trajectory planning problem.

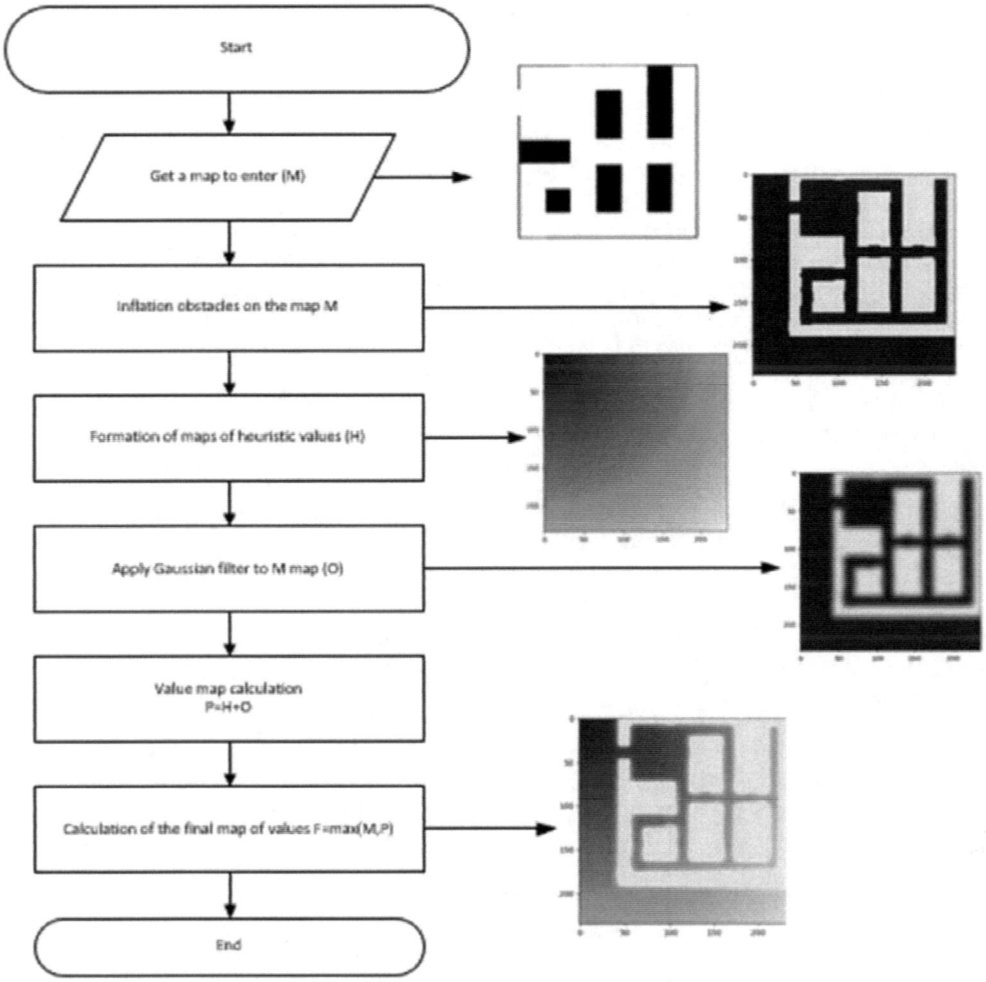

Figure 4. Algorithm for creating a cost map

Initially, the type in which data from measuring devices will be sent for processing was determined. When using a laser rangefinder and the ultrasonic sensors, a two-dimensional obstacle map, subsequently processed, is obtained.

The next step will be inflation (increase) of obstacles on the input map M, this is necessary so that the unmanned vehicle can stop at some distance from the obstacles. This stage is implemented by means of the Open CV library [3] (computer vision library as well as image processing), since, in fact, an image as an input is received.

This map is then presented as a cost map. The first is to fill an empty array with heuristic values. These costs are formed taking into account the distance from the center to the specified segment (cell). This is necessary so that in the future the path cost for the goal can be calculated.

Next is the input map M processing; it is necessary to apply a Gaussian filter (blur filter), which is most often used when working with convolution matrices. This step is also carried out using the Open CV library and is necessary to determine the cost for cells near obstacles. Then a summary map of costs consisting of a heuristic costs map and a map with determining the cost near obstacles is calculated. And the final step in processing the value map is the final map calculation with the calculation of the maximum values.

And in the end, applying all the data, it is necessary to train the model applying one of the reinforcement learning methods - AACA.

To implement this algorithm, it is necessary to use two neural networks, but at the same time, each of them will use the functions and improvements that are in other reinforcement learning algorithms. AACA algorithm (Figure 5) combines the best parts of both methods, i.e. the algorithm predicts both the cost function V (s) and the optimal policy function π (s) (s-state). The training agent uses the Value (Critic) to update the optimal policy (Actor) function. It should be noted here that in this case the policy function means the probabilistic distribution of the action space. The training agent determines the conditional probability P (a | s) i.e. the parameterized probability that the agent will choose the action a (action) in the current state s.

In AACA, there is a global network and several working agents, each with its own set of network parameters. Each of these agents interacts with its own copy of the environment at the same time as other agents interact with their environment. The reason of its performing better than neural networks with a single agent (besides making more work faster) is that each agent's experience is independent of the experience of others. Thus, the overall experience available for learning becomes more diverse. [5.6]

Understanding the use of benefit estimates, not just discounted rewards (formula 1.1 where $\gamma(r)$ is a discount factor indicating to the agent which of its actions were rewarded and which were penalized) allows agents to determine not only how good its actions were, but also how much better they were than expected (formula 1.2 where Q(s, a) is the value of action in state s, V(s) shows how good the current state is without considering actions). Intuitively, this allows the algorithm to focus on where network predictions were missing.

Total reward: $R = \gamma\,(r)$ (1.1)

Advantage: $A = Q\,(s, a) - V\,(s)$ (1.2)

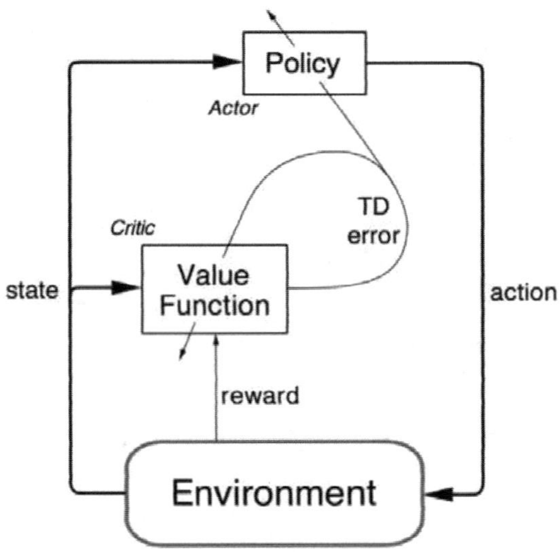

Figure 5. Actor Critic method Algorithm

The ResNet18 convolutional neural network is used to process the map and a multilayer perceptron is used to process the kinematic parameters. At the output, two vectors are obtained and then they are combined into one. Subsequently, after the AACA algorithm two values are obtained: state estimation and action estimation.

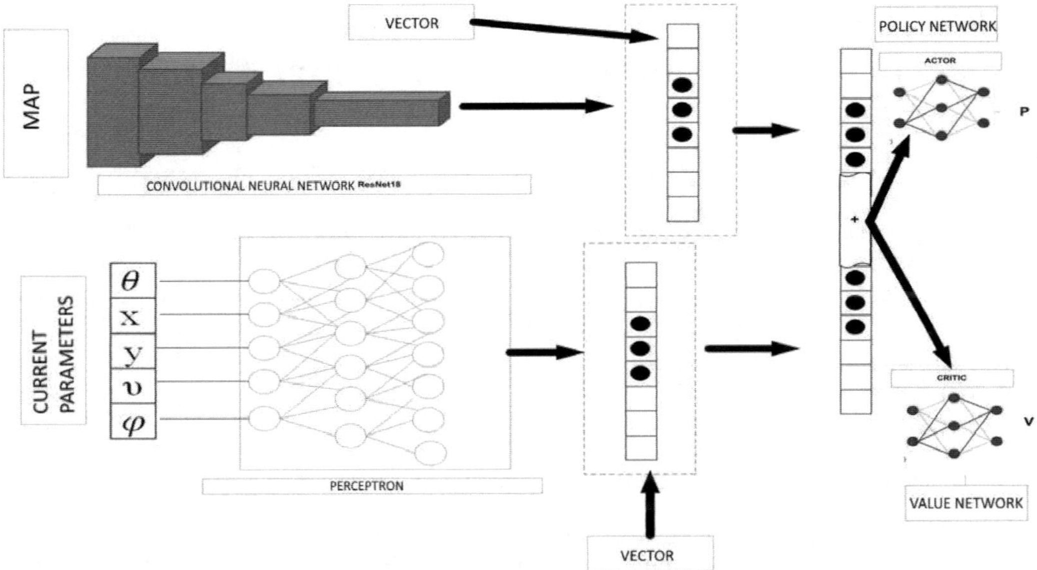

Figure 6. Actor Critic algorithm diagram in terms of solving the trajectory planning problem

4. Conclusion

As a result of the work, a reinforcement learning algorithm was developed. It showed that for the most correct operation and the algorithm learnability, it is necessary to put the system in a large number of situations. The advantages are that these algorithms have the ability to learn and they will allow considering more contingencies in the future and solve them.

A number of problems remain that can be identified as further directions of the algorithm development. One of the problems is that this system cannot be fully implemented in any vehicle, but if to train the system on one model, the entire series of these models can be implemented with the developed system. There are also requirements for increasing the operation speed and learning of algorithms, as well as the use of more complex algorithms for developing local trajectories for the comparative function. The research this direction will be continued. Tests were carried out in a simulation environment; the next step will be to introduce this system into an unmanned vehicle prototype.

5 References

[1] Dalal N and Triggs B 2005 Histograms of oriented gradients for human detection *CVPR* pp 886–893
[2] *Convolutional neural network* URL: https://intellect.ml/svertochnayanejronnaya-set-convolutional-neural- network-cnn-6013 (application date 30.05.2018)
[3] Bradski G and Kaehler A 2008 *Learning OpenCV* O`Reilly Media p 580
[4] Rosten E and Drummond T 2006 Machine learning for high-speed corner detection *European Conference on Computer Vision* **1** pp 430–443
[5] Konda V R and Tsitsiklis J N 1999 Actor-Critic Algorithms *NIPS* **13** pp 1008-1014
[6] Bhatnagar S. et al. 2009 Natural actor-critic algorithms *Automatica* **45** 11
[7] Shariq Iqbal and Fei Sha 2019 *Actor-Attention-Critic for Multi-Agent Reinforcement Learning*

IOP Conf. Series: Materials Science and Engineering 1086 (2021) 012016 doi:10.1088/1757-899X/1086/1/012016

Development and study of ways to extend the multifuel capability of ICE

S A Plotnikov[1,3], A N Kartashevich[1], P Ya Kantor[1,2] and M N Vtyurina[2]

[1] Vyatka State University, 36, Moskovskaya Str., Kirov, 610000, Russian Federation
[2] Vyatka State Agricultural Academy, 133, Octyabrsky Avenue, Kirov, 610017, Russian Federation

E-mail: [3] plotnikovsa@bk.ru

Abstract. In the recent years, no considerable growth of use of alternative fuels by vehicle fleets has been observed, in spite of the availability of a vast volume of research data and recommendations. The identified reason is the reluctance of manufacturers to produce ICE modifications for operation with fuels different from DF. The problem could be solved by development and study of ways to extend the multifuel capability of ICE. Basic ways were identified to extend the limits of the usability of alternative fuels with different physical and chemical properties. Regressions were suggested allowing for determining the chemical nature of additives and their degree of influence on the decreasing of the kinematic viscosity of the fuel. A mathematical solution was obtained describing conditions of physical existence of a fuel emulsion. An analysis of a number of combustion inhibitors was carried out. A mechanism of inhibition of additives of new generation was analyzed. Subcaloric fuel combustion process (generator gas) was studied, ways to improve its efficiency were suggested. Based on the theoretical analysis and experimental data, generalized assessment criteria were formulated for the usability limits of non-conventional energy materials. The obtained results could help to essentially extend the fuel base of engines in real operation environments.

1. Introduction

In the recent years, no considerable growth of use of alternative fuels (AF) by vehicle fleets and in other branches of the national economy has been observed. Simultaneously, there is a vast volume of various research data in the sphere of alternative energy [1, 2, 3]. Published sources report on developed recommendations and experimental ICE designs, allowing for prompt introduction of a broad range of non-conventional energy materials in land-based vehicles.

The analysis of the causes of this juncture leads to the following main conclusion. It is known that the operational properties of non-conventional fuels are considerably different from those of the conventional DF (Table 1). Manufacturers are reluctant to manufacture different individual ICE series for specific alternative fuel types, which, in addition, differ in on-board storage facility designs, feed and (or) ignition methods, control packages and other operational limitations.

For problem solution it is highly advisable to develop and study the ways to extend the multifuel capability of ICE's by extending the usability limits of alternative energy sources. These are, first of all, fuels with physical and chemical properties deviating from the requirements of the standards currently in effect.

Table 1. Distinctive properties of alternative fuels

Alternative fuel	Differences of properties	Influence on engine performance
1. Vegetable oils	High kinematic viscosity, tendency to deposit	Increased loads on engine parts, increased deposits, appreciably reduced reliability and durability
2. Lowest aliphatic alcohols	Low values: - Cetane number - Heat of combustion - Physical stability of the mix	Reduced output power and fuel efficiency, increased cost of design manufacture
3. Compressed natural gas	Molecular state	Increased cost of design and manufacture, appreciably reduced reliability and durability
4. Industrially produced gases (biogas, generator gas)	Low lower heating value, molecular state	Reduced output power and fuel efficiency, increased cost of design manufacture
5. Water	Poor lubrication properties, very high corrosion activity	Reduced output power and fuel efficiency, appreciably reduced reliability and durability
6. Liquefied ammonia	Low lower heating value, appreciable corrosion activity	Reduced output power and fuel efficiency, appreciably reduced reliability and durability

The range is rather broad – heavy fuels, fuel emulsions and compositions with low stability, fuels with low self-combustion ability, fuels with poor low-temperature properties, etc. [4].

2. Literature Review

In paper [5], based on the experimental results of paper [6] on the mechanical properties of mixtures of diesel fuel with rape seed and sunflower oils, empirical formulas for viscosity calculation of such mixtures were obtained. For approximation, two approaches were used, named by the authors algebraic and logarithmic. The divergence between the experimental and the calculated results reached up to 27% in case of rape seed oil.

The results of rather recent studies of fuel mixtures containing ethanol are reported in paper [7] and thesis [8]. The studies have demonstrated the efficiency of use as motor fuel of a mixture containing 96% of diesel fuel and 4% of absolute ethanol. Beside that, not solved problems are identified associated with increased emissions of carbon monoxide and unburnt hydrocarbons, as well as with the stability of the mixtures containing high ethanol percentage.

Paper [9] reports on test results of a K6S310DR locomotive diesel engine operating with diesel fuel mixed with natural gas in several power output modes. It was found out that at 5% methane addition the smoke and carbon monoxide emissions, thereby, the most positive effect is reached in the rated mode of operation. A problem was identified, associated with the feeding of heterogeneous fuel mix into the cylinders of a diesel engine.

The problems associated with the stability and the viscosity of rape seed oil based fuel emulsions were considered in [10, 11]. Emulsions were studied with up to 40% methanol content, stabilized with alkenyl succinimide. It was found out that the methanol concentration growth is followed by decreasing mixture viscosity, thereby, the stability time of the emulsion also gets bad. Thus, in this situation, important operation factors contradict each other.

The maximum stability time of 51 min was achieved for an emulsion containing 10% wt. of methanol; thereby, the emulsifier concentration was extremely high – about 9%. The components ratio considered by the authors to be optimum – 29% of methanol and 5.4% of the emulsifier - has a very small time of emulsion stability of 14.5 min. Measurement data [11] demonstrate also insignificant viscosity of the fuel mixture along with the concentration growth of the emulsifier.

Papers [12, 13] suggest variants of the multifuel capability of diesel engines for areas with tropical climate using palm oil as alternative fuel. The authors note that the addition of palm oil to diesel fuel lead due to higher superficial tension coefficient and viscosity of the additive to the growth of the average droplet diameter of the pulverized fuel and to a narrower spray angle. These factors, in their turn, are followed by an increase of the ignition delay, combustion duration and, as a consequence, by a decrease of the engine efficiency. For compensation of the specified negative effects, it is recommended to additionally introduce hydrogen peroxide into the fuel mixture. In accordance with authors' statement, free radicals generated during the peroxide degradation will act as inchoative centers of chain reactions intensifying the oxidation of the fuel. Unfortunately, the theoretical calculations are not supported by any experimental data.

Paper [14] reports on studied of diesel engine operation with transesterified mustard oil with addition of vol.15 ppm titanium dioxide TiO_2 powdered to average particle size of 50 nm. It is stated that the titanium dioxide presence leads to delay time reduction of the ignition, accelerates the fuel vaporization grade, lowers the carbon activation temperature which stimulates its completer combustion, and accelerates the carburation. There is also a statement, that TiO_2 additive leads to reduced emission of carbon and nitrogen oxides and reduced smoke content in the exhaust.

3. Research objectives

The objective of the study was the extension of the multifuel capability of ICE by wider usability limits of liquid alternative fuels.

The mission of the study was the development of adaptation ways of physical and chemical properties of alternative fuels to DF properties.

4. Methods and materials

The investigation methods were theoretical studies, laboratory tests, bench testing of diesel and spark-ignition ICE's.

Table 2. Devices and equipment as a part of experimental installation

Type of test	Equipment	Designation	Note
Kinematic viscosity	Viscosimeter	VPZh-2	Divergence ± 1 %
Stability	Stop-watch	SOP Pr-2a-2-010 "AGAT"	Accuracy ± 0.1 sec.
Limiting temperature of filtration capability	Temperature measurement instrument for filtration capability	ITF	No regular error
Bench tests	Internal combustion engine	D-245.S2, DDE UP168-1	Power 70 kW Power 2.2 kW
Toxicity of exhaust gases	Gas analyzer	MGT-5	Accuracy ± 3 %

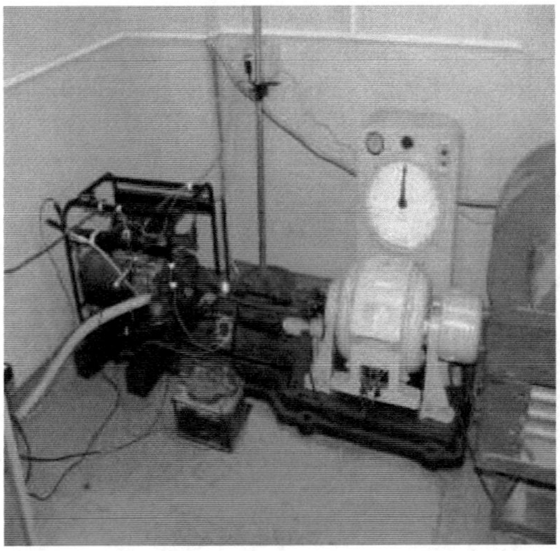

a)

b)

Figure 1. General view: a) spark-ignition engine DDE UP168-1; b) diesel engine D-245

General appearance and list of equipment and instruments, see figure 1 and Table 2.

The theoretical analysis and laboratory tests of physical and chemical properties of motor fuels were conducted by FGBOU VO Vyatka State University. The bench tests of the ICE's were carried out in UO Bielorussian State Agricultural Academy and FGBOU VO Vologda State Dairy Farming Academy named after N.V. Vereschagin.

The processing of the experimental data and the graph plotting were carried out in Microsoft Exel and Corel Draw applications [15].

5. Result and discussion.

1. One of the most critical properties restricting alternative fuel use is the kinematic viscosity. Appropriate accuracy is vitally important for determining of the chemical nature of the additives allowing for required decreasing of viscosity of the mixed fuel.

Absence of a reliable theoretical molecule interaction model of different high-molecular substances for calculation of the mixture viscosity makes it vital to construe particular empirical formulas. To describe the dependency of the viscosity of the mixture versus component concentration, exponential approximation is used sometimes [22]. Such approximation could be well consistent with experimental results in a certain range of concentrations, however, it is not backed by any objective physical reasons, being in this sense merely a formal expedient. Mixture viscosity calculations via the average balanced flowability of the components often misleads to results differing by several times from the experimental data.

The processing of the results of our measurements of the kinematic viscosity of mixtures of diesel fuel with rape seed oil containing 0 to 50% of the oil [16] demonstrated that the dependency of viscosity ν versus oil content c is rather satisfactorily described by quadratic regression

$$\nu = \nu_0 + ac + bc^2, \tag{1}$$

where $\nu_0 = 4{,}524$, $a = 96{,}54 \cdot 10^{-3}$ and $b = 3{,}407 \cdot 10^{-3}$ (ν in cSt, c in %).

If the mole fraction of the rape seed oil χ in the mixture is taken the argument, then, the measurement results fit well into a linear function

$$\nu = 4{,}17 + 69{,}24\,\chi\text{, cSt.} \tag{2}$$

The deviation of the experimental data from the results of the calculation results from the above-referred formulas do not exceed 3%.

In order to reveal the degree of influence of different physical factors on the mixture viscosity, we have undertaken for experimental evaluation of the viscous flow energy E. For that, viscosity versus temperature dependency was studied. In accordance with the generally acknowledged theory of activation, this dependency shall appear as follows:

$$\nu = Ce^{E/(kT)}, \tag{3}$$

where k is the Bolzmann constant; T is the absolute temperature. The value C has a weak dependence on T, and it can be assumed constant in a narrow temperature interval.

The obtained experimental evaluations of the activation energy, see Table 3.

The data of Table 3 make a certain trend to the growth of the activation energy along with the growth of the concentration of the admixtures in the solution visible, which could be associated with the presence of an additional dipole-dipole attraction of polar fragments of the oil and the additive molecules. However, the specified differences never reach beyond the experimental error, and cannot obviously be a reason for significant variations of the mixtures' viscosity values observed in the reported measurements.

Table 3. Activation energy of viscous flow in mixtures of diesel fuel with rapeseed oil

Mixture composition	Activation energy E, eV
100% diesel fuel	0.120 ± 0.005
80% diesel fuel + 20% rape seed oil	0.101 ± 0.010
60% diesel fuel + 40% rape seed oil	0.135 ± 0.009
50% diesel fuel + 50% rape seed oil	0.128 ± 0.008
49% diesel fuel + 49% rape seed oil + 2% DIFRON H372 additive	0.132 ± 0.009

We have also studied the influence of DIFRON H372 cetane number increasing additive on the fuel mixture viscosity, with 2-ethyl hexyl nitrate as effective substance. It was found out that adding of 2% (wt.) of the additive to the fuel mixture leads to a decrease of the kinematic viscosity by 10% and above.

The data of the theoretical analysis and the laboratory tests [16] confirm the indicated extension way of the usability limits of high-viscosity and weight-increased fuels.

2. The use of fuel emulsions and compositions goes hand in hand with the other not less important problem – the insufficient physical stability.

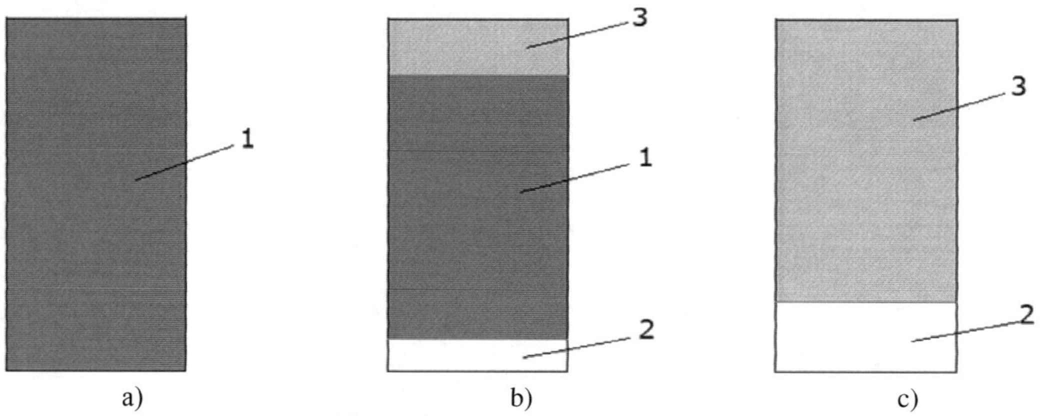

Figure 2. Three emulsion states:
a) homogeneous emulsion; b) partial stratification; c) full stratification

Figure 2 shows in terms of time the three states of the emulsion of diesel fuel, distilled water +and ethanol. Definition of the volumes: 1 – $V_э$ (emulsion volume), 2 - $V_{вс}$ (volume of the distilled water + ethanol solution), 3 – $V_{дт}$ (volume of the diesel fuel (DF)), V_0 is the total volume of the entire emulsion composition.

Then, in any moment of time, the following equation is true:

$$V_{дт} + V_{вс} + V_э = V_0 \quad (4)$$

We assume that the volume changes of the phases in the emulsion will develop in the course of time exponentially, since most of natural processes in the nature and in the engineering develop exponentially.

We set the change rate of the emulsion with the time:

$$(V_э)'_t = \mathrm{c}e^{-\mathrm{k}_3\mathrm{t}} \quad (5)$$

Similarly, we write now the change rates of the of the volumes of the diesel fuel and of the water-ethanol solution. Then, equation (1) can take the following appearance:

$$ae^{-k_1 t}\mathrm{dt} + be^{-k_2 t}\mathrm{dt} + \mathrm{c}e^{-\mathrm{k}_3\mathrm{t}}\mathrm{dt} = 0. \quad (6)$$

Having integrated equation (3) we obtain the following solution:

$$\frac{a}{k_1} - \frac{ae^{-k_1 t}}{k_1} + \frac{b}{k_2} - \frac{be^{-k_2 t}}{k_2} + c\frac{e^{-k_3 t}}{k_3} = V_0. \quad (7)$$

The values a, b, c correspond to the initial volume change rates of the respective phases, whereas the coefficients k_1, k_2, k_3 define the the respective changes of the respective change rates of the respective volumes. Along with the dropping down of the water-ethanol drops, the density of the remaining emulsion will get back to the DF density on the top, и and the water-ethanol mix density on the bottom. Thus, it can be easily calculated, by which moment of time the emulsion volume will be zero, that is, the stability time can be obtained [17, 18].

3. Most of alternative fuels have the cetane number (CN) lower than that of the DF.

The analysis of references shows that the combustion process can be influenced by means of special additives.

It is known that the combustion process is distinguished by the radical mechanism. In the initial stage, free atoms, radicals and other instable compounds are generated which have high chemical activity. Subsequently, these compounds react with the initial substance, the chain furcates, and multiple active centers are built. Thereby, the reaction significantly accelerates. The interactions of the active centers between one another, with molecules of other compounds, with the combustion chamber (CC) walls, on the other hand, decelerate the reaction rate - a chain break occurs. With reference to the operation of a diesel ICE, it is vitally important to inhibit the chain furcation stage, which can affect the uniformity of the combustion process.

The main drawback of ethanol as alternative fuel is its low flammability. Quantitatively, this fact is illustrated by a significant decrease of the cetane number of the alcohol-DF mixture compared with pure DF. The cetane number of pure ethanol is 8, whereas the optimum cetane number for normal diesel engine operation must be 40…50. The reduced cetane number of the mixture is stipulated by the following circumstances. The specific heat of evaporation of ethanol is 3.5 times higher than that of DF, consequently, the temperature of the fuel-air mixture in the cylinder is reduces, and the delay time of its self-ignition grows, accordingly. On the other hand, due to the presence of an oxygen atom in the alcohol molecule, the combustion time of the fuel mixture is reduced followed by a rougher diesel engine operation with ethanol-fuel mixture.

An improvement of the properties of a fuel mixture can be achieved with additives which, as set forth above, either lower the self-ignition temperature or prolong the combustion time

As combustion inhibitor increasing the cetane number, we suggest using molybdenum disulfide MoS_2 [19].

The assumed inhibition mechanism of the additive looks as follows: At interaction of the initial hydrocarbon with oxygen, free radicals are generated in the first oxidation stage:

$$RH + O_2 \rightarrow R\cdot + HOO\cdot. \quad (8)$$

Then, radical R· joins an oxygen molecule and becomes a peroxide radical $RO_2\cdot$:

$$R\cdot + O_2 \rightarrow RO_2\cdot, \quad (9)$$

In its turn, the latter detaches a hydrogen atom from another hydrocarbon molecule:

$$RO_2\cdot + RH \rightarrow ROOH + R\cdot, \quad (10)$$

building a hydroperoxide and a new free radical R· which continues the chain. In presence of MoS_2, the kinetic chain can be broken. The catalyst activity of molybdenum disulfide is associated with the presence of anionic vacancies on the lateral crystallite faces. In MoS_2 structure (Figure 3), there are three sulfur atom types with different co-ordination by molybdenum. A part of Mo ions is localized in partially unfinished trigonal prisms – on end faces and layer corners. It is supposed that the kinetic chains are broken by the reactions of peroxide radicals with molybdenum disulfide on the surface of its particles expressed by radicals' joining the "protruding" molybdenum atoms marked in Figure 3. As a result, the fuel oxidation process is slowed down. The combustion process is extended in time, the temperature and pressure growth gets smoother which is equivalent to a higher cetane number.

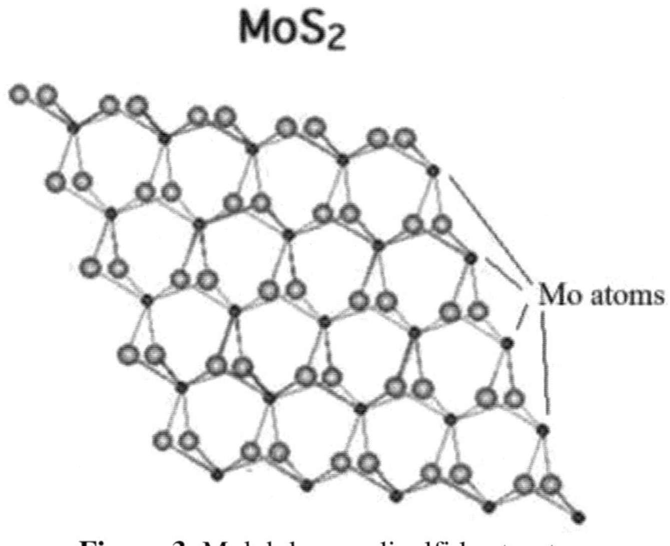

Figure 3. Molybdenum disulfide structure

One more, a bit less probable but generally possible influence mechanism of molybdenum disulfide on the combustion process of the fuel mixture could be as follows. At temperatures $t \geq 650^oC$ characteristic for air-fuel mixture prior to self-ignition, dehydration of the hydrocarbons of the hydrocarbons could take place, e.g., by the following reaction:

$$C_nH_{2n+2} \rightarrow C_nH_{2n+1}\cdot + H\cdot, \quad H\cdot + H\cdot \rightarrow H_2. \quad (11)$$

Along with that, in ethanol is present in the fuel mixture, it could also be dehydrated as follows:

$$CH_3 - CH_2 - OH \rightarrow CH_3 - CHO + H_2, \quad (12)$$

since the characteristic energies of C – H and O – H links are close to each other (appr. 430 kJ/mole). The resulting hydrogen enters into a hydrocracking reaction with cyclic (naphtenetic) hydrocarbons, which are mostly homologue to cyclohexane and cyclopentane. In these processes, molybdenum disulfide acts as a catalyst hydrocracking which can occur as follows:

IOP Conf. Series: Materials Science and Engineering 1086 (2021) 012016 doi:10.1088/1757-899X/1086/1/012016

$$C_nH_{2n} + H_2 \rightarrow C_nH_{2n+2}; \tag{13}$$

for example, cyclopentane hydrocracking leads to the opening of the ring to n-pentane.

It is known that at equal C number, the cetane number of a cyclic hydrocarbon is, as average, by 20…40 less than that of the respective alkane, being a specific case of the general tendency of cetane number growth with the growth of the relative number of hydrogen atoms in a hydrocarbon molecule. Thus, the result of the hydrocracking process is the increased cetane number of the fuel, which, in its turn, leads to a shorter delay of the self-ignition and a smoother combustion process. Since the cyclic hydrocarbons content in the diesel fuel is rather big (20% to 60%) the influence of the assumed mechanism could be significant.

The results of the reported theoretical excerpts and the analysis data of the diesel engine performance parameters [19, 20] allow to consider the indicated extension way of the usability limits of low-cetane fuels vital and promising.

4. An essential problem of use of generator gas as a fuel are its low heat value and low combustion velocity.

The combustion processes of the generator gas, as well as those of the generator gas mixture with gasoline are for the time being poorly studied. That is why, a theoretical treatment of the operation conditions of an ICE fueled with a mixture of generator gas and gasoline is required.

From literature references, it is known that the basic combustible components of the generator gas are hydrogen H_2 and carbon monoxide CO, with respective volume fractions of 12…20% vol. and 17…32 % vol. The reaction of their oxidation, as per Arrhenius equation, is determined as follows:

$$\frac{dC}{dt} = C_1 C_{O_2} k = C_1 C_{O_2} A e^{-\frac{E}{RT}}, \tag{14}$$

where k is the oxidation reaction constant; E is the activation energy; T is the absolute temperature; R is the universal gas constant; A is a pre-exponential factor characterizing the efficiency of the collision of the molecules of the reacting substances; C_1 is the fuel (hydrogen of carbon monoxide) concentration; C_{O_2} is the oxygen concentration.

The activation energy for H_2 and CO molecules, in accordance with literature reference, is 242 and 201 kJ/mole, respectively. The known activation energy value of gasoline vapors is 78.5…80 kJ/mole, that is, 2.5 times less than the activation energy of carbon monoxide. As to the pre-exponential factor A in (14), determined by the effective sections of the interacting molecules, for combustion reactions of hydrogen, carbon monoxide and gasoline vapors, it must be of the same numerical order. However, no significant growth of the gasoline combustion rate relative to the generator gas combustion is observed, which for good reasons could be anticipated due to the exponential dependency on the activation energy. The absence of such growth is obviously stipulated by the limitation of the gasoline combustion rate by the evaporation rate of its droplets, being quite a slow process. On the average, the decrease rate of the square of the gasoline droplet diameter for gasoline combustion temperatures will be, in accordance with the available reference data, as follows:

$$K = \left|\frac{d(d^2)}{dt}\right| = 4{,}3 \cdot 10^{-6} \text{ m}^2/\text{s}. \tag{15}$$

Then, the preliminary calculation of the combustion time could be made based on the following formula:

$$\tau = \pi d_0^2/K, \tag{16}$$

where d_0 is the initial gasoline droplet diameter being, on average, 0.065 mm. The calculation shows that the combustion time will be, for gasoline, the combustion time $\tau = 0.0031$ s.

Calculation example of combustion time in an ICE with spark ignition operating with pure gasoline, pure generator gas and with a mixture of 30% gasoline and 70% GG (Table 4).

Table 4. Combustion time calculation data

Frequency of rotation, min^{-1}	Structure of a mix	Time of combustion, sec
3,000	Gasoline – 100%	0.0031
3,000	Gasoline – 30% + Generating gas - 70%	0.0036
3,000	Generating gas - 100%	0.0039

Based on the data of Table 4, the time required for combustion of generator gas under given conditions is longer than that of gasoline by approximately 0.001 seconds. In case of the mixture containing 30% of gasoline and 70% of generator gas, the time growth will be 0.0005 seconds.

Comparative data of bench tests of an ICE with spark plugs at operation with pure gasoline, pure generator gas and their mixture completely confirm the results of the theoretical studies [21].

Thus, the problem of use of low-calorie alternative fuels can also be partially coped with.

Conclusions

1. Based on the data of the studies, generalized criteria of the assessment of the usability limits of alternative fuels were identified.
2. Sufficient proof was provided of real possibilities of operation of serial ICE's with alternative fuels with certain distinctive engine properties.
3. The extension of the usability limits of alternative fuels by adaptation of their motor properties improves the multifuel capability and extends the fuel base of internal combustion engines.

References

[1] Markov V A, Devyanin S N, Zykov S A and Gajdar S M 2016 *Biofuels for Internal Combustion Engines* (Moscow: SIC «Engineer») p 292
[2] Plotnikov S A 2011 Improving the Performance of Diesel Engines by Creating New Alternative Fuels and Improving Fuel Supply Equipment *Author's abstract diss. doct. tech. sciences* (Nizhny Novgorod)
[3] L'otko V, Lukanin V N and Hachiyan A S 2000 *Use of Alternative Fuels in Internal Combustion Engines* (Moscow: Publishing house of MADI) p 311
[4] Terent'ev G A, Tyukov V M and Smal' F V 1989 Motor Fuels from Alternative Raw Materials (Moscow: Chemistry) p 272
[5] Markov V A, Chainov N D and Loboda S S 2018 Physicochemical Properties of Petroleum Motor Fuels with Vegetable Oil Additives and Their Effect on Diesel Performance *Vestnik MGTU named after Bauman N E (Ser. Mechanical Engineering)* **5** pp 108–122.
[6] Vasil'ev I P 2009 Influence of Vegetable Fuels on Environmental and Economic Indicators of Diesel (Lugansk: Publishing house of VNU named after Dal' V p 240
[7] Markov V A, Vallejo Maldonado P R and Biryukov V V 2015 Alcohol Fuels for Diesel Engines *Proceedings of Higher Educational Institutions. Mechanical Engineering* **11** pp 39-52
[8] Biryukov V V 2017 Methods for Improving the Efficiency of Diesel Operation when Using Ethanol as an Ecological Additive to Diesel Fuel *Diss. cand. tech. sciences* (Moscow)
[9] Kurmanova L S 2018 Methods of Organizing the Working Cycle in Heat Engines for Operation on a Mixture of Diesel Fuel and Natural Gas *Volga Transport Bulletin* **6** (72) pp 108-114

[10] Shemyakin A V, Terent'ev V V, Panov Y A and Ivanov A A 2017 Selection of the Composition of the Methanol-Rapeseed Emulsion for Its Use as Diesel Fuel *Tractors and agricultural machinery* **11** pp 10-14.
[11] Ivanov A A and Panov Y A 2016 Selection of the Composition of the Methanol-Rapeseed Emulsion for Its Use as a Fuel for the D-242 Diesel *Bulletin of the St. Petersburg State Agrarian University* (St. Petersburg: Publishing house of SPbSAU) pp 152-155.
[12] Oshchepkov P P, and Adegbenro S A 2017 Alternative Fuel for Vehicles in Nigeria Based on Palm Oil *RUDN Bulletin. Series: Engineering Studies* **18** (4) pp 437-444
[13] Oshchepkov P P, Zaev I A, Smirnov S V and Bizhaev A V 2019 Research on Biodiesel with Palm Oil and Hydrogen Peroxide Additives *Agricultural machines and technologies* **13** (3) pp 48-53
[14] Pandian A K, Munuswamy D B, Radhakrishana S, Ramesh Bapu B R, Beemkumar N and Devarajan Y 2017 Influence of an Oxygenated Additive on Emission of an Engine Fueled with Neat Biodiesel *Pet. Sci.* **14** pp 791–797
[15] Burlakov M 2001 Corel DRAW 9 Directory (St. Petersburg: Piter) p 496
[16] Plotnikov S A, Kantor P Ya, Kozlov I S and Vtyurina M N 2018 Investigation of the Motor Properties of Mixtures of Diesel Fuel with Rapeseed Oil *Proceedings of NSTU named after Alekseev R E* (Nizhny Novgorod) **2** (121) pp 169-174.
[17] Plotnikov S A, Shyshkin G P and Smol'nikov M V 2019 Calculation of the Stability of Ethanol-Fuel Emulsion for Use in Diesel Engines *Engine building* **1** pp 24–27
[18] Vtyurina M N and Plyago A V 2017 Study of the Properties of Ethanol-Fuel Emulsions with Additives *Transport Systems* **2**
[19] Kantor P Ya, Plotnikov S A, Plyago A V and Vtyurina M N 2018 The Effect of Molybdenum Disulfide on the Combustion of Ethanol-Fuel Emulsion *Proceedings of NSTU named after Alekseev R E* (Nizhny Novgorod) **4** (123) pp 215-220.
[20] Plotnikov S A, Kuimov E A, Plyago A V, Kartashevich A N and Vtyurina M N 2018 RF patent № 2668225, MPK C10L 1/32
[21] Plotnikov S A, Kantor P Ya, Zubakin A S and Vtyurina M N 2019 Some Features of the Engine Running on Generator Gas *Proceedings of NSTU named after Alekseev R E* (Nizhny Novgorod) **2** (125) pp 192-196
[22] Kuvykin V I and Kuvykina E V 2016 Viscosity of a Mixture of Hydrocarbons *Natural and Mathematical Sciences in the Modern World* **1** (36) pp 46–51.

Development of KORSAK 8x8 small-size all-terrain vehicle design project

A G Kapustin[1], A V Papunin and A M Belyaev

Nizhny Novgorod State Technical University n.a. R.E. Alekseev, Minin str., 24, 603950, Nizhny Novgorod, Russia

E-mail: [1]kapustinalexandr2017@yandex.ru

Abstract. The paper presents KORSAK 8x8 Small-Size All-Terrain Vehicle Design Project developed in Nizhny Novgorod State Technical University n.a. R.E. Alekseev. Performance specifications of the vehicle non-inferior to those of the existing similar vehicles. This all-terrain vehicle has better ergonomics and visibility from the driver seat. The cabin design of the vehicle complies with the requirements of the applicable Codes and can be certified for use in the Russian Federation. For inspection of the aesthetic properties of this all-terrain vehicle, a 1:5-scale dummy was made.

Introduction

For the time being, there is a considerable interest, in terms of scientific research and practical use, in manufacturing of highly-efficient small-size light land-based vehicles. The use of light all-terrain vehicles allows for lower operation costs and a possibility of easy driving over heavy-going terrain impassable for heavy all-terrain vehicles. Korsak 8x8 snow and swamp-going vehicle currently under design in Nizhny Novgorod State Technical University n.a. R.E. Alekseev (NNSTU) has a high level of all-terrain capacity and environmental friendliness of the propulsion unit due to low ground pressure, being intended for carriage of passengers and cargo in absence of roads, including on virgin snow, swamped terrain, open water basins and quick sand. Previously, the authors hereof have developed Korsak 6x6 All-Terrain Vehicle [1, 2]. Tests were made for flotation and drive path profile capacity, as well as for swimming [3-6]. Under recognition of the available backlog, the all-terrain vehicle presented in the paper has better mobility performance.

Project description

Conducted polling and its subsequent analysis have shown that would-be buyers of all-terrain vehicles of this class want to possess a vehicle complying with the following requirements: an engine above 25 kW; ride comfort; availability of useful cargo space; availability of a safety framework; availability of side rear-view mirrors; the vehicle shall feature individually-shaped body; a multimedia system shall be installed in the basic configuration.

Figure 1 shows the exterior appearance of Korsak 8x8 All-Terrain Vehicle meeting the declared requirements.

IOP Conf. Series: Materials Science and Engineering 1086 (2021) 012017 doi:10.1088/1757-899X/1086/1/012017

Figure 1. KORSAK 8x8 Small-Size All-Terrain Vehicle

For concise technical specifications of KORSAK 8x8 All-Terrain Vehicle see Table 1.

Table 1. Basic specifications of KORSAK 8x8 All-Terrain Vehicle

Specifications	Value
Gross vehicle weight (GVW), kg	1200
Load carrying capacity, kg	400
Design version:	1) passenger version; 2) cargo version; 3) cargo-and-passenger version
cabin	front section
engine compartment	middle section
Number of passenger seats	2-7 (dependent on the design version)
Average ground pressure with full load, MPa (kgf/cm^2)	0.025 (0.25)
Engine	Lifan 2v90 (37 h.p.)
Overall length/ width/ height, mm	3375/ 1600/ 1900
Track width, mm	1320
Wheel base, mm	2250
Road clearance, mm	300
Weight of towed trailer, kg	400
Max. travel speed, km/h	
on-road/ swim	50/ 5
Road clearance, mm	300
Obstacles bridged:	
ascend (descend)/ flank slope	35° / 25°
Transmission:	1) mechanical; 2) hydrostatic
Turning control	Motorcycle-type handlebar, power turn
Propulsion unit:	8x8 wheel formula, low-pressure tires, tire dimensions 25”-12,5”-10”
Body	Carrier frame - self-carrying, steel, water-tight. Body - glass-fiber reinforced plastic. Safety framework.

The vehicle configuration can be different and depends on the required number of passengers and cargo quantity. Figure 2 shows different arrangement layouts of the passenger seats.

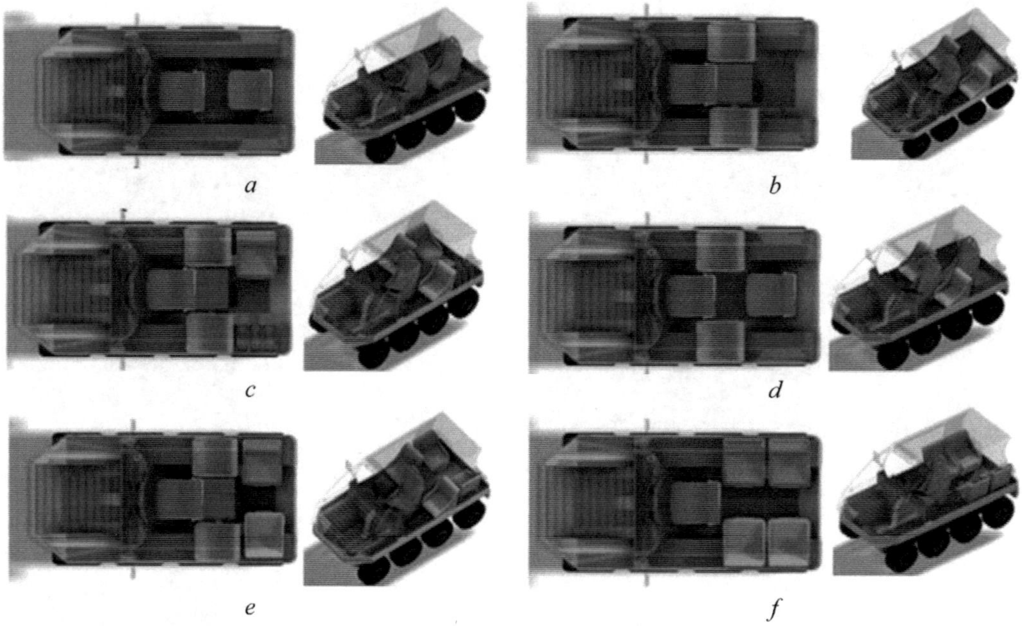
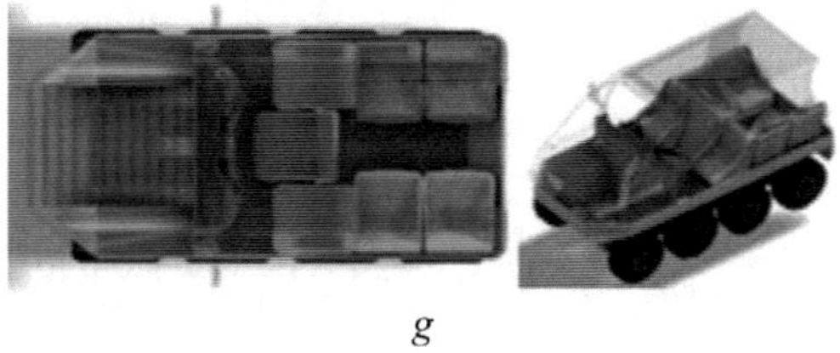

Figure 2. Configuration variants of KORSAK 8x8 All-Terrain Vehicle: *a* – two-seat, *b* – three-seat, *c, d* – four-seat, *e, f* – five-seat, *g* – seven-seat.

This all-terrain vehicle has better ergonomics and visibility from the driver seat than the existing similar vehicles. Fig. 3 shows the visibility from the driver and the passenger seats during boarding, as shown in Figure 2 *c*. The boarding of the driver and the body frame of the all-terrain vehicle is designed in a manner to meet the requirements of the visibility in accordance with the applicable codes.

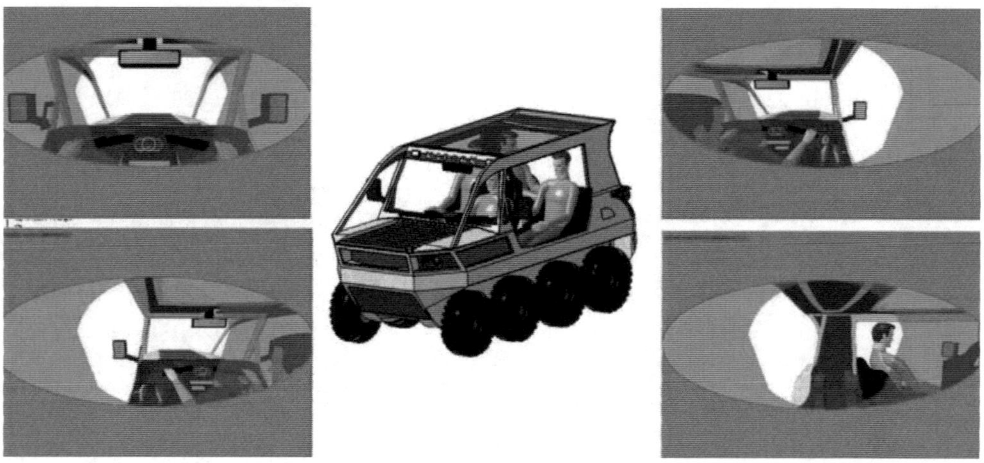

Figure 3. Visibility form driver and passenger seats

Also, safety framework calculations were made for compliance with GOST R 50943-2011. The structure shall withstand the design impacts imitating the capsizing of the snow and swamp-going vehicle. It is exactly the loading of the front and side edge of the cabin roof with a test loading device at displacements of 5° in longitudinal, and 25° in transversal directions. The load produced by the loading device makes 1.5 of the vehicle weight. Fig. 4 shows the framework of the vehicle in the deformed state.

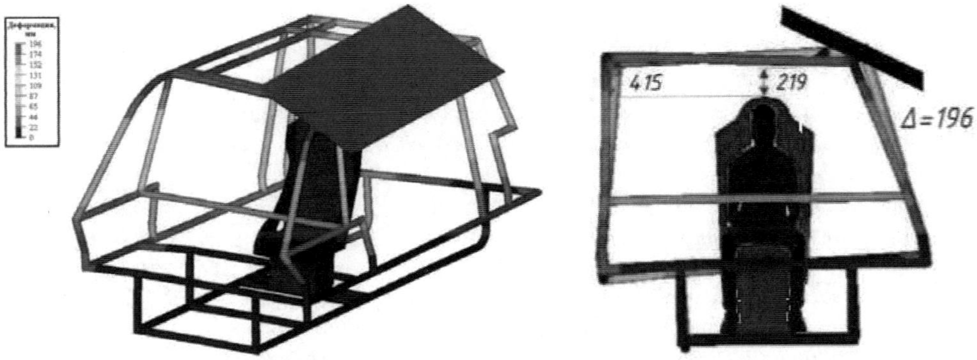

Figure 4. Load accommodation diagram by the design model and apperarance of the deformed framework

The obtained calculation results demonstrate, that the framework deformation in the vertical direction is 196 mm and does not exceed admissible values and may be certified for use in the Russian Federation.

Result and discussion

The paper presents a description of a design project developed in Nizhny Novgorod State Technical University n.a. R.Ye. Alekseyev of KORSAK 8x8 Small-Size All-Terrain Vehicle being a descendant of previously designed 6x6 chassis. It is demonstrated that the new vehicle has better specifications than existing similar vehicles and complies with the anticipations of would-be buyers of all-terrain vehicles of such class. Different configuration variants are presented dependent on the arrangement and the number of the passenger seats. The visibility from the driver and the passenger seats is shown. Calculations demonstrate that the developed safety frame complies with GOST R 50943-2011, and the all-terrain vehicle may be certified in Russia.

The following design stage will be creation of mathematical models for 8x8 vehicles with low-pressure tires [5-8], calculation of mobility parameters over virgin snow [7-11], and the assessment of the efficiency of their use for carriage of goods and passengers in the given conditions.

Acknowledgment

This study was conducted in continuation of the research conducted by "Nizhny Novgorod scientific and practical school of snow vehicles" and with financial support of the grant of the President of the Russian Federation No. MD-226.2020.8.

References

[1] Kapustin A G et al. 2019 Development of design project of mini cross-country vehicle KORSAK *IOP Conference Series: Materials Science and Engineering* **695(1)** 012026 doi: 10.1088/1757-899X/695/1/012026

[2] Papunin A V et al. 2018 A dynamic model of unsupported pit traversal by a vehicle with 6×6 wheel arrangement *IOP Conference Series: Materials Science and Engineering* **386(1)** 012001 doi: 10.1088/1757-899X/386/1/012001

[3] Papunin A V et al. 2018 Calculation of trench width to be overcome by multi-axis wheeled vehicle *IOP Conference Series: Earth and Environmental Science* **194(6)** 062027 doi: 10.1088/1755-1315/194/6/062027

[4] Papunin A V et al. 2019 Field research of profile trafficability of 6×6 wheel assembly KORSAK vehicle *Journal of Physics: Conference Series* **1177(1)** 012051 doi: 10.1088/1742-6596/1177/1/012051

[5] Belyakov V et al. 2015 Multifunctional vehicle for coastal areas *12th International Conference on the Mediterranean Coastal Environment, MEDCOAST 2015* pp 945-951

[6] Belyaev A et al. 2019 Development of 8×8 all-terrain vehicle with individual wheel drive *VEHITS 2019 - Proceedings of the 5th International Conference on Vehicle Technology and Intelligent Transport Systems* pp 556-561

[7] Anikin A et al. 2017 Calculation of traction capabilities of wheeled vehicles on low-pressure tires on snow *19th International and 14th European-African Regional Conference of the ISTVS*

[8] Makarov V, Zeziulin D and Belyakov V 2014 Prediction of all-terrain vehicles mobility in snowscape scenes *Paper presented at the18th International Conference of the ISTVS*

[9] Porubov D et al. 2017 Study of efficiency of using all-terrain vehicles during the winter period *19th International and 14th European-African Regional Conference of the ISTVS*

[10] Kurkin A et al. 2017 Ground vehicle for ice conditions monitoring 13th International MEDCOAST Congress on Coastal and Marine Sciences, Engineering, Management and Conservation, MEDCOAST 2017 **2** pp 827-838

[11] Goncharov K O, Belyakov V V and Makarov V S 2018 Estimating method of efficiency of wheel chassis driving on snow *IOP Conference Series: Earth and Environmental Science* **194(3)** 032005 doi: 10.1088/1755-1315/194/3/032005

IOP Conf. Series: Materials Science and Engineering 1086 (2021) 012018 doi:10.1088/1757-899X/1086/1/012018

Efficiency of the simulation model of a light commercial vehicle

D A Butin[1], A V Tumasov, A S Vashurin, M E Gnenik

Nizhniy Novgorod State Technical University, 24 Minin street, Nizhniy Novgorod, Russian Federation

E-mail: [1]danila.butin@nntu.ru

Abstract. The paper describes the results of the efficiency analysis of the simulation model of light commercial vehicle GAZ GAZelle NEXT. The efficiency of the model is confirmed by comparing the results of field test and calculation studies of controllability. The model is built in the MSC ADAMS/Car software package. It accounts for the mass-inertial characteristics of sprung and unsprung masses; the model of the suspension system simulates the kinematics and elasto-kinematics of the suspension and steering control, as well as elastic and damping elements. The tire model simulates vertical stiffness and damping, the slip stiffness depends on the vertical load, the longitudinal slip coefficient and the lateral angle of tire relative to the contact surface. The assessment was made by comparing the data obtained experimentally and by calculation. The test is conducted using the Fishhook method applied by NHTS for the vehicle stability analysis. The resulting errors are analyzed based on relative and absolute error values. When analyzing lateral acceleration and yaw rate deviations, the mathematical statistical tools are employed: average deviation, dispersion, root-mean-square deviation. The discrepancy between the results and the deviation of the average values does not exceed 10%. Minor deviations and scatter of results shows the efficiency of the simulation model and that this model can be employed for simulating the curvilinear motion of light commercial vehicles and assessment of their controllability and stability characteristics.

1. Introduction

The efficiency analysis of the model is required to assess the degree of similarity of the model [1]. The model efficiency can only be proved by experiment. For confirmation of the model efficiency the model reactions are assessed for consistency with the model object. The reactions are analyzed based on average responses values and dispersion of the response deviations. The model is characterized by the efficiency zone which is a set of situations when the model is efficient. The model accuracy is characterized by mismatch between relevant parameter of the model and the subject of study.

2. Object of study

The subject of research and modeling is GAZ GAZelle NEXT model A21R22 with a gross weight of 3500 kg, load on the front axle 1325 kg, load on rear axle – 2175 kg; curb weight 2040 kg. The vehicle is equipped with Cordiant Business tires 185/65R16C. It is designed with independent front suspension on double wishbones, independent rear suspension on semi-elliptical springs. The suspension is equipped with telescopic oil shock absorbers and anti-roll bars. The transmission is mechanical with a five-speed gearbox. The vehicle features rear axle drive and open type differential.

The simulation of the motor vehicle is carried out in the object-oriented software package MSC ADAMS Car. The simulation model features the mobility in all 6 degrees of freedom. It takes into account mass-inertial characteristics of sprung and unsprung masses; the suspension model simulates the kinematics and elasto-kinematics of the suspension and steering control, as well as elastic and damping elements. The tire model is built on the basis of the H.B. Pacejka theory. It takes into consideration the vertical stiffness and damping, the slip stiffness depends on the vertical load, the longitudinal slip coefficient and the lateral angle of tire relative to the contact surface.

3. Experimental and theoretical research

The methodology and results of the vehicle field tests the under study are published in papers [2-9]. The controllability and reactions of the subject of interest is subjected to the Fishhook test used by NHTSA [10] for the dynamic rollover stability analysis of the vehicle. A steering robot is used for this test. The computer-aided analysis of the vehicle significantly improves the reliability and repeatability of results. The model efficiency assessment is based on results of one test run. The experimental research is conducted using measurement equipment of the 'Transport Systems' Shared Facilities Centre of NNSTU.

The following parameters were recorded during the tests: steering wheel angle, longitudinal vehicle speed, yaw rate, lateral and longitudinal acceleration, as well as the displacement of the center of gravity in the plane of the road. Overlay graphs are shown in the figures for the purpose of clarity when comparing field tests and calculation results. The field tests and simulation results were overlaid within the time domain with reference to the common starting point of the steering wheel rotation. Relative and absolute deviations were used for the model efficiency analysis. The average deviation relative of the calculation results vs the experiment was computed by formula 1.

$$\bar{x} = \frac{1}{n}\sum_{i=1}^{n}\left(\frac{X^{i}_{\text{experiment}} - X^{i}_{simulation}}{X^{i}_{\text{experiment}}}\right) \quad (1)$$

Figure 1 shows the steering wheel angle curves during the experiment and simulation. The model employed the dependence of the steering wheel angle on field test results, therefore the graphs coincide.

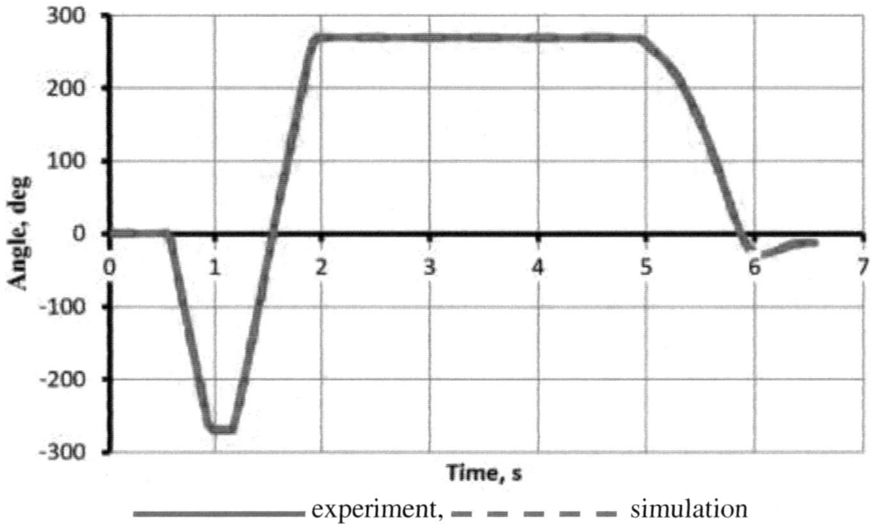

Figure 1. Steering wheel angle

The speed of motion exerts rather strong influence on how the vehicle reaction to the steering wheel. Figure 2 shows the graphs of longitudinal motion speed during the test and simulation.

The average deviation of the longitudinal motion speed of the model is 1.6 km/h. The maximum motion speed deviation is 3.87 km/h. The average relative deviation of the model speed is 3%, but not exceeding 6%.

Figure 3 shows the yaw rate graph. The yaw rate is one of the important indicators for the vehicle controllability analysis.

The average relative deviation of yaw rate is 6%. The yaw rate deviation during the simulation and testing is not constant and can be represented as distribution of yaw rates as shown in Figure 4. It is evident from the histogram that the distribution of the yaw rates is normal. The absolute average deviation of yaw rates is -1.0234 deg/s. The dispersion of yaw rate deviation is D=4.464 mm^2/s^4, root-mean-square deviation is σ=2.11mm/s^2. The maximum yaw rate deviation is 7.05 deg/s.

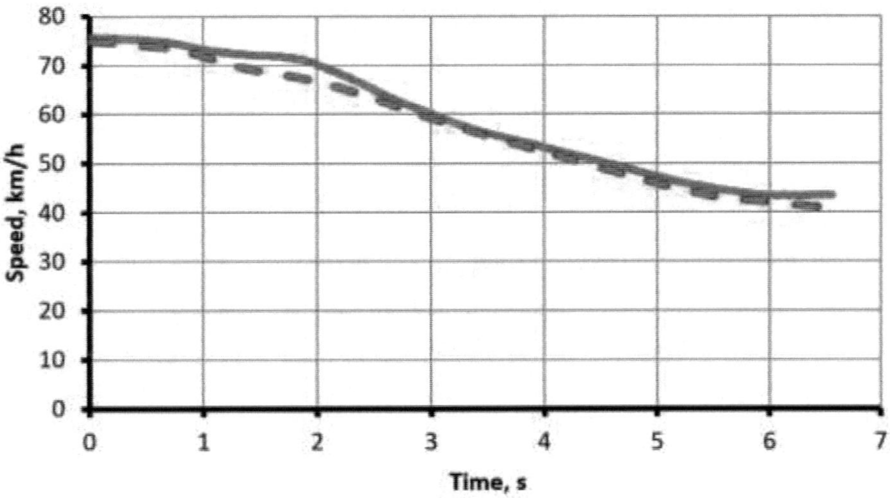

experiment, simulation

Figure 2. Longitudinal travel speed

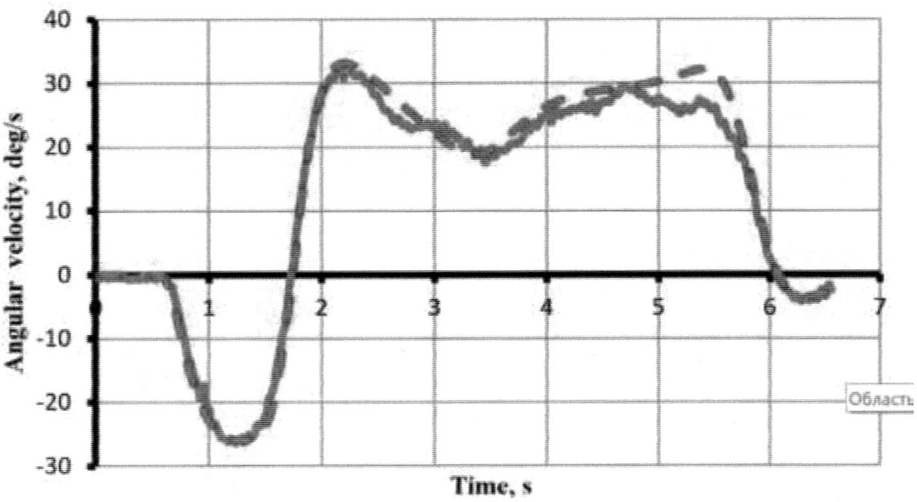

experiment, simulation

Figure 3. Yaw rate

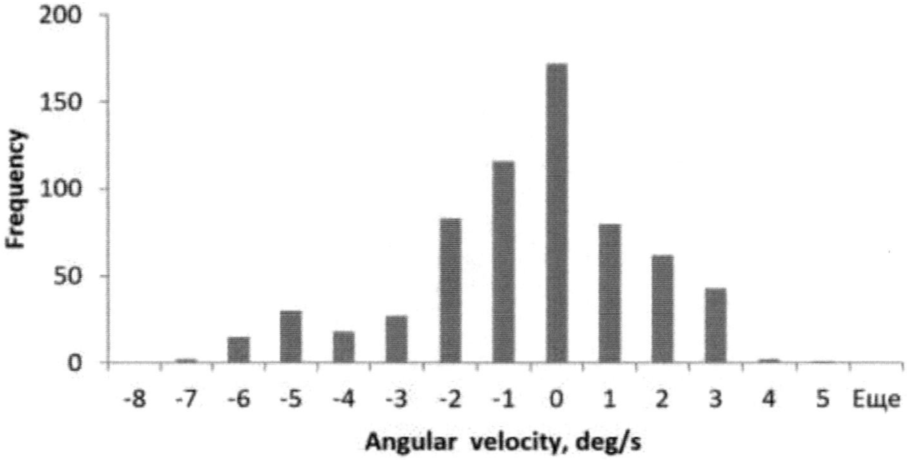

Figure 4. Distribution of yaw rate deviations

Figure 5 shows the lateral acceleration graphs.

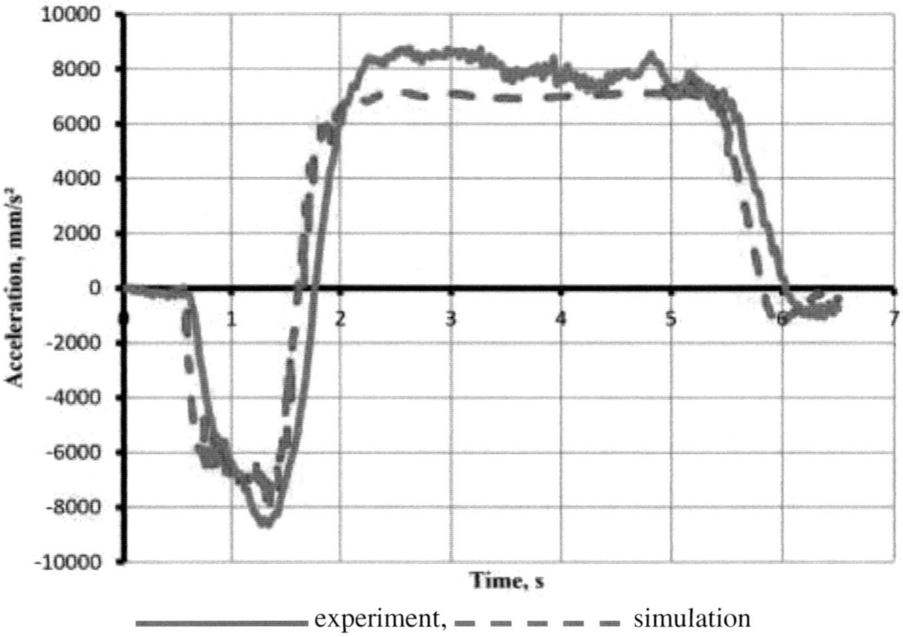

Figure 5. Lateral acceleration

Average relative deviation of lateral acceleration 7%. The difference in lateral accelerations can be represented as a distribution of lateral accelerations as shown in Figure 6. The average deviation of lateral accelerations 391.92 mm/s^2. Dispersion of lateral acceleration deviations is D=2483002 mm^2/s^4, root-mean-square deviation σ=1575.7 mm/s^2. Maximum lateral acceleration deviation is 5814 mm/s^2.

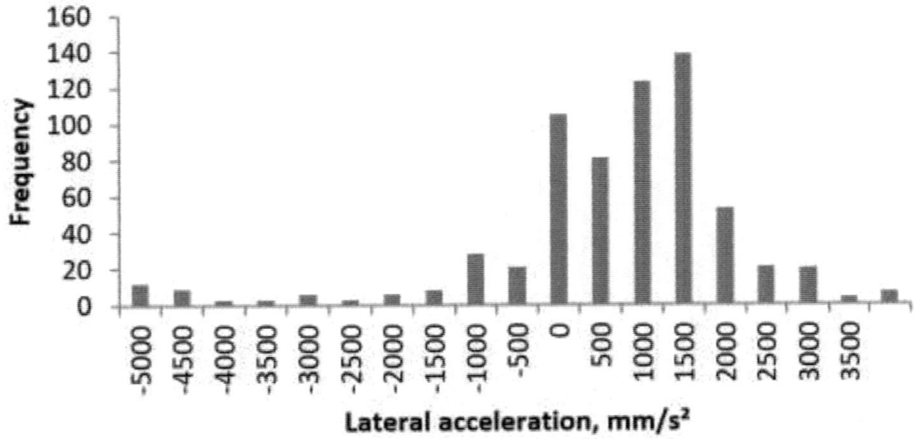

Figure 6. Distribution of lateral acceleration deviations

Since the investigations were conducted in different coordinate systems, for trajectory error calculation some transformations and alignment of the trajectories are required. For the purpose of alignment, the trajectories were shifted and turned in such a way that initial conditions of motion would coincide. The straight-line motion of the vehicle at the beginning of the test is taken as the common axis, while the common point of the trajectories is determined by the time when the steering wheel starts turning. Overlaid trajectories of the vehicle center of gravity, field studies and simulation modeling are shown in Figure 7.

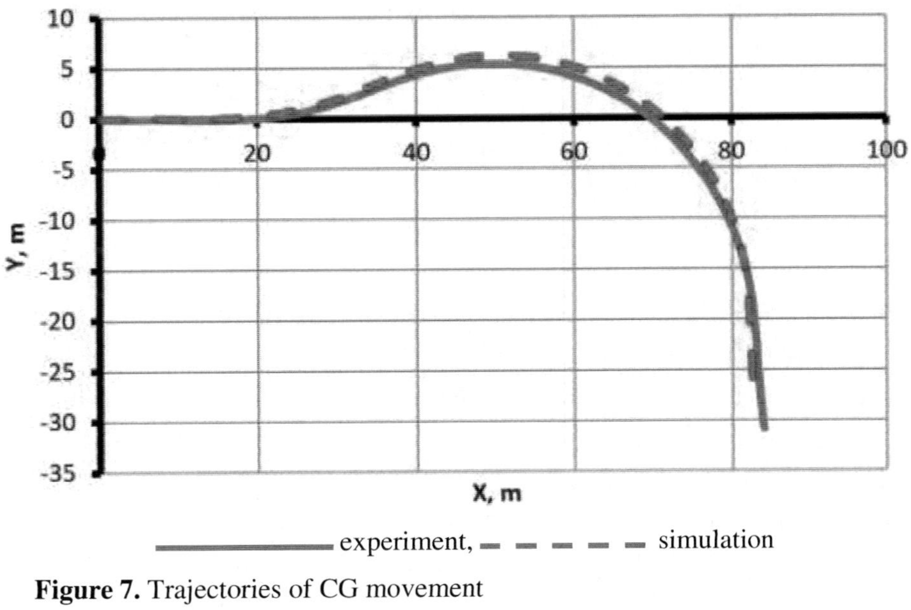

Figure 7. Trajectories of CG movement

The difference in trajectories is the distance between the center of gravity position of the model and that of a vehicle in the physical world during the test, at the same point in time that is calculated by formula (2).

$$\Delta^t = \sqrt{\left(X^t_{simulation} - X^t_{experiment}\right)^2 + \left(Y^t_{simulation} - Y^t_{experiment}\right)^2} \quad (2)$$

The variation of the distance between the centers of gravity relative to the time of the vehicle maneuver is shown in the form of a graph in Figure 8.

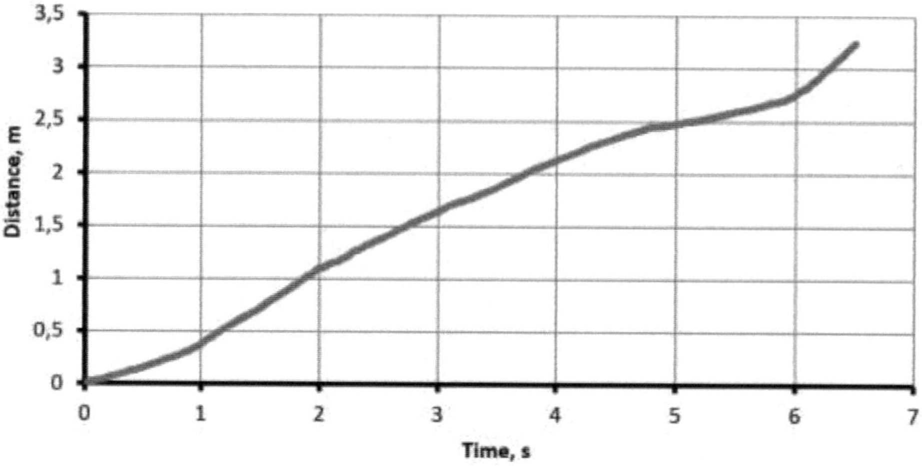

Figure 8. Trajectory deviation

The maximum deviation of the trajectories at the end of the test was 3.24 m, with the total distance traveled 106.93 m. The relative trajectory deviation from the path is 3%.

Conclusions

The efficiency of the simulation mathematical model was analyzed by comparing the results of a full-scale experiment with the simulation results. The efficiency and accuracy analysis of the model was based on the Fish Hook test results. The model efficiency was assessed by the parameters characterizing controllability and stability of the vehicle motion: longitudinal speed, yaw rate, lateral acceleration, trajectory of motion. The comparison of the experiment and simulation modeling results proves the model efficiency. The discrepancy between the results and the deviation of average values does not exceed 10%. The absolute deviation of the longitudinal speed of motion does not exceed - 6%, that of the trajectory – 3%. The average deviation of lateral acceleration is 7%, yaw rate - 6%. The average yaw rate deviation is -1.0234 deg/s, that of lateral acceleration – 391.92 mm/s^2. The standard deviation of yaw rate is 2.11 mm/s^2, that of lateral acceleration – 1575.7 mm/s^2. Minor deviations and scatter of results shows the efficiency of the simulation model, and that this model can be employed for simulating the curvilinear motion of light commercial vehicles and assessing their controllability and stability characteristics.

References

[1] Skotnikov G I 2016 Efficiency analysis of the mathematical model for curvilinear motion of light commercial vehicle road train *News bulletin of higher educational institutions, Mechanical engineering* **8(677)** pp 29-37

[2] Beresnev P et al. 2017 Simulation of road impacts for forecasting of durability of vehicle's suspension elements and supporting system *19th International and 14th European-African Regional Conference of the ISTVS*

[3] Groshev A et al. 2018 Estimation of light commercial vehicles dynamics by means of HIL-test bench simulation *IOP Conference Series: Materials Science and Engineering* **315(1)** doi: 10.1088/1757-899X/315/1/012010

[4] Leliovsky K Y et al. 2019 Vibration load of transmission units at vehicle's motion over different roads. *Journal of Physics: Conference Series* **1177** 012053 doi:10.1088/1742-6596/1177/1/012053

[5] Vashurin A S et al. 2015 Impact of the electronic dynamic stability control system of a light commercial vehicle on safety of the vehicle curvilinear motion *Fundamental studies* **12-4** pp 665-670

[6] Toropov Ye I et al. 2017 Full-scale tests for debugging of the software and hardware package designed for analysis of controllability of light commercial vehicles equipped with electronic stability control systems. Specifics of the vehicle maneuvering on dry and wet asphalt *Transactions of NNSTU n.a. R.E. Alekseev* **4 (119)** pp 172-177
[7] Toropov Ye I et al. 2019 Verification of the methodology for virtual physical studies of the vehicle curvilinear movement dynamics based on road test results *Transactions of NNSTU n.a. R.E. Alekseev* **2 (125)** pp 210-216
[8] Tumasov A V et al. 2016 Estimation of Influence of ESP on LCV Active Safety in Condition of Curvilinear Movement *International Conference on Vehicle Technology and Intelligent Transport Systems* pp118-123 doi10.5220/0005894401180123
[9] Zezyulin D et al. 2014 Modeling of roads impacts for life prediction of light commercial vehicles parts *FISITA 2014 World Automotive Congress*
[10] Federal Motor Vehicle Safety Standards; Electronic Stability Control Systems; Controls and Displays *National Highway Traffic Safety Administration* URL: https://www.nhtsa.gov/fmvss/federal-motor-vehicle-safety-standards-electronic-stability-control-systems-0 (data of application: 03.12.2020).

IOP Conf. Series: Materials Science and Engineering 1086 (2021) 012019 doi:10.1088/1757-899X/1086/1/012019

Elastomer elastic element visco-elastic properties' impact on damping in a vehicle suspension

E V Stepanov[1], S M Ogorodnov, L N Orlov and M K Chegurov

[1]Nizhny Novgorod State Technical University n.a. R.E. Alekseev, Minin st., 24, 603950, Nizhny Novgorod, Russia.

E-mail: [1] evgeniystepanov1991@nntu.ru

Abstract. Suspension structure, its elastic elements made of elastomers, is described in the paper. This allows to considerably decrease a vehicle's unsprung weight improving its ride quality. The research is aimed at determining elastomer elastic element's damping impact on a vehicle's smooth ride quality. The impact estimate was carried out by comparing simulation modelling results of GAZelle NEXT vehicle's front suspension and an equivalent suspension with an elastomer elastic element. Dual-mass single-axle suspension model was chosen as a calculation model. The modelling was performed in MATLAB/Simulink software package. Standard suspension key parameters, that is spring rate and shock absorber resistance characteristic, were experimentally obtained for each vehicle. Elastomer element elastic response was obtained at static necking and averaged with respect to loading and unloading characteristics. Damping capacity was estimated basing on the damped oscillations process realized on a special pendulum impact testing machine. "Relative" elastomer damping coefficient was calculated. All the measured parameters were reduced "awheel" by grapho-analytical methods. Vibration acceleration root-meant-square values (RMS) on the car frame and normal awheel response of a vehicle's motion under various speed conditions over smooth cobbled road served as assessment criteria. Roadbed microprofile mathematic model was realized basing on superposition harmonic function method with various oscillations frequency and amplitude. Analytical findings reveal that a standard suspension with a cylindrical spiral spring and a hydraulic shock absorber outperforms the elastomer elastic element structure in all the given motion modes with respect to the ride quality criteria. The research findings bring about the conclusion that elastomers lack visco-elastic characteristics necessary to provide better ride quality compared to the standard suspension structure.

1. Introduction

Non-metallic elastic elements in vehicles' suspension are used as extra ones, allowing to appropriately adjust the elastic response, e.g. making it non-linear and increasing its energy capacity. Elastomer elastic elements, Figure 1, when certain requirements to the suspension's elastic response observed, considerably decrease the vehicle's unsprung weight and metal expenditure as well as improve the ride quality.

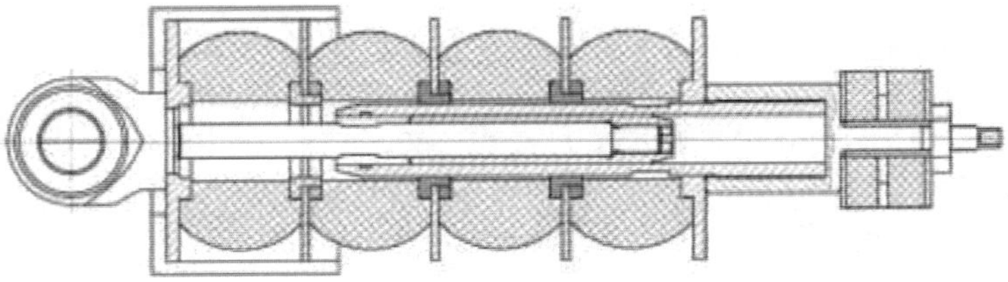

Figure 1. Flowchart of elastic elastomer element in a vehicle's suspension

2. Experimental research

Elastomer has viscous friction and its elastic response is of non-linear character [1]. Figure 2 shows experimentally obtained real elastomer elastic response allowing to determine viscous friction value ($F_{тр}$) and rigidity. Elastomer viscous friction in its physical sense is close to «dry» friction. Elastomer viscous friction impact on the nature of dynamic processes in suspension is difficult to estimate as the type of friction is not determined by clear-cut physical boundaries. This is actually «dry» friction, but its value remains non-constant and depends on elastic element deformation f and force F acting on the elastic element.

Moreover, elastomer viscous friction affects oscillatory processes damping rate in suspension and can be determined by "relative" damping coefficient «$k_э$», similar to the damping coefficient of hydraulic shock absorber. An experiment was carried out to determine the elastomer "relative" damping coefficient. The experiment allowed to calculate damping logarithmic decrement in elastomer and the "relative" damping coefficient value.

The damping decrement allows to estimate the oscillations damping rate and is mathematically determined as a ratio of two subsequent maximum deviations of the system oscillations damping process, one way from the equilibrium state, Figure 3.

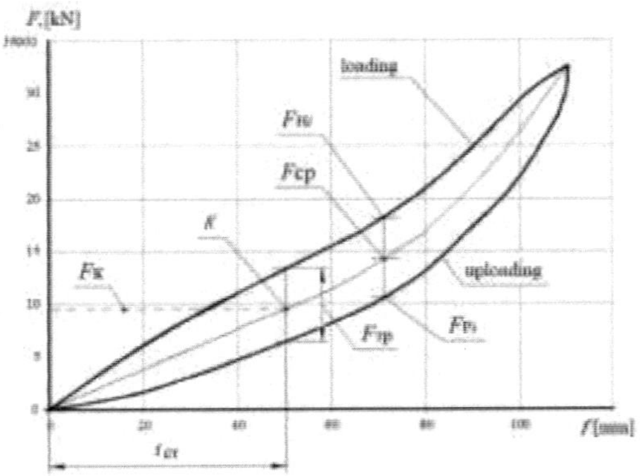

Figure 2. Elastomer viscous friction characteristics

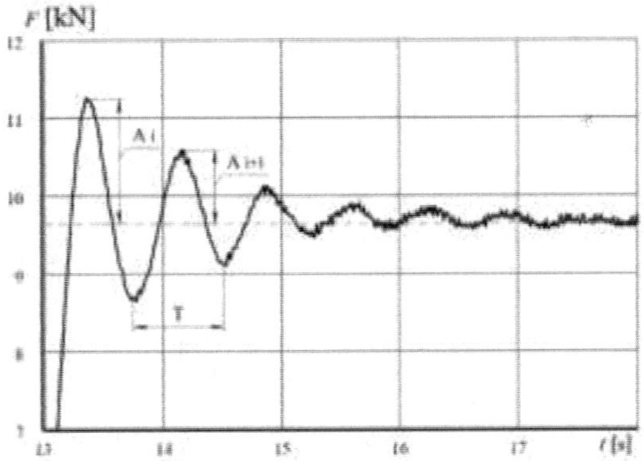

Figure 3. Oscillations damping process in elastomer elastic element

Relative elastomer resistance coefficient equals to

$$k_э = \frac{2*lnd}{T} M \quad (1)$$

where: M – the vehicle's sprung weight, coming onto the elastomer; d – damping decrement; T – oscillation period. For the given elastomer element structure relative resistance coefficient equals to $k_э$ = 1592 N s/m.

It would be practical to analyze elastomer friction impact on the suspension critical performance characteristics – vertical acceleration root-meant-square values (RMS) $\sigma_{\ddot{z}}$, passed on to the sprung weight and normal response forces RMS σ_F, affecting the wheel in the point of tyre-road contact [6, 7].

IOP Conf. Series: Materials Science and Engineering 1086 (2021) 012019 doi:10.1088/1757-899X/1086/1/012019

3. Theoretical research

Mathematical models are applied to solve such tasks. A simpler mathematical (calculation) model is usually used when researching dynamic processes. Calculation models can become more sophisticated depending on the requirements to applicability and accuracy of analytical experiment result rates. The choice of the vehicle's calculation model is based on certain assumptions [2-11], about oscillations independence in longitudinal and transverse planes. The vehicle's transverse and longitudinal oscillations amplitudes are considerably lower than heave oscillations amplitudes, so they can be disregarded for a first approximation and the vehicle's mathematical model can be linear.

At sprung weight distribution coefficient value ε of each suspension close to unity [3], weight fluctuations are described by independent set of equations, and design diagrams of a vehicle's front and rear suspensions heave oscillations can be presented by single-mass and dual-mass calculation models.

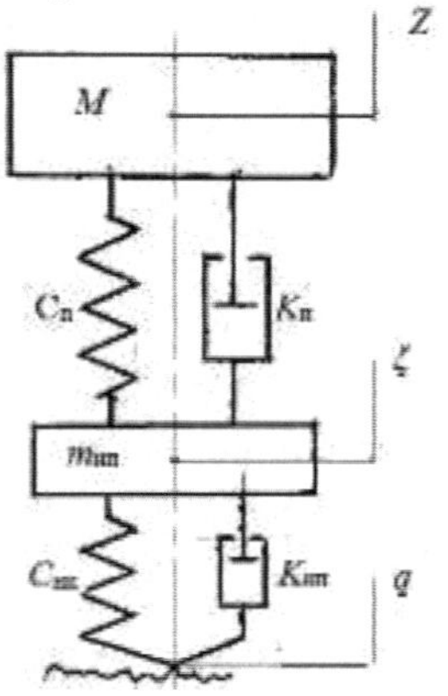

Figure 4. Calculation model of a suspension with elastomer

Dual-mass single-axle calculation model [4], Figure 4 is suggested for preliminary analysis of elastomer friction impact on the suspension damping properties.

Dual-mass single-axle calculation model allows to obtain consistent results at relatively simple initial motion equations. Sprung weight M, Fig. 4, is considered as a solid object incorporating all the weights elastically connected with it. Unsprung suspension weight $m_{нп}$ is connected with the sprung part by an elastic linkage with $C_п$ rigidity and a smoother with $K_п$ damping coefficient. $K_п$ damping coefficient can take on $k_э$ or K_a, values corresponding to the hydraulic shock absorber damping coefficient.

Tyre impact on oscillating processes nature is considered $C_{нп}$ by tyre stiffness reduced values and $K_{нп}$ tyre damping coefficient.

Differential equation sets describing calculation model weight motion with respect to viscous friction elements are given below [4, 5]:

$$\begin{cases} M\ddot{z} + K_п(\dot{z} - \dot{\xi}) + C_п(z - \xi) = 0 \\ m_{нп}\ddot{\xi} - K_п(\dot{z} - \dot{\xi}) - C_п(z - \xi) + K_{нп}(\dot{\xi} - \dot{q}) + C_{нп}(\xi - q) = 0. \end{cases} \quad (2)$$

where: z, $\dot{z}$, $\ddot{z}$ – sprung weight motion, speed and acceleration respectively; ξ, $\dot{\xi}$, $\ddot{\xi}$ – unsprung weight motion, speed and acceleration respectively; $q, \dot{q}$ – microprofile ordinate and rate of ordinate change respectively; $C_п$ – suspension stiffness equal to $C_Э$ elastomer rigidity; $K_п$ – shock absorber resistance coefficient (if equipped with elastomer it is considered to be equal to $k_э$ relative resistance coefficient).

Shock absorber resistance coefficient should be calculated taking into consideration the suspension kinematics – reduced awheel.

Analytical methods of solving dynamics tasks imply the choice of disturbance forming methods passed on from the bearing surface to the vehicle's wheels. When a vehicle is in motion, the disturbance affecting the wheel is formed by the microprofile. The road microprofile ordinates change randomly and presenting a law of their change in the form of a mathematical function is a difficult task. There are several disturbance forming methods [3, 4].

The simplest methods of road disturbance modelling are based on deterministic views and realized by analytical dependencies. A tyre is considered to smoothen outlines of irregularities assuming the irregularity profile to have sinusoidal form, symmetrical with respect to its mean value. *q(s)* microprofile function within each S_0 period can be analytically presented as a sum of trigonometric Fourier's series, comprising an infinite set of the functions' harmonic components. A possibility to present a continuous function in the form of Fourier integral allows to realize the task of forming microprofile function as a random process along the spectral density of its dispersions. The most significant normative document, classifying roadbeds is considered ISO 8608.

In some papers [3, 4] spectral density dependence of microprofile irregularities distribution is set as angular frequency function while transfer function is formed by a special filter. Forming random microprofile disturbance as a sum of harmonic constituents is an adequate method based on well-known views of the possibility to present a random process with set characteristics by a sum of harmonic constituents with frequency and amplitude selected in the appropriate manner. When solving a practical task of forming random continuous disturbance in the form of harmonic constituents, correlation of amplitudes of road irregularities waves and their length is taken into account, identified by spectral density characteristic of microprofile dispersion. When modelling a disturbance, spectral microprofile characteristics are used which are obtained by any of the known methods. Harmonic constituents' frequencies of a random disturbance process can be selected on a linear scale with the preset discretization rate or corresponding central frequencies of octave or 1/3-octave bands. A harmonic constituent amplitude is determined according to the dispersion value for the preset frequency band. Superposition harmonic function method with various oscillations frequency and amplitude is applied in the paper. These make up the random process spectrum (modelling with a polyharmonic function) [12]. Figure 5 shows the method flowchart.

Integral sum of spectral density function is the dispersions sum of interval ordinates of the frequency band in question. We consider that in the case of microprofile being presented by a superposition of periodical functions with various irregularity length (route frequency) and amplitudes, overall dispersion of microprofile spectral density can be determined as a sum of dispersions of the microprofile route frequency assigned slots, Figure 6. Spectral density overall dispersion constituents are regarded as harmonic process dispersions with the assigned slot medium frequency.

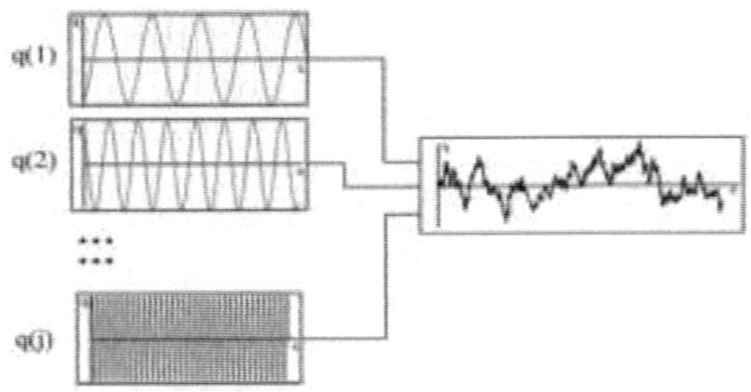

Figure 5. Microprofile mathematical representation

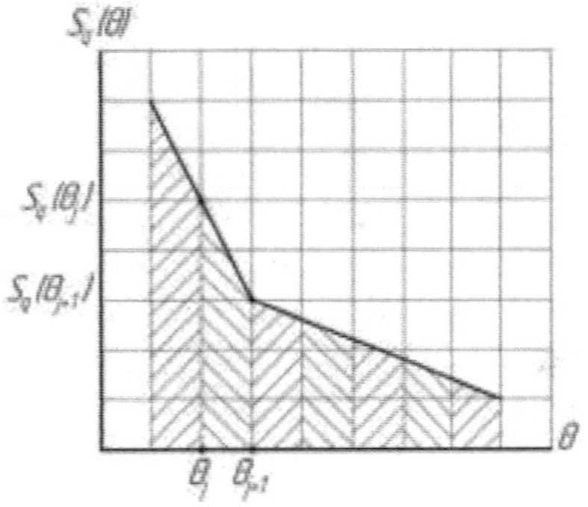

Figure 6. Microprofile spectral density dispersion

Dispersion corresponding to the harmonic process with the interval medium frequency $\{\theta_j, \theta_{j+1}\}$ is equal to the space limited by a function curve within the preset frequency band

$$D_j = \frac{S_q(\theta_{j+1})+S_q(\theta_j)}{2}(\theta_{j+1} - \theta_j) \quad (3)$$

where: D_j – function dispersion of microprofile spectral density for the route frequency interval $\{\theta_j, \theta_{j+1}\}$; $S_q(\theta_j)$, $S_q(\theta_{j+1})$ – spectral function values at the route frequency values equal to θ_j, θ_{j+1}.

Interval frequency mean value is accepted as equal to

$$\theta_{j\text{cp}} = \frac{\theta_j+\theta_{j+1}}{2} \quad (4)$$

where: $\theta_{j\text{cp}}$ – interval medium frequency $\{\theta_j, \theta_{j+1}\}$.

The road microprofile contains irregularities of different length, which can cause various frequency disturbances. The given parameters are related by ratios

$$T_\omega = \frac{2\pi}{\omega} = \frac{s}{\text{v}} \quad (5)$$

where: T_ω – time of overpassing an irregularity [c]; $\omega=2\pi f$ – road irregularities frequency, [rad/s]; f – cyclic frequency, $\text{f} = \frac{\text{v}}{\text{s}}$, [$\text{c}^{-1}$]; v – vehicle speed, [m/s]; s – irregularity length, m.

Differential equations are solved by numerical methods in MATLAB/Simulink programme. Figure 7 shows a microprofile model realized according to expressions (3) – (5). Process modelling algorithm allows for the disturbance effect frequency diversity at different (constant) vehicle speed, e.g. (40, 60, 80) km/h. Processes of changing vertical acceleration and normal response are time function. Selected process realization should be stationary. Elastomer friction is taken into account by "relative" elastomer damping coefficient.

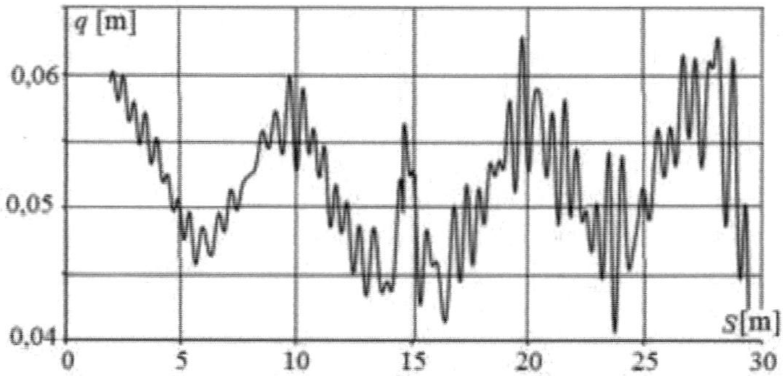

Figure 7. Smooth cobbled road microprofile model

Oscillation damping efficiency by elastomer friction is compared to the oscillation damping efficiency by a hydraulic shock absorber of a "standard" suspension. Mathematical model of dual-mass elastomer suspension was performed in MATLAB/Simulink software package.

4. Results and conclusions

Figure 8 and 9 show analytics results – vibration accelerations (RMS) dependences $\sigma_{\ddot{z}}$, of normal response in a wheel contact patch σ_F for elastomer suspension and the same dependences for hydraulic shock absorber suspension.

Comparative analysis of vibration acceleration RMS and normal responses for a standard hydraulic shock absorber suspension and elastomer suspension brings about the conclusion that elastomer elements lack oscillation damping efficiency under various speed conditions. Therefore, in order to obtain better smooth ride quality performance characteristics it is necessary to combine elastomer elements and hydraulic shock absorber. However, such structures have perspective advantages over sprung suspensions from the point of view of energy capacity and need further research and development.

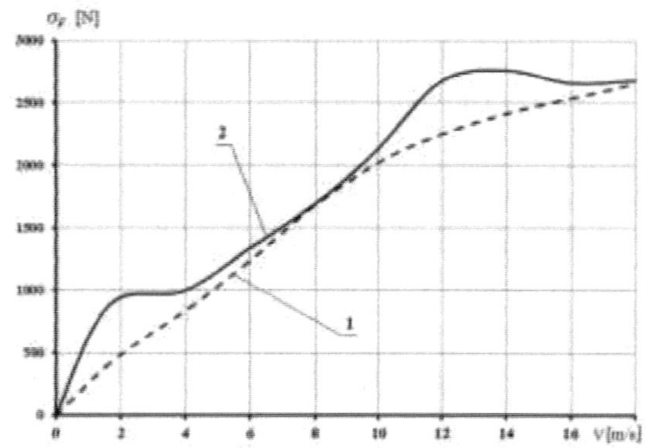

Figure 8. Normal response RMS awheel: 1 – for hydraulic shock absorber suspension; 2 – for «relative» elastomer damping coefficient.

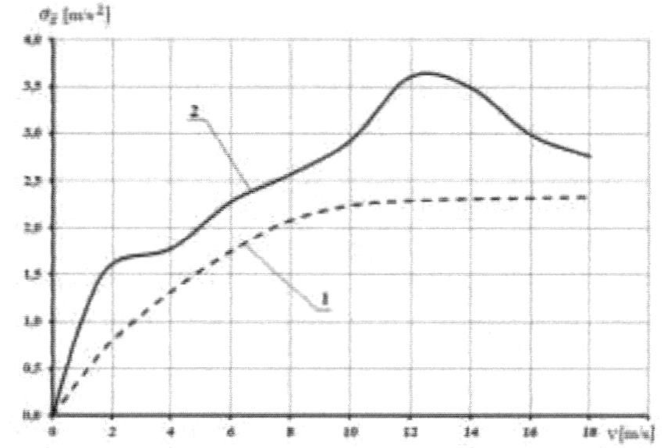

Figure 9. Vibration acceleration RMS on the car frame: 1 – for hydraulic shock absorber suspension; 2 – for «relative» elastomer damping coefficient

Acknowledgment
The research was funded by Ministry of Education and Science of the Russian Federation under the project "Creating high-technology manufacturing of GAZelle Next model car series with new electronic architecture of electronic systems " according to the Agreement № 075-11-2019-027 dated 29.11.2019 (Decree of the Russian Government dated 09 April 2010 №218).

Experimental research was carried out with Shared Use Center equipment of NNSTU «Transport Systems».

References:
[1] Christensen R 1974 *Introduction to viscoelasticity theory* Mir p 339
[2] Yatsenko N N and Prutchikov O K 1968 *Lorry smooth ride quality* M.: Mashinostroenie p 220
[3] Rotenberg R V 1972 *Vehicle suspension* M.: Mashinostroenie p 392
[4] Khachaturov A A 1976 *Road-tyre-vehicle-driver system dynamics* M.: Mashinostroenie p 535
[5] Musarsky R A 2008 *Mathematical models of wheeled vehicles* Nizhny Novgorod. Lobachevsky State University p 164
[6] *GOST 12.1.012-2004 Occupational safety standards system (OSSS). Vibration safety. General requirements* "National Research Center of Engineering Systems Control and Diagnostics" p 46
[7] *Technical guidelines of wheeled vehicles safety* Approved by Decree of the Russian Government №720 dated 10.09.2009
[8] Beresnev P et al. 2017 Simulation of road impacts for forecasting of durability of VE-hcle's suspension elements and supporting system *19th International and 14th European-African Regional Conference of the ISTVS*
[9] Leliovsky K Y, Makarov V S and Belyakov, V V 2019 Vibration load of transmission units at vehicle's motion over different roads *Journal of Physics: Conference Series* **1177(1)** 012053 doi:10.1088/1742-6596/1177/1/012053
[10] Makarov V S et al. 2019 Study of uneven surfaces distribution on forestry roads *Journal of Physics: Conference Series* **1177(1)** 012041 doi:10.1088/1742-6596/1177/1/012041
[11] Zezyulin D et al. 2014 Modeling of roads impacts for life prediction of light commercial vehicles parts *FISITA 2014 World Automotive Congress*
[12] Maleev S.I et al. 2019 Analytical investigation of suspension dynamics *Journal «Modern high-end technologies»* **6** p 88-95

Electronic component base for controlling the engine, mechatronic transmission and adaptive brake force limitation system as part of the electronic platform of the Formula SAE class sports vehicle

A Kulagin[1] and V Khozerov

Nizhny Novgorod State Technical University n.a. R.E. Alekseev, Minin str., 24, 603950, Nizhny Novgorod, Russia

E-mail:[1] kulagin@nntu.ru

Abstract. The article presents a description of the research and development work carried out by NSTU employees as part of their participation in the international project Formula Student . The paper deals with the process of creating an electronic platform for a sports car of the Formula SAE class, which includes electronic control units for the engine, mechatronic transmission and brake system, united into a common data bus. The process of selection and adaptation of the engine control unit based on the required parameters, the prototyping process of mechatronic transmission systems is described, the concept of the development of the brake control unit is reflected

1. Introduction

Among the most important parts of a modern sports car, as the suspension, frame, braking system, there are electronics and mechatronics controlled by it. Over the past decades, there has been a tendency in the modern automotive industry to transfer control over a number of vehicle components to mechatronic systems that can ensure accurate, coordinated and error-free operation of vehicle components. A number of vehicle systems are shifting from the control of the driver to the control of electronic systems, those systems make it possible to concentrate on solving a wide range of tasks, reducing driver fatigue and increasing the safety of the vehicle as a whole. The elements included in the electronic control system must meet the high requirements of safety, reliability and resistance to environmental influences. In the process of solving the problems of creating a control system for the internal combustion engine, transmission, and information output to the driver, there is a need to create an electronic platform This solution minimizes the laboriousness of setting and adjusting the operating parameters of the executive systems.

2. Creation of an engine management system

The engine of the sports car is a four-stroke internal combustion engine of the Honda CBR 600 RR motorcycle with a modified air-fuel intake system. To be able to control the systems of the engine, with the above changes, it is possible to install the following control units: Mega Squirt ECU, Speduino ECU, RusEFI Frankenso ECU (Figure 1). A comparative analysis of the characteristics of these control units can be seen in Table 1.

Table 1. Characteristics of the control units

Models / Parameters	Mega Squirt ECU	Speduino ECU	RusEFI Frankenso ECU
Microprocessor	Motorolla HC9S12C64	ATmega2560	STM32F407
Number of fuel injectors connected to the ECU	8 with phased fuel injection and 12 with non-phased injection	4	12 with phased fuel injection
CAN-BUS support	+	+	+
Source	Closed	Open	Open
Electronic circuit	Closed	Open	Open
Types of shafts position sensors which are possible to use	Hall, VR	Hall, VR	Hall, VR
System upgrade options	No	Yes. It is possible to upgrade the ECU with modules: Throttle - By - Wire , GDI , etc.	Yes. It is possible to upgrade the ECU with modules: Throttle - By - Wire , GDI , etc.

Based on the data in Table 1 and the technical characteristics of the ECU, the RusEFI Frankenso ECU electronic control unit is a universal solution with the possibility of expanding the electronic platform within the framework of the project being developed. The open source code of this ECU and publicly available options for expansion allow you to upgrade the control unit for specific tasks. With CAN bus support and electronic throttle control, this ECU can be used even in driverless vehicles [1].

Figure 1. Engine electronic control unit RusEFI Frankenso ECU

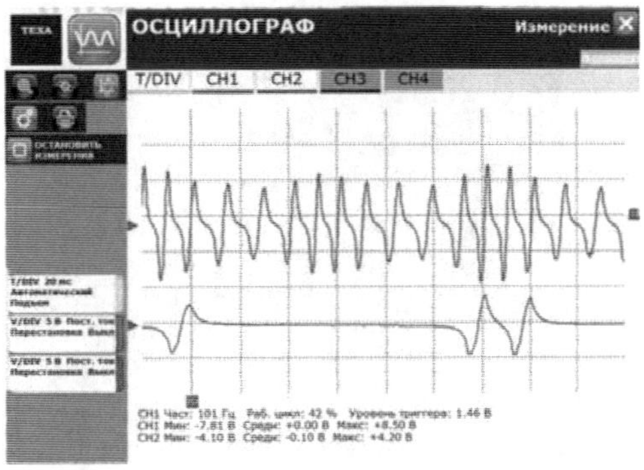

Figure 2. Oscillogram of signals from crankshaft and camshaft position sensors

During adapting the control unit, the problem of reading the signal parameters from the crankshaft and camshaft position sensors was discovered, because both of the sensors have a VR (variable reluctance)

signal with an analog signal generation. Figure 2 shows an oscillogram of the signals from the crankshaft and camshaft sensors with the correlation between the pulse tops. These pulse peaks can be used to track the strokes of each cylinder.

During test measurements with the connection of the crankshaft and camshaft sensors to the RusEFI Frankenso ECU, the presence of distortions in the read signal parameter was detected. Figure 3 shows an oscillogram obtained directly from the engine control unit. In this signal image, there is no correlation between the sensor signals.

Figure 3. Signals recognized by the control unit before applying the noise reduction circuit

To minimize the influence of distortion of signal parameters, when reading these parameters through the RusEFI Frankenso ECU, a noise reduction circuit was developed. Based on the application of the noise reduction circuit, an oscillogram was obtained without signal distortion (Figure 4).

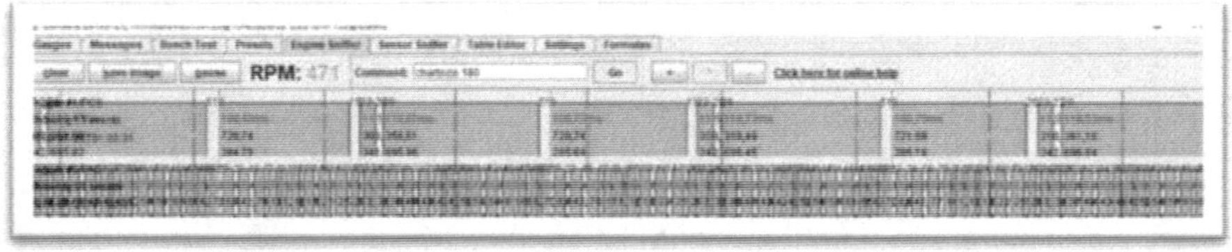

Figure 4. Signals recognized by the control unit after applying the noise reduction circuit

The ignition system, controlled by the RusEFI Frankenso ECU, required the manufacture of control modules for Denso OEM coils installed on the engine by the manufacturer. These modules have been developed due to the lack of proposals on the market that satisfy in terms of dimensions and characteristics. The ignition modules (Figure 5) are based on an IGBT transistor that allows a signal to be applied to the primary coil of the selected ignition coil.

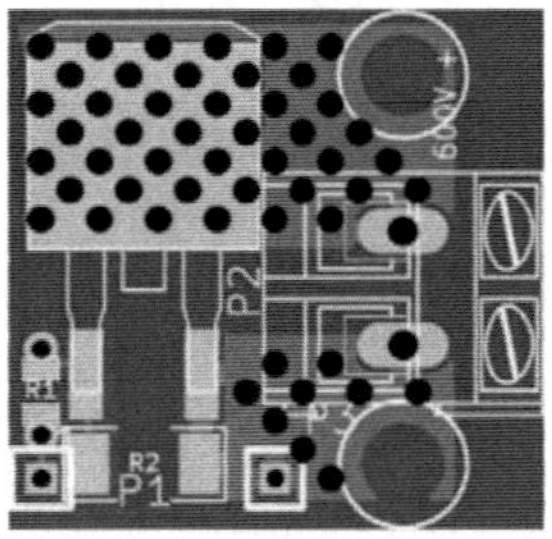

Figure 5 Ignition module for OEM coils

Figure 6. Transmission ECU of a FORMULA STUDENT sports car

OEM sensors installed on the engine of the Honda CBR 600 RR motorcycle are part of a group of sensors that can interact with the RusEFI Frankenso ECU. Tests have shown that there is no need to adapt Frankenso to read signal parameters routinely.

3. Creation of a mechatronic transmission control system.

In terms of creating mechatronic control systems for shifting the gearbox and clutch stages, the key parameters are the speed and time of switching, the force of the actuators, as well as the synchronization of these parameters to ensure the operability of the system [2].

IOP Conf. Series: Materials Science and Engineering 1086 (2021) 012020 doi:10.1088/1757-899X/1086/1/012020

The creation of a prototype of the mechatronic system is based on the use of the Arduino Mega 2560 microcontroller (Figure 6). The system being developed is capable of working not only in stand-alone mode, but also in the mode of an electronic assistant, displaying recommendations on gear shifting on the dashboard screen.

A key feature of the development of the mechatronic gear shifting system is the selection of actuators for the implementation of gear shifting, as well as the implementation of the clutch drive.

In the course of the work, a group of actuators were tested, consisting of a stepper motor (Figure 7), a linear actuator (Figure 8) and a DC motor (Figure 9).

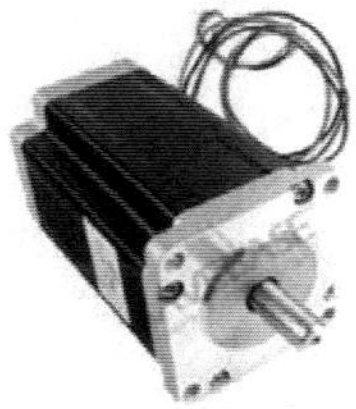

Figure 7. Stepper motor

Figure 8. Linear actuator with hall sensor

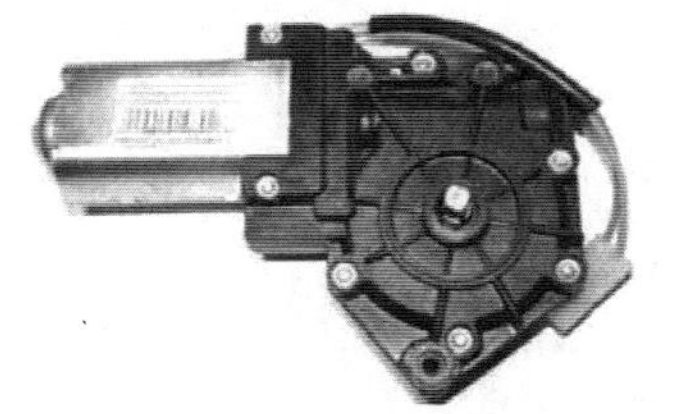

Figure 9. DC motor

The actuators shown have been tested down from the following key parameters:

1. Switching time
2. Maximum force/torque
3. Power consumption
4. Element weight
5. Overall dimensions

In order to test the actuators, unique mounting systems were created to interact with the transmission mechanisms. The development was carried out using 3D scanning, rapid prototyping using additive technologies and CNC machines [3].

Table 2 shows the comparative characteristics of the actuators based on the tests performed.

Table 2. Comparative characteristics of actuators

		Stepper motor	Linear actuator with hall sensor	DC motor
1.	Torque, N / m	3, 3	-	3
2.	Maximum effort, N	-	3500	-
3.	Shaft rotation frequency, rpm	60	-	65
4.	Starting torque, N / m	3 , 3		fourteen
5.	Maximum speed of movement of the slider, mm / sec		sixteen	
6.	Weight, kg	2	0 , 64	0.95

Tests of the stepper motor show the possibility of gear changes in a cycle of 0.36 seconds when integrated into the gearbox system of a Formula Student sports car. This time indicator can be optimized downward due to the use of a complicated operation algorithm and the expansion of the hardware complex for collecting and processing information.

During the tests of the stepper motor, positive operation as an actuator was recorded in the entire range of shifting gears in the transmission of the Honda CBR 600 RR engine.

However, the following disadvantages were identified that limit the use of this element in the mechatronic control system:

1. During testing this electric motor as a clutch release actuator, the inability of this electric motor to release the clutch was revealed.

2. The mass of the actuator is - 2 kg, which, taking into account the use of the gearbox as a gear change element and clutch drive, doubles the weight.

3. The time of the complete switching cycle of one gearbox stage is 0.71 seconds.

4. Overall dimensions in combination with the weight impose additional requirements on the load-bearing elements of these components in terms of high strength.

5. Significant overall dimensions of the stepper motor control unit.

6. The presence of additional electromagnetic fields caused by the operation of the electric motor, which can have a negative impact on the complex of operation of all electronic control systems of the engine.

Tests of the linear actuator have shown that the maximum angle of rotation of the gearbox shaft when changing gears is within 24 degrees from the initial mark. That is, an angle of 24 degrees is the range for up and down gears. To implement gear shifting based on the use of a linear actuator, within the entire angular range, it is necessary to comply with the requirements for maintaining maximum force. To realize the maximum force, the shoulder of the impact on the gearbox shaft must be at least 10 mm when measured from the center of the shaft. With this leverage, the effort required to shift gears is 658.8 N.

Based on additional criteria for the shift time and comparing this parameter with a force of 658.8 N, the actuator speed is 12 mm / s. At this travel speed, it will take no more than 0.35 seconds to change gear.

When evaluating a linear actuator for the ability to disengage the clutch, it was found that the time required for the linear actuator to release the clutch is 0.78 seconds. At the same time, the same period of time is required to return the clutch to its original position, in view of this, the time for the working process doubles.

The advantages of using a linear actuator are:

1. The mass of the actuator is 640 grams.

2. The time of the complete switching cycle of one gearbox stage is 0.35 seconds.

3. Availability of feedback systems for position determination. The linear actuator design includes a Hall sensor. The presence of this element excludes the use of additional sensors on the gear shift shaft

The disadvantages of using a linear actuator are:

1. The parameters of the overall dimensions of this type of actuator are associated with the need to use massive and geometrically developed fastening components, as well as the integration of these fasteners into the design of the engine and gearbox.

2. The need to change the design of the linear actuator to adapt to the gearbox drive mechanisms.

In the DC motor test, the time to complete a gear change cycle was 0.35 seconds, which is comparable to the changeover time using a linear actuator. The time taken to disengage the clutch was 0.76 seconds.

Advantages of using DC motor:

1. Mass of the actuator - 950 grams.

2. The time of the complete switching cycle of one gearbox stage is 0.35 seconds.

The disadvantages of using a DC motor are:

1. The need to adapt the design of the electric motor to the standard gearbox system.

2. Lack of feedback systems for position determination.

To solve the problem of the lack of systems for tracking the position of the DC motor shaft, the accelerator pedal position sensors of a motorcar were used. A key feature of the accelerator pedal position sensor is the presence of a second channel for position sensing and troubleshooting when determining the application of the shift shaft and clutch lever.

4. ECU of the brake system.

To increase the level of safety, control precision of a sports car, as well as control of skids, it is necessary to use driver assistance systems capable of proactively locking and unlocking the wheels, as well as controlling the torque. For this, a brake system control unit is being developed, the basis of which will include the tasks of controlling the steering wheel angle, acceleration along each of the three spatial axes, the angular speed of each of the four wheels, pressure in the brake circuits and comparison and processing of the obtained data to control the position car on the track.

Having a communication bus, the control unit can not only influence the operation of the brake circuits, but also send signals to the engine and gearbox control units. These operational capabilities will allow system to control the torque on the drive wheels. Important criteria for the development of the entire system are its modularity, compatibility with different types of vehicles, safety, reliability, efficiency and low cost. Figure 10 shows a schematic diagram of the developed brake control unit.

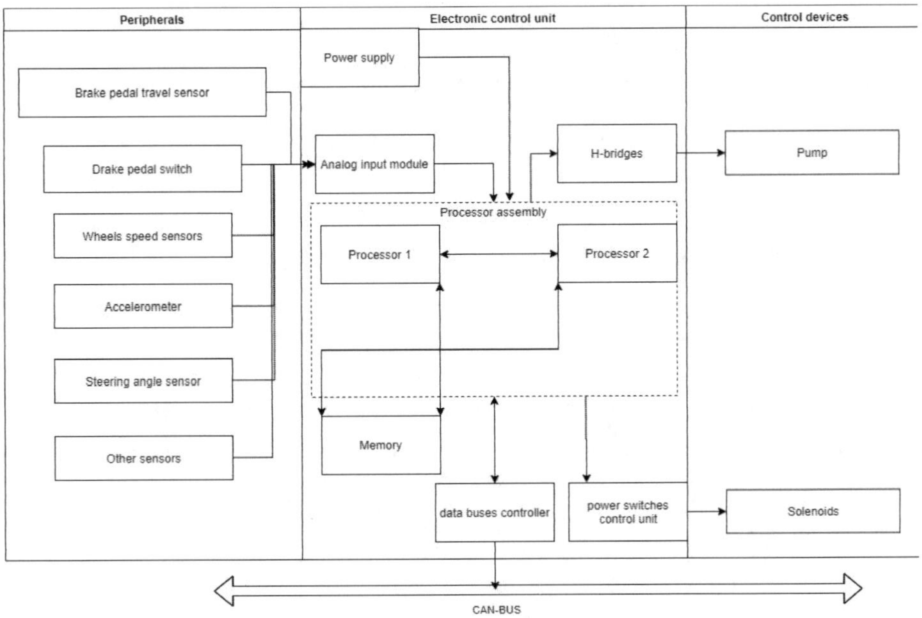

Figure 10. Schematic diagram of the ECU of the brake system

5. Conclusion

At the current stage, two components have been fully implemented: the mechatronic transmission ECU and the engine ECU. The created control systems, united in a data bus supporting the Controller Area Network standard, become part of the modified electronic platform. The implementation of extensions to this platform, including the electronic brake control system, will make it possible to adapt this system to an unmanned racing car platform.

References

[1] Khozerov V S and Kulagin A L 2020 Development of an electronic platform of the Formula SAE sports car *Collection of materials of the XIX all-Russian youth scientific and technical conference "the Future of technical science"* / NSTU named of R. E. Alekseev. N. Novgorod

[2] Khozerov V S 2019 On the question of control systems for a AMT gearbox using the Arduino UNO microcontroller *Proceedings of the X V III All-Russian Youth Scientific and Technical Conference "The Future of Technical Science"* NSTU n.a. R.E. Alekseev. N.Novgorod p 145

[3] Khozerov V S and Kulagin A L 2020 Optimization of the development process of embedded systems of racing cars *Collection of materials of the XIX all-Russian youth scientific and technical conference "the Future of technical science"* NSTU named of R. E. Alekseev. N. Novgorod

Ensuring reliability of electronic control of engine fuel supply subsystem in operation

Y Bazhenov[1,3], M Bazhenov[1] and V Kalyonov[2]

[1]Vladimir State University named after Alexander and Nikolay Stoletovs, Gorky St. 87, Vladimir, 600000, Russia
[2]Dealer center Peugeot, Kuibyshev St. 24A, Vladimir, 600035, Russia

E-mail: [1] bagenovyv@mail.ru

Abstract. The paper presents results of a study of operational reliability of electronic control of an automotive engine fuel supply subsystem. Impacts on the structural elements of the subject subsystem stipulated by physical and chemical processes of heat transfer and dissipation, absorption of magnetic and electric fields, wear and tear, etc., leads to occurrence of various kinds of damage in them. Causes of the subsystem getting inoperable and the effect of failures of its elements on the engine operation are revealed. Considerable attention is paid in the paper to the diagnostics of the technical state of order of the subsystem which must be performed in scheduled car maintenance intervals in accordance with the developed method and algorithm. The deviation of the fuel pressure in the fuel rail from the setpoint in the electronic control unit was taken as the diagnostic parameter for the evaluation of the technical condition of the subsystem. Using Microsoft Excel, an analytical equation was selected that describes the dependency of the diagnostic parameter deviation versus the running time of the vehicle allowing for the substantiation of its rated values. The results of the performed studies were tested and introduced into the technological processes of maintenance and repair of Peugeot cars of "Auto Tract Peugeot" dealership, Vladimir City.

Introduction

The operation of a modern internal combustion engine is controlled by an electronic engine control system (EECS), which ensures high technical and economic performance and compliance with environmental requirements to exhaust gas toxicity. During the operation, various types of damage and faults inevitably occur in the EECS (violation of adjustments, change of electrical characteristics, corrosion failure of contacts, insulation damage, etc.). The on-board self-diagnostics system of the electronic engine control system cannot provide its reliable operation, since the electronic control unit stores the fault code only when any of the diagnostic parameters exceeds the rated value set in the software, i.e. when the fault has already occurred. Untimely detection and elimination of damages leads to engine malfunction, full or partial loss of serviceability (failure). To reduce the number of failures, to detect hidden damage and to ensure the required level of reliability of the electronic control systems, it is necessary to monitor the technical condition of their elements during routine maintenance of motor vehicles at car service workshops.

Overview of previous works

The issues of ensuring the operational reliability of the electronic control of the fuel supply (FS) subsystem are poorly studied. The performed studies are mostly aimed at the development of diagnostic methods for individual EECS elements, studying the causes of malfunctions and their impact on the technical and economic performance of the engine, maintenance and repair methods of internal combustion engines equipped with control systems [1-6]. In the adopted maintenance regulations of cars, no monitoring of the technical condition of the elements of the FS subsystem is envisaged. This leads to concealed unsolved failures in a number of vehicles after routine maintenance, which entails an increase in the number of operational failures. The analysis of literary sources shows that 30 to 38% of failures of gasoline engines occur exactly due to damages of EECS structural elements [7-10]. Considering that, studies aimed at ensuring the reliable operation of electronic engine control systems are relevant and of great practical importance.

Methods

The EECS is a complex technical system of a motor vehicle containing several dozen structural elements, which, in terms of their functional purpose and diagnostical convenience, are advisable to de subdivides into several subsystems. The number of subsystems is based on the available number of individual diagnostic parameters depending on the complexity of the EECS used [9, 11].

In the present paper, the selected DUT, 1.6 THP Turbo Tiptronic (110 kW) EECS of Peugeot 408, is divided into 4 subsystems: fuel supply control, air supply control, valve timing phase control and exhaust gas toxicity control. Studies carried out in the maintenance shop of Peugeot dealership in Vladimir during maintenance, repair and diagnostics of electronic control systems of gasoline engines have shown that the *fuel supply subsystem* is the least reliable subsystem of EECS. The results of the studies of the operational reliability of this subsystem are given in Table 1 and in the form of bar graphs and their theoretical smoothing distribution curves of running time distribution before failure in Figure1.

The shapes of the curves, as well as the calculated values of the coefficient of variation v show that the distribution of the running times before failure of the parts of the FS subsystem are adequately described by Weibull distribution. Testing the hypothesis for fitting of the experimental data into this distribution using Pearson's chi-squared test confirmed its validity.

Table 1. Statistical estimates of the numerical reliability characteristics of the fuel supply subsystem

Structural element	Mean time to failure, thousand km	Mean square deviation	Coefficient of variation
1. Electric gasoline pump	119.3	68.35	0.57
2. High pressure fuel pump (with fuel pressure regulator)	90.6	56.84	0.63
3. Fuel injector	125.8	46.39	0.37
4. Fuel pressure sensor	185.3	117.06	0.63

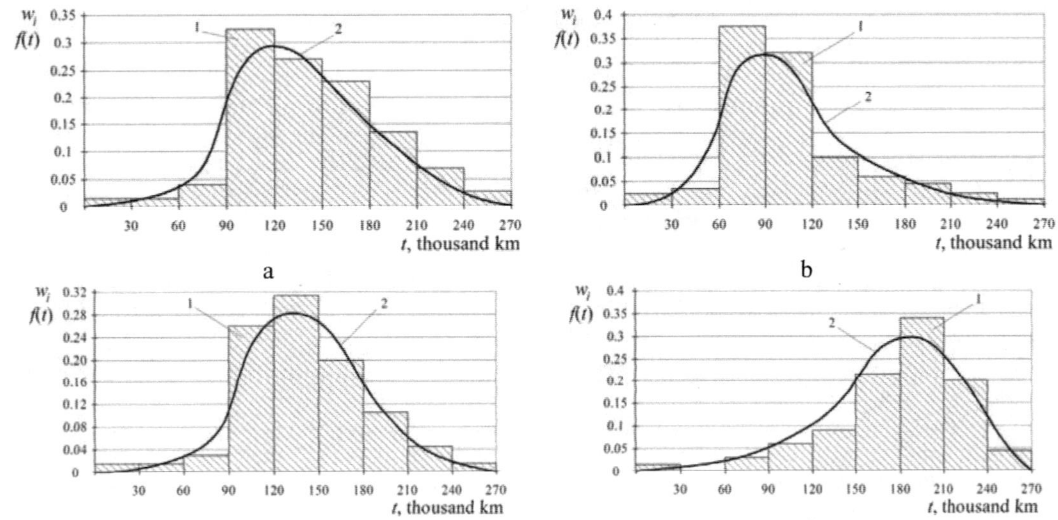

Figure 1. Bar graphs (1) and theoretical curves (2) of running time to failure distributions of structural elements of the fuel supply subsystem: a) electric gasoline pump; b) high pressure fuel pump; c) fuel injector; d) fuel pressure sensor

The operation of the elements of the FS subsystem is based on the use of electrical, electromagnetic, electronic, mechanical and other processes. In the parts of the subsystem, there are processes of heat transfer and dissipation, absorption of magnetic and electric fields, wear, etc. The impact of the listed processes leads to the occurrence of various kinds of damage in the structural elements of the FS subsystem and, in case of untimely troubleshooting, to a loss of operability (Table 2).

Table 2. Causes of FS subsystem failures and their impact on engine operation

Subsystem element	Causes of element failure	Failure effect
1	2	3
1. Electric gasoline pump	Stall of the gasoline pump motor armature, hanging and wear of brushes, destruction of insulation and inter-turn short circuit in the pump motor winding, collector wear	The engine cannot be started because the fuel supply to the engine cylinders is blocked
2. High pressure fuel pump	Failure of tightness of the shut-off valve of the pressure regulator due to its contamination, destruction of insulation and inter-turn short circuit in the pressure regulator winding, wear of pump plunger pairs	Reduction of engine power down to 80%, increase of fuel consumption by 18-20% and increase of emissions of harmful substances with exhaust gases by 8-10%
3. Fuel injector	Insulation destruction and inter-turn short circuit in electromagnetic valve winding, accumulation of lacquer deposits and coking of nozzle metering elements reducing the section of calibration channel	Decrease in power by 20-25%, increase in fuel consumption by 17-19%, increase in emissions of harmful substances by 12-15%, unstable operation of ICE in idle mode
4. Fuel pressure sensor	Contamination of sensor's sensing element, sensor solder connection contact violation	Reduction of engine power by 25-30%, increase of fuel consumption by 12-15% and emissions of harmful substances by 5-7%

As follows from the table, a significant part of the failures of the structural elements of the FS subsystem having the excitation winding is associated with the destruction of its insulation, which undergoes natural wear and aging due to temperature changes, vibrations, moisture, etc. Short-circuits of the winding to the housing occur due to local deterioration of insulation which occurs due to friction of winding turns between each other or against the housing, as well as to the insulation fouling.

The analysis of the obtained results given in the table shows that the technical condition of the structural elements of the FS subsystem has a serious influence on the engine operation. In case of a failure, for example, of the electric gasoline pump, the engine gets completely inoperable, since the fuel supply to the cylinders is blocked. The technical condition of the high-pressure fuel pump has a significant impact on the performance of the engine. When various kinds of mechanical damages occur in it, the engine power goes down to 80%, the fuel consumption and the emissions of harmful substances with the exhaust gases (carbon monoxide (CO), hydrocarbons (CH), nitrogen oxides (NO_X)) are significantly increased.

The main reliability indicators of any technical object, along with the mean time to failure, are the probability of its failure-free operation $P(t)$ and the failure rate $\lambda(t)$ which clearly show the dependence of the failures on the running time [12, 13]. Statistical estimates of these indicators for the elements under test are given in Table 3.

While analyzing the data given in the table, it is possible to draw certain conclusions on the operational reliability of the parts of the FS subsystem under test. For the initial running time intervals 0 to 90 thousand km, the probability of the parts failure, with the exception of the high-pressure fuel pump, is at a sufficiently high level. By the running time interval of 120-150 thousand km, the probability of failure-free operation of almost all elements of the FS subsystem is significantly reduced. Up to that running time, operational damages occur and accumulate in these elements (wear, lacquer deposits and coking of metering elements, deterioration of insulation in the windings, etc.) which reduce the efficiency of their operation and, in the long run, the inevitable loss of operability. Almost all parts of the FS subsystem get completely worn out within a running time of 270 thousand km.

Table 3. Indicators of fuel supply subsystem elements failure-free operation by running time intervals

Indicator	Running time interval, thousand km.								
	0-30	30-60	60-90	90-120	120-150	150-180	180-210	210-240	240-270
Electric gasoline pump									
$P(t)$	0.98	0.97	0.94	0.63	0.37	0.15	0.06	0.03	0.01
$\lambda(t)\cdot 10^{-4}$	0.40	0.41	1.26	12.19	15.11	22.25	27.78	37.04	66.67
High pressure fuel pump									
$P(t)$	0.98	0.94	0.56	0.25	0.14	0.08	0.04	0.03	0.01
$\lambda(t)\cdot 10^{-4}$	0.79	1.22	16.66	26.08	18.18	17.54	26.66	33.32	64.87
Fuel injector									
$P(t)$	0.98	0.97	0.94	0.68	0.36	0.17	0.06	0.02	0.01
$\lambda(t)\cdot 10^{-4}$	0.51	0.52	1.06	10.59	20.29	24.76	31.11	40.07	67.98
Fuel pressure sensor									
$P(t)$	0.98	0.98	0.95	0.89	0.80	0.58	0.25	0.05	0.01
$\lambda(t)\cdot 10^{-4}$	0.52	0.00	1.06	2.22	3.64	10.37	27.16	45.61	65.67

The graphical interpretation of the change in the statistical estimates of the failure intensity $\lambda(t)$ and the probability of failure-free operation $P(t)$ by t running time is shown based on the example of the electric gasoline pump and the fuel pressure sensor (Figure 2).

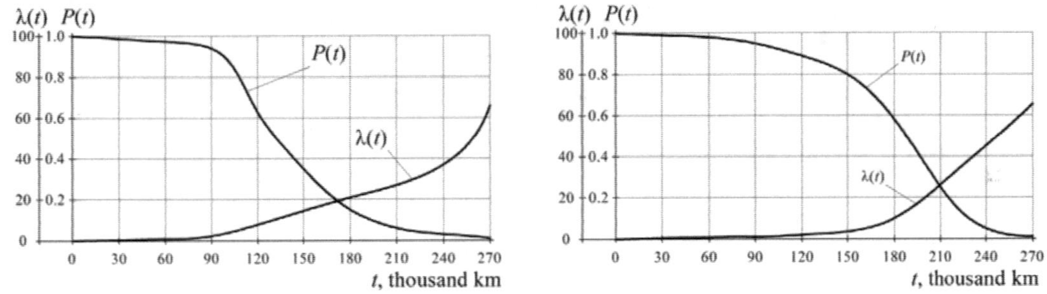

a b

Figure 2. Probability change curves of failure-free operation and failure rate: a) electric gasoline pump; b) fuel pressure sensor

The completed studies of the operational reliability of the structural elements of the subsystem show that the process of their technical condition change depending on the running time of the car is smooth and monotonous, leading to the occurrence of so-called gradual failures. For such failures, the change of a technical condition parameter is analytically well described by an entire rational function of the nth order [14-16].

Based on the results of processing of the statistical data of the operational reliability of the components of subsystem under analysis, the following analytical dependencies of the change in the intensity of their failures $\lambda(t)$ versus the running time were obtained:

$$\begin{aligned} \lambda_{gp}(t) &= 0.199t^3 - 1.877t^2 + 8.820t - 9.035; \\ \lambda_{fp}(t) &= 0.521t^3 - 7.139t^2 + 32.210t - 29.460; \\ \lambda_{fi}(t) &= 0.126t^3 - 0.873t^2 + 5.521t - 6.392; \\ \lambda_{ps}(t) &= 0.152t^3 - 0.478t^2 - 0.750t + 2.139, \end{aligned} \quad (1)$$

where $\lambda_{gp}(t)$, $\lambda_{fp}(t)$, $\lambda_{fi}(t)$, $\lambda_{ps}(t)$ are the failure rates of the electric gasoline pump, high-pressure fuel pump, fuel injector, fuel pressure sensor, respectively; t is the vehicle running time.

In order to ensure reliable operation of the FS subsystem and to reduce the cost of repairs after failures, it is necessary that most of them should be prevented during routine maintenance of motor vehicles with a preliminary assessment of the technical condition of the subsystem using diagnostic methods.

The deviation of the fuel pressure in the fuel rail P_{fp} from the setpoint in electronic control unit (ECU) was assumed as the diagnostic parameter characterizing the technical condition of the subsystem. The limit value P_{lfp} and the rated value P_{nfp} of this parameter are set on based on studies of the engine operation processes during the design of EECS by manufacturers and are given in the technical regulations for after-sales maintenance and repair of Peugeot motor vehicles [17]: P_{lfp} = 9 bar; P_{nfp} = 3 bar.

It is advisable to use the *permissible* value of P_{pfp} parameter rather than the limit value, as a standard, when performing control and diagnostic works to evaluate the technical condition of the FS subsystem in automotive service workshops. To determine it, it is necessary to find out the regularity of change of the fuel pressure deviation from the ECU setpoint versus vehicle running time.

The expiry of the service life assigned to the subsystem by its design is stipulated by the gradual accumulation of various damages in their elements (wear, corrosion, etc.). The development of such damages depending in the course of time or along with the running time is smooth, monotonous, leading to occurrence of gradual failures, therefore, with some probability, the change in the technical condition parameter by running time t can be described by the power function [18, 19, 20]:

$$y(t) = y_n + \upsilon t^{\alpha}, \quad (2)$$

where y_n is the rated value of the technical condition parameter; υ is the intensity of the parameter change along with the running time; t is the running time of the DUT; α is the exponent determining the dependence of parameter y on running time t.

The change in the analog diagnostic parameter versus the running time for most units and assemblies of a motor vehicle is described by the same functions as the parameters of the technical condition. The analytical equation selected using Microsoft Excel, describing the change in the diagnostic parameter P_{fp} versus the running time, will have the following appearance:

$$P_{fp}(t) = P_{nfp}(t) + \upsilon t^{\alpha} = 3 + 0.044t^{1.03}, \quad (3)$$

where $P_{nfp}(t)$, $P_{fp}(t)$ are, respectively, the rated and the current values of diagnostic parameter P_{fp}.

The exponent α and intensity of change of the diagnostic parameter υ versus the running time are determined experimentally, based on the processing of the experimental data on the operational reliability. The graphical presentation of the analytical equation is given in Figure 3.

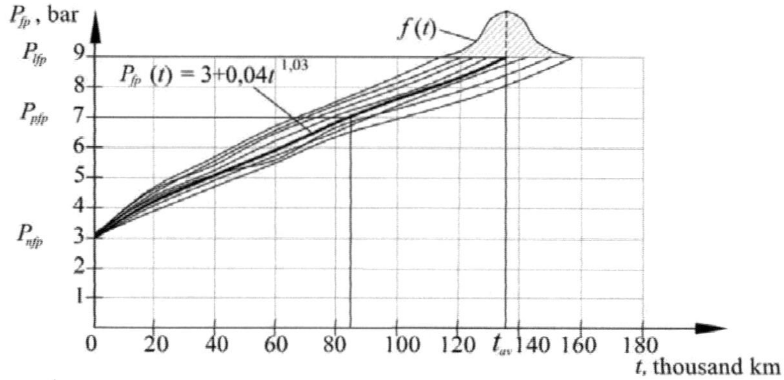

Figure 3. Graph of P_{fp} diagnostic parameter change by operating time: P_{nfp}, P_{lfp}, P_{pfp} – nominal, limit and permissible values of diagnostic parameter; t_{av} – average time between failures

Knowing the periodicity of the vehicle diagnostics (for Peugeot, t_m = 20 thousand km), the permissible value of the diagnostic parameter P_{pfp} is determined from the expression

$$P_{pfp} = v\left(\sqrt[\alpha]{\frac{P_{lfp}}{v}} - t_m\right) = 0.044\left(\sqrt[1.03]{\frac{9}{0.044}} - 20\right) = 7 \text{ bar}.$$

Fault detection and localization in the fuel supply subsystem is performed in accordance with the algorithm developed and tested in the Peugeot dealership of Vladimir (Figure 4).

After the elimination of faults detected as a result of diagnostics, the fuel pressure P_{fp} deviation in the rail is checked again. If its value does not exceed the permissible standard (7 bar), the technical condition of the FS subsystem is deemed restored guaranteeing its failure-free operation till the next maintenance is carried out.

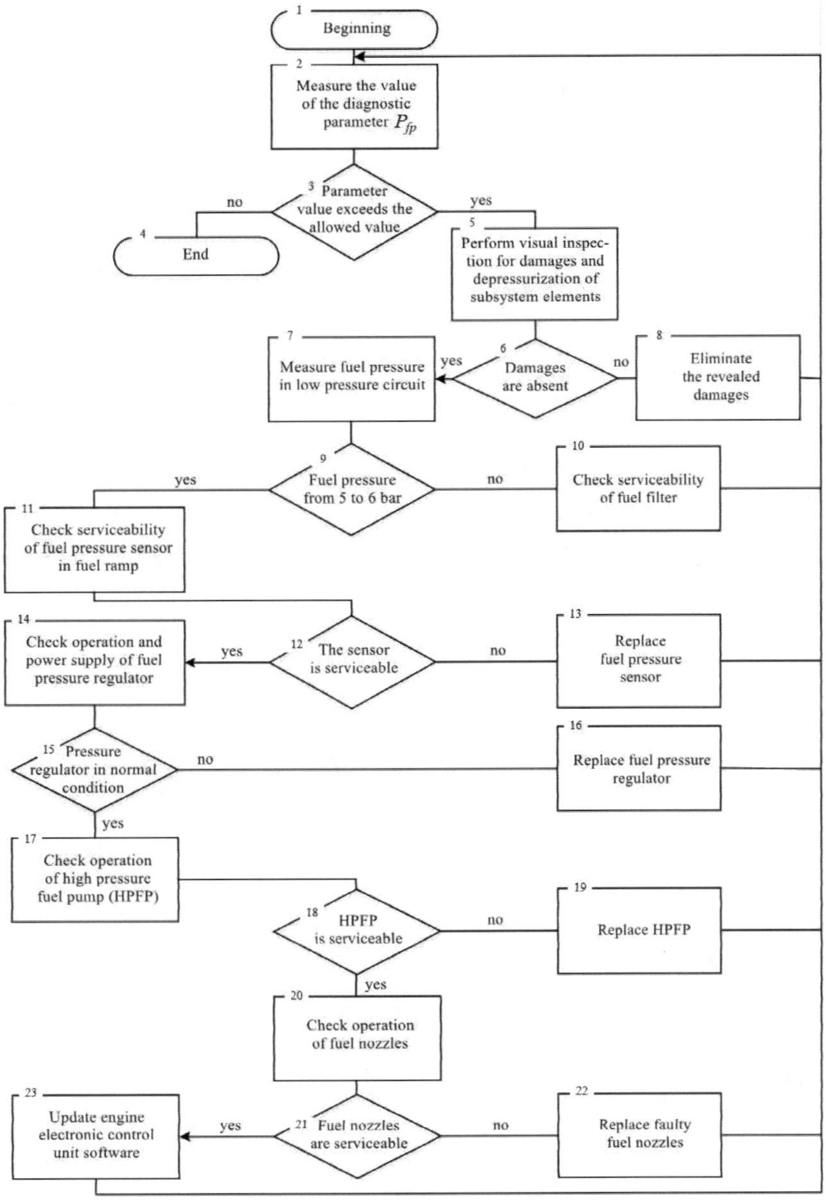

Figure 4. Algorithm of fuel supply subsystem diagnostics

IOP Conf. Series: Materials Science and Engineering 1086 (2021) 012021 doi:10.1088/1757-899X/1086/1/012021

Conclusions

The results of the study demonstrate that failures of the fuel supply subsystem lead to a significant worsening of the engine performance and an increase in emissions of harmful substances into the environment. To ensure reliable operation of the structural elements of this subsystem, it is necessary to carry out diagnostics of their technical condition with the periodicity of the scheduled vehicle maintenance. The FS subsystem diagnostics technology proposed in the paper makes it possible to significantly reduce the number of its failures during the operation due to the timely detection of damage occurring in it which remain concealed to the on-board diagnostics system of the vehicle.

References

[1] Goncharov A A 2004 *Improvement of the technology for electronic control systems diagnostics of automotive engines* PhD Eng. thesis. Orenburg p 96
[2] Hedicks E 1995 *Event based engine control: practical problems and solutions* SAE paper No 950008
[3] Herner A and Riel H 2013 *Automotive electrics and electronics* Moscow. Za Rulyom p 625
[4] Sarkisov A A 2004 Smart diagnostics of failures of automobile sensors *Electronics and electrical equipment of transport* **3-4** p 46-52
[5] *Diagnostics of electronic systems of motor vehicles with NTS devices 9th edn. NPP "NTS"* 2007 Samara p 262
[6] Tweg R 2003 *Diagnostics of the electronic control system of the automotive engine* Astrel. Moscow p144
[7] Rendall M 2008 *Electrical and electronic equipment of automobiles* Alfamer Publishing. Moscow p 284
[8] Palaguta K and Kuznetsov A 2007 System for injection ICE diagnostics by its running speed characteristics *Automotive industry* **4** p 21-24
[9] Bazhenov Yu and Kalyonov V 2016 System for ensuring the operability of EECS in operation *World of transport and technological machines* **4 (55)** p 24 - 33
[10] Bazhenov Yu and Bazhenov M 2018 Study of the operational reliability of automotive engines *Reliability* **18-4** p 22- 27
[11] Kalyonov V 2019 *Development of a system for ensuring the operability of automotive engine control systems of a motor vehicle in operation* PhD Eng. Thesis. Moscow p157
[12] Bazhenov Y V 2016 *Basics of machine reliability theory* Forum. Moscow p 320
[13] Kovalenko T 2016 *Experimental data processing* M. National Open University "INTUIT" p 179
[14] Gnedenko B V, Belyaev Y K and Soloviev A D 2018 *Mathematical methods in reliability theory. Main reliability characteristics and their statistical analysis* Publication "URSS" p 584
[15] Zubarev Yu 2017 *Fundamentals of reliability of machines and complex systems* St. Petersburg. Lan p 180
[16] *RD 50-690-89. Methodological instructions. Reliability in engineering. Methods of evaluation of reliability indicators based on experimental data* 1990 Publishing House of Standards. Moscow p 132
[17] *Electronic portal of car manufacturer PEUGEOT. SERVICE BOX* 2019 https://servicebox.peugeot.com
[18] Tunin A A 2007 *Diagnostics of electronic control systems of passenger car engines* Solon-Press. Moscow p 352
[19] Bigus G, Daniev Yu, Bystrova I and Galkin D 2014 *Diagnostics of technical devices* Moscow. Publishing house of MSTU n.a. N.E. Bauman p 615
[20] Stukanov V A and Boyaryshnikov A A 2019 *Reliability and technical diagnostics of motor vehicles* Moscow. FORUM: INFRA-M p 420

Evolution of requirements to state of order of automotive tires for traffic safety

I A Yerasov[1,3], D N Proshin[1] and D V Glazunov[2]

[1]Nizhny Novgrod State Technical University n.a. R.E. Alexeev, Russia, 603950, Nizhny Novgorod, ul. Minina, 24
[2]Kirghiz-Russian Slavic University, Kirghizia 720000, Bishkek City, ul. Kiyevskaya, 44

E-mail: [3] erasoff2013@yandex.ru

Abstract The paper presents the results of a study of the influence of the amendment of the requirements to the technical state of order of motor vehicles, specifically, to the design of the tires used on the front and rear axles of motor vehicles on the motor traffic safety. It has been determined that newly developed requirements aim on the reduction of the difference of the traction between the front and the rear axle for more prompt response of the braking mechanisms and improved braking efficiency. It has been demonstrated that the implementation of the design requirements to vehicles developed in 2010 an 2013 allows for reduction of the accident rates of motor vehicles using different tire types by 3%. Further tightening of the operational requirements to motor vehicles in this direction has a reserve allowing for max. 1% of further reduction of the accident rate whereafter the reserve of the traffic safety improvement will be deemed exhausted.

Introduction

For the time being, injuries and mortality of people in traffic accidents constitute one of the major negative development factors of our society. Scientific studies evaluate economic losses due to traffic accidents up to 5% of GDP, thereby, alone the mechanical repair works on the damaged vehicles weigh well 2.5%. During 2020 1.5 million people became COVID-19 victims, whereas in traffic accidents 1.25 million casualties were reported. The average global reduction of gross domestic products in 2020 makes up 6% [1]. That is, we confront comparable figures of the global economy losses due to COVID-19 pandemics and due to traffic accident, thereby, the priority in the countermeasures of the humankind is definitely shifted towards the struggle against COVID-19, whereas the matters of the traffic accident rates remain undeserved in the shadow of the struggle against the new threat. Along with that, it is worth saying that the effect of the improvement of the motor traffic safety will be no less important than the victory over COVID-19, that is why solving of problems relevant to the reduction of traffic accident rates is a vital research subject.

Overview of previous studies

The beginning of the studies on provisions for traffic safety is now associated with Ralph Nader's Unsafe at any Speed: The Designed-In Dangers of the American Automobile [2], thereby, in connection with the highest relevance of this problem for highly motorized countries, the majority of the studies have been made by American scientists [3,4,5]. In Europe, Swedish scientists score the leadership [5,6,7]. Individual interesting studies can be encountered in other countries, too [8]. In Russia, the improvement of the motor traffic safety has been the subject of studies by such scientists, as: V.V.Abartsumyan, O.P.Gudjoyan [9], V.I.Konoplyanko, O.P.Gudjoyan,

V.V.Zyryanov [10]. An important contribution was made by scientists of Nizhny Novgorod Technical University, such as A.Groshev., A.Vashurin, P.Musarskiy, A.Tumasov [11]

The results of the specified studies allow for conclusions that, for the time being, the problem is rather completely investigated with wide utilization of mathematical statistics for determining of the influence of any specific factor on the automotive traffic safety. However, the issue of the influence of combined use of different tire designs on the front and the rear axle of a motor vehicle on the automotive traffic safety has not enjoyed considerable interest, since it is deemed during the design of a motor vehicle that it should normally bear all identical tires, be tested on all identical tires and be delivered to the consumer on all identical tires. There are only few studies dedicated to problems of motor vehicle driving on a combination different design tires, whereas they touch upon mostly the operation of special mission vehicles [12]. The replacement of normally carried tires by tires of a different design is within the competence of technicians in charge of operation matters whose actions are guided by basic provisions on admission of motor vehicles to the operation being an integral part of Russian Traffic Rules [13,14].

Based on the document, the requirements to the carried tires were reviewed five times in the last 27 years, however, no papers substantiating the necessity for such revisions are publicly available.

Methods

A conducted analysis of amendments to the design requirements of a motor vehicle in terms of combined utilization of different tire types has revealed that, in 1993, it was forbidden to utilize tires with different tread patterns on one axle to prevent derapage of the vehicle at emergency braking in absence of a lateral force. In 2002, the tread pattern difference clause was completed by the prohibition of combining of studded and non-studded tires, tires with new and restored tread, frost-resistant and frost-sensitive tires on one axle. All these interdicts were also to prevent derapage at emergency braking in absence of lateral forces. In 2010, a requirement was developed concerning combining of different tire types between the front and the rear axle prohibiting any combinations of studded and non-studded tires in motor vehicles. In 2013, an interdict was adopted against the use of winter and all-season tires with less than 4 mm residual tread depth. In the same year, a technical regulation on safety of wheel-type motor vehicles forbidding combining of winter and summer tires, as well as the use of studded tires in summer and summer tires in winter was adopted. Beside that, banned were tires with twice restored tread pattern and tires with once restored tread pattern on the front axle. Thus, the evolution of the requirements to the technical state of order of motor vehicles aims, for the time being, uniform tiring in terms of the uniformity of their properties on all wheels of a motor vehicle. Thereby, the control properties of motor vehicles [15] are mostly influenced by changes in the stiffness of the tire structure, rather than by tire type and tread pattern. The limitation of the traction difference between the wheels is mostly intended for ensuring the vehicle stability during emergency braking.

The results of traction property studies of different tire models can be found in specialized automotive journals like "Autoreview", "Za Rulyom", "Auto-Bild", "Auto Moto und Sport", etc. [16-19]. Analyzing the regulatory requirements to the tiring, it should be noted , that till 2010 it was admissible to mount new studded tires onto the front axle with worn summer tires on the rear axle. Such a combination could have a traction difference of up to 0.19 between the front and the rear axis on ice, up to 0.3 on snow, up to 0.3 on wet asphalt, and up to 0.4 on dry asphalt. The ban of combined use of studded and non-studded tires allowed for a reduction of the traction difference down to 0.13 between the front and the rear axis on ice, down to 0.2 on snow, down to 0.2 on wet asphalt, and down to 0.3 on dry asphalt. The prohibition on combined use of winter and summer tires did not alter the obtained traction differences since the specified difference in the parameters of the interaction of the wheel with the road can occur due to different tire manufacturers. A further improvement of the automotive traffic can be achieved by use of tires by a single manufacturer. In this case, the traction difference could be reduced, due to non-uniform wear, down to 0.05 between the front and the rear axis on ice, down to 0.1 on snow, down to 0.1 on wet asphalt, and down to 0.15 on dry asphalt.

The derapage of a vehicle with a combination of different tires on the front and the rear axle is stipulated by early onset of wheel blocking of the rear axle wheels in respect to the front axle. Thereby, the lateral force leads to a drift of the rear axle. Thus, in order to prevent the derapage of the vehicle, it is important to strive for early front wheel blocking which is most difficult to realize in motor vehicles without automatic braking control, e.g., only with brake bias valves. It should be noted, that a braking action check should be carried out in full outfit when the blocking of the rear wheels is most probable. Driving with full payload forms big load on the vehicle axles preventing the blocking of the wheels, that is why we consider hereinafter only the travel parameters of vehicles in full outfit only for studies of emergency situations of derapage.

The author [20] has obtained for vehicles with such design dependencies for determining of the wheel blocking range reach time:

$$\begin{cases} t_{11} = \dfrac{\dfrac{G_1\varphi}{B_1} + \dfrac{h}{L}\varphi\left[k(1-W)\dfrac{M_\alpha g\varphi}{B_1+B_2}\right]}{B_1\left[1-\dfrac{h}{L}\varphi(1+k)K_0\right]} \\ t_{12} = \dfrac{G_2\varphi k}{B_2\left[1-\dfrac{h}{L}\varphi(1+k)K_0\right]} \end{cases}$$

where t_{11} and t_{12} are the moments of the brake system tripping of the front and the rear wheels, respectively; φ is the traction value of the wheels; L is the wheel base of the vehicle; a is the distance of the front axle of the vehicle to the center of gravity; h is the height of the center of gravity; G_1 is the partial vehicle weight accommodated by the front axle; G_2 is the partial vehicle weight accommodated by the rear axle; k is the braking force distribution ration by the vehicle axles determined as follows:

$$\frac{a}{L} + k - k = \varphi\frac{h}{L} + \varphi k \rightarrow \mathrm{k}\left(1+\varphi-\frac{a}{L}\right) = \rightarrow k = \frac{a-h}{L\left(1+\varphi-\dfrac{a}{L}\right)}$$

Hence, considering the statutory limitation of the maximum duration of the deceleration growth in the brake mechanism, admissible limit values of B_1 and B_2 parameters can be obtained which represent the ratio of the pressure in the brake system of the front and rear wheels versus the brake force exerted on the respective wheels. The minimum values of these parameters can be determined based on the following equations:

$$t_{12} = \frac{G_2\varphi k}{B_2\left[1-\dfrac{h}{L}\varphi(1+k)K_0\right]} \rightarrow B_2 > \frac{G_2 k}{t_{12}\left[\dfrac{1}{\varphi}-\dfrac{h}{L}(1+k)K_0\right]}$$

$$\frac{t_{11}}{t_{12}} = \frac{B_2}{B_1}\frac{G_1}{G_2 k} < 1 \rightarrow B_1 > \frac{B_2 G_1}{G_2 k}$$

$$t_{11} = \frac{G_1\varphi}{B_1\left[1-\dfrac{h}{L}\varphi(1+k)K_0\right]} \rightarrow t_{11}-t_{12} = \left(\frac{G_1}{B_1\left[\dfrac{1}{\varphi+\Delta\varphi}-\dfrac{h}{L}(1+k)K_0\right]} - \frac{G_1}{B_1\left[\dfrac{1}{\varphi}-\dfrac{h}{L}(1+k)K_0\right]}\right)$$

$$\mathrm{t}_{11}-\mathrm{t}_{12} = \left(\mathrm{t}_{11}-\mathrm{t}_{12}\frac{\left[\dfrac{1}{\varphi}-\dfrac{h}{L}(1+k)K_0\right]}{\left[\dfrac{1}{\varphi+\Delta\varphi}-\dfrac{h}{L}(1+k)K_0\right]}\right)$$

That is, on dry asphalt, at 0.4 traction difference between the front and the rear wheels, with the rear wheels with less traction, the response time of the of the wheel brake system shall be increased by 1.9, whereas at 0.3 traction difference it should be increased by 1.6. On ice, at 0.19 traction difference between the front and the rear wheels, with the rear wheels with less traction, the response time of the of the wheel brake system shall be increased by 2.2, whereas at 0.13 traction difference it should be increased by 1.6. On snow and wet asphalt, at 0.3 traction difference between the front and the rear wheels, with the rear wheels with less traction, the response time of the of the wheel brake system shall be increased by 2.3, whereas at 0.2 traction difference it should be increased by 1.8. While using uniform tire model of a single manufacturer, the response time difference of the brake system would not exceed 1.3x on ice, 1.4x on snow, 1.3x on wet asphalt, and 1.3x on dry asphalt.

The necessity to ensure the stability of the vehicle under lateral acceleration leads to a necessity to consider that a part of the traction reserve of the wheels will be consumed by the compensation of the lateral force. Then, the magnitude of the required response time difference of the wheel brake system will, for this specific case, have the following appearance:

$$\delta T = \frac{\left[\frac{1}{\sqrt{\varphi^2 - \left(\frac{a_y}{g}\right)^2}} - \frac{h}{L}(1+k)K_0\right]}{\left[\frac{1}{\sqrt{(\varphi+\Delta\varphi)^2 - \left(\frac{a_y}{g}\right)^2}} - \frac{h}{L}(1+k)K_0\right]}$$

That is, on dry asphalt, at 0.4 traction difference between the front and the rear wheels, with the rear wheels with less traction, the response time of the of the wheel brake system at a lateral acceleration of 2.5 m/s^2 should be increased by 1.9, whereas at 0.3 traction difference it should be increased by 1.7. On snow and wet asphalt, at 0.3 traction difference between the front and the rear wheels, with the rear wheels with less traction, the response time of the of the wheel brake system shall be increased by 4, whereas at 0.2 traction difference it should be increased by 3. While using a uniform tire model of a single manufacturer, the response time difference of the brake system would not exceed 2x on snow, 1.3x on wet asphalt, and 1.3x on dry asphalt.

Considering that automotive vehicle designers envisage uniform tiring providing for even dry asphalt road conditions a response time of the wheel brakes equal to 0.5 s, then, combining of different tires will, for the time being, increase the response time of the brake system up to 0.8 s, whereas this value was equal to 1.0 s before 2010. Thus, a combination of tires with different traction between the front and the rear axle on dry asphalt will, for the time being, lead to an increase of the collision speed by 2.1 m/s (or 7.5 km/h), whereas, before 2010, this value was equal to 3.5 m/s (or 12,6 km/h).

Thereby, substituting in equation different traction difference values of the wheels, it was found out that the lateral acceleration does not essentially interfere this parameter.

Thus, the safety of the traffic in a vehicle with combined tiring of the front and the rear axle, as well as at driving in mixed cycle can only be provided for by incomplete use of the available brake force range. The influence of the traction difference between the wheels of the front and the rear axle of a vehicle on the accident rate of the automotive traffic can have the following appearance:

$$U = \Psi U_\Sigma; \qquad \Psi = 0{,}5\left\{ e^{\left(6+ {}^{3}\!/_{K\varphi^2}\right)\left(0{,}06-{}^{0{,}01}\!/_{K\varphi}\right)^{-1{,}32}} + \frac{0{,}04}{0{,}06 - {}^{0{,}01}\!/_{K\varphi}} \right\},$$

where Kφ is the utilization rate of the brake force equal to the ratio of the braking distance of the vehicle with different tiring of all the axles to the same vehicle with uniform tiring:

$$K\varphi = \frac{S_0}{S\Delta\varphi} = \left(\frac{1}{1 + \frac{2\varphi g \Delta t}{v}}\right)$$

where Δt is the surplus time of the deceleration growth required to provide for stabilization of the vehicle with different tiring of the front and the rear axle.

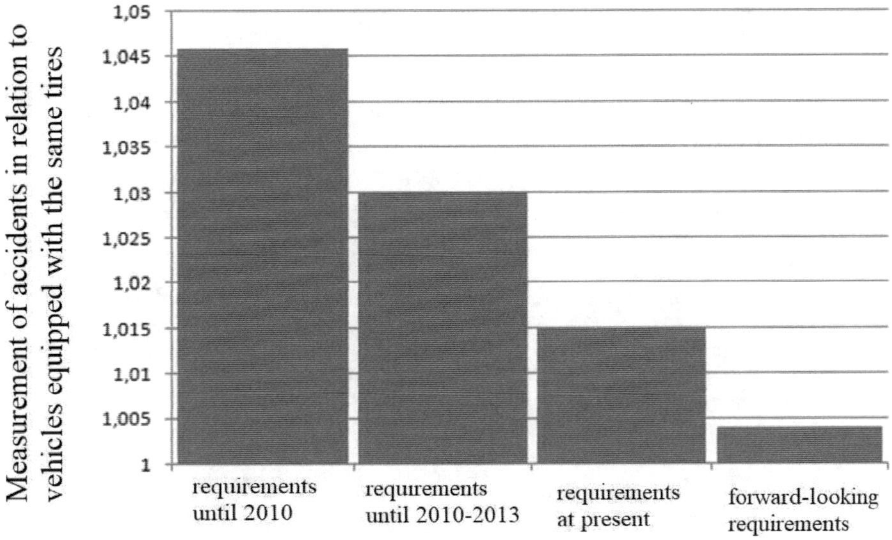

Figure1. Influence of amendments of technical state-of-order requirements to motor vehicles on improvement of accident rates of motor vehicles with different tire designs on the front and the rear axle

Conclusions

The obtained results show unambiguous correlation of the reduction of traffic accidents by 3% form 2010 caused by motor vehicles combining different tire types on the front and the rear axle with the tightening of the requirements to the technical state of order of automotive vehicles participating in the motor traffic in terms of uniform tiring of the front and the rear axle. Additional measures of surveillance for prevention of non-uniform tiring of motor vehicles could reduce the number of traffic accidents by further 1%. Thereby, it will be the ultimate possible achievement in the traffic safety improvement since even tires of one batch have different wear after a certain operation period, stipulating a certain inevitable difference of traction between the wheels which could be compensated only by braking efficiency reduction by increased response time of the rear axle brakes.

References

[1] *World's first road death. London, RoadPeace* 2003 http://www.roadpeace.org/articles/ WorldFirstDeath.html, accessed 17 November 2003

[2] Nader R 1965 *Unsafe at any speed* New York: Grossman Publishers p 365

[3] Winnett M A, Ireland D J and Herrod R T 1991 *Risk assessment by the measurement of driver acceleration signatures* Technical report

[4] Almqvist S and Hyden C 1994 *Methods for assessing traffic safety in developing countries* Lund p 20

[5] Committee S H 1997 The Automobile: A Century of Progress *Society of Automotive Engineers* **48**

[6] Jansson J 1994 Government and Transport Infrastructure – Investment *Swedish national road and transport research* p 221-243

[7] Hydén C and Linderholm L 1984 The Swedish Traffic Conflict *Technique International Calibration Study of Traffic Conflict Techniques. NATO ASI Series* pp 133-139

[8] *Wypadki drogowe - element analizy technicznej i opiniowania. Project funded by the European Community under the 'Competitive and Sustainable Growth' Programme (1998- 2002)* Ernst Tomasch Graz University of Technology 2004 p 136

[9] Ambartsumyan V V, Babanin V N, Gudjoyan O P and Petridis A V 1998 *Automotive traffic safety* M. Mashinostroyeniye p 304

[10] Konoplyanko V I, Gudjoyan O P, Zyryanov V V, Kosolapov A V 1998 *Automotive traffic organization and safety* Educat. guidance for higher education institutions Kemerovo: Kuzbassvyaz'isdat p 236

IOP Conf. Series: Materials Science and Engineering 1086 (2021) 012022 doi:10.1088/1757-899X/1086/1/012022

[11] Tumasov A, Groshev A, Musarsky R,.Trusov Y and Kostin S 2014 Estimation of light commercial vehicles dynamics by results of road tests and simulation *Proceedings of the ASME Design Engineering Technical Conference* **3**

[12] Manyanin S E and Cherevastov M G 2020 A study on the issue of controllability of a vehicle with super-balloons *Transactions of NNSTU n.a. R.E. Alekseev* **3 (130)**

[13] Vashurin A, Molev Y, Proshin D and Trusov Y 2020 Developments of requirements to stability of motor vehicles at emergency braking *Transactions of NNSTU n.a. R.E. Alekseev* **2 (129)**

[14] Molev Y and Proshin D 2010 Determining of maximum admissible wheel traction difference between front and rear axle providing for vehicle stability during braking *Transactions of NNSTU n.a. R.E. Alekseev* **2 (81)**

[15] Makarov V, Molev Y, Proshin D and Cherevastov M 2020 Response of Motor Vehicles, Front and Rear Axles of Which Are Fitted with Different Tires. Main Aspects *MATEC Web of Conferences* **329** 0101 https://doi.org/10.1051/matecconf/202032901012

[16] https://www.autozeitung.de

[17] https://autoreview.ru

[18] https://www.auto-motor-und-sport.de

[19] https://www.zr.ru

[20] Sotskov D A 1988 *Enhancement of active safety of motor vehicles during braking*. Dr. Eng. thesis p 547

Experimental and calculation method for determining the parameters of a car suspension with an elastic elastomeric element

E V Stepanov[1], S I Maleev, R A Musarsky and S M Ogorodnov

[1]Nizhny Novgorod State Technical University n.a. R.E. Alekseev, Minin st., 24, 603950, Nizhny Novgorod, Russia.

E-mail: [1] evgeniystepanov1991@nntu.ru

Abstract. We present a suspension design the elastic elements of which are made of elastomers. This allows us to reduce the weight of unsprung parts of the vehicle significantly, which affects the quality of ride in a positive way. The aim of the study is to determine the elastic characteristic, the damping coefficient, the natural oscillation frequency of the sprung masses and the specific dynamic energy intensity of the suspension with elastomeric elastic elements. The determination of these parameters is carried out on the basis of experimental data got as a results of static and dynamic loading of a package of elastomeric elements. The elastic characteristic of the elastomeric element is obtained by static reduction and is not linear. Stiffness estimation of an elastomeric element for a specific point of characteristic or for a limited area is suggested. Also, there are significant friction losses during deformation, it is quite difficult to take them into account as a separate parameter when evaluating the elastomer damping properties. It is recommended to estimate their absorption capacity using a "conditional" damping coefficient. Its value is determined by the nature of the damped oscillations process under dynamic loading of an elastomeric elements package implemented on a special pendulum copra. The method of reducing the elastic characteristic and the damping coefficient of the elastomer to the "wheel" using graphoanalytic techniques is described in detail. The values of natural oscillation frequency and the energy intensity of the suspension with elastomeric elastic element are calculated based on the values of the suspension static and dynamic deflections and the forces in the tire contact with the supporting surface. Relying on experience in analyzing the standard car suspension parameters, it is concluded that using elastomeric elastic elements in the suspension design of trucks, buses and all-terrain vehicles is advisable.

1. Introduction

The elastic suspension element reduces the level of vibration loads transmitted from the support surface to the wheels and unsprung masses (body, frame) of a car (vehicle). Despite the fact that suspensions with non-metallic (mainly rubber) elastic elements have been known for at least 100 years, they have not been widely used in any country.

Being made of synthetic materials "rubber" elastic elements used at present have a relatively low durability, but which is more important do not give the necessary elastic response and the suspension energy intensity to the vehicle. In addition, the temperature range of their operation is limited due to the properties of the materials.

Elastomeric elastic elements are partially devoid of these disadvantages. Getting a package of elastomeric elements installed into the suspension (Figure 1), it is possible to reduce the weight of elastic elements and unsprung parts of the vehicle significantly, which affects the quality of ride in a positive way, reduces metal consumption and provides a number of other advantages. Using elastomeric elements is currently hindered by the fact that only certain techniques for calculating the mechanical properties and choosing the size of the elastomer have been elaborated [1], these techniques do not allow us to evaluate the operation of this object as a part of a car suspension. It is necessary to develop some additional techniques to determine all the key parameters of the suspension. These parameters are: elastic characteristics of the suspension (stiffness, static, dynamic and maximum deflections), natural oscillation frequency of the sprung masses of the car, deformation work and suspension capacity, a number of other parameters.

The elastic characteristic of the suspension represents the dependence of the normal force (reaction) $N\kappa$ acting on the wheel at the point of contact with the support surface from the suspension deflection. Therefore, all forces acting in the suspension (resistances of the shock absorber, elastic element, and others) must be pre-determined and reduced to the "wheel". To determine the suspension elastic characteristic, it is necessary to get a kinematic diagram of the guiding device and the characteristic of the elastic element.

2. Experimental determination of the elastic characteristics

Tests to determine the elastic characteristic of the elastomeric elements package were carried out on a special bench-testing unit under elastic element static loading. The maximum load can be set in accordance with the elastomer strength or the kinematic features of the suspension guiding device (the maximum possible deformation of the elastic element). The design of the package of elastomeric elements eliminates friction between the elements and the guiding device they are assembled on. During the tests, it was found that the elastomeric elastic element has significant internal friction (hysteresis losses). As a result of the tests, the load and discharge curves of the elastic element were determined and represented graphically (figure 2).

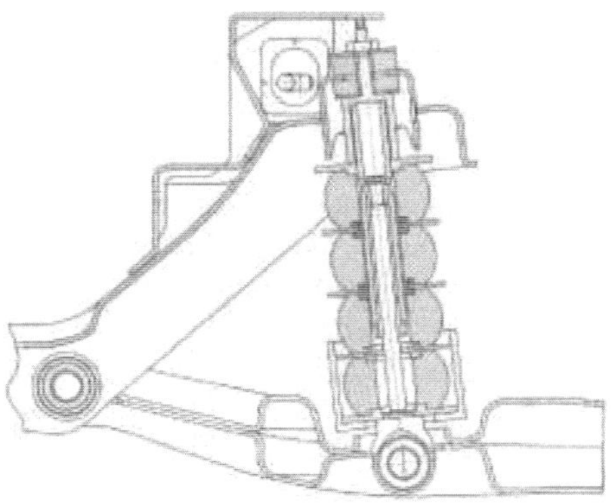

Figure 1. Scheme of the elastomeric elastic element installation into the car suspension

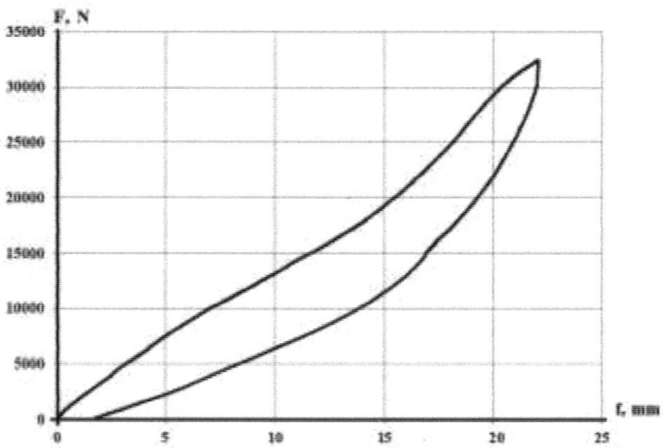

Figure 2. The elastic characteristic of the elastomeric element

A discrete set of experimentally obtained points is approximated by continuous curves using the least mean squares method. When testing elastic elements with internal friction, the elastic characteristic (the dependence of the elastic force F from the deformation value of the elastomeric element f) can be determined as the dependence of the average value of the force Fcp from the elastic element deformation f.

The elastic characteristic of an elastomeric element is nonlinear, so its stiffness can be determined only at a point or for a limited area considered linear. The average value of the stiffness at the point *Ccp i* of the nonlinear elastic characteristic can be determined in accordance with the expression.

$$C_{\text{cp}\,i} = \frac{F_{\text{cp}\,i}}{f_i} \tag{1}$$

where: $C_{\text{cp}\,i}$ is the stiffness value at point i of the nonlinear elastic characteristic; $F_{\text{cp i}}$, f_i, are the average value of the force and the value of the deformation at point i of the nonlinear elastic characteristic, respectively.

$$F_{\text{cp}\,i} = \frac{F_{\text{н}\,i}+F_{\text{p}\,i}}{2} \tag{2}$$

where: $F_{\text{н i}}$, $F_{\text{p i}}$ are the value of the force at point i in the loading and unloading section of the elastic characteristic, respectively (Fig. 2).

The average value of stiffness for a limited section of the nonlinear elastic characteristic is determined in accordance with the expression

$$C_{\text{пр}\,i} = \frac{F_{i+1}-F_i}{f_{i+1}-f_i} \tag{3}$$

where: $C_{\text{пр i}}$ is the reduced stiffness on the i-th section of the elastic characteristic; F_i, F_{i+1} are the previous and the subsequent values of the vertical force on the elastic characteristic acting on the elastomeric elastic element; f_i, f_{i+1} are the previous and the subsequent values of the reduced vertical deformation of the elastic element.

The elastic characteristic of the elastomeric element provides the scheduling of the suspension elastic characteristic, followed by the calculation of the reduced values of stiffness and damping coefficient.

3. Experimental determination of the damping coefficient

Since the elastomeric element design eliminates the friction of the sleeve material against the outer and inner guiders, the only source of friction can be the friction caused by the loss of energy in the sleeve material during their deformation. To solve the problem of the amount of friction and its sufficiency for oscillation suppression in the car suspension when driving, tests were carried out to assess the elastomeric elastic element damping properties.

To determine the damping coefficient of an elastomeric elastic element we use method based on measuring the parameters of the system vertical damped vibrations.

The scheme of the bench-testing unit for determining the damping coefficient is shown in figure 3.

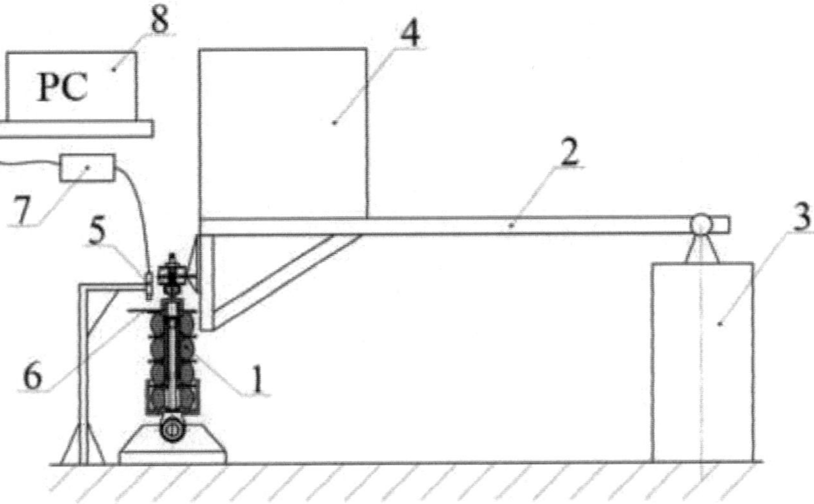

Figure 3. Bench-testing unit for determining the damping ratio

The elastomer package 1 is attached to the pendulum copra 2, which is connected to the rack 3 by a hinge. The rack 3 is fixed rigidly on the foundation. To provide a load on the elastomer, a load 4 is attached to the pendulum 2, the mass " M " of which must correspond to the reduced load acting on the elastomer package. To create damped vibrations, the pendulum of the raised is raised to a small height and released using a "quick reset" device. Vibrations are recorded with the help of a displacement sensor 5, 6, which registers a damped process by changing the coordinate " z ", movement in the vertical direction. The signal from the sensor is processed by the ADC 7 and recorded using a personal computer 8. Figure 4 shows the result of the experiment, which is the process of damping vibrations in an elastomeric element.

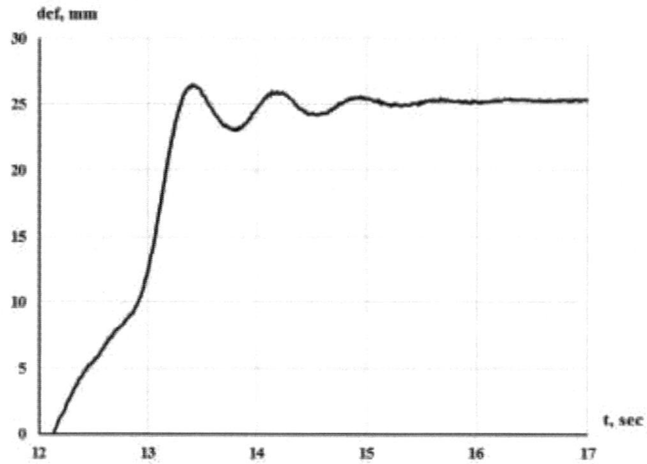

Figure 4. The process of damped vibrations in an elastomeric elastic element

The internal friction of the elastomer affects the damping properties of the suspension in a certain way. It is difficult to account how internal elastomer friction affects the nature of dynamic suspension processes because the type of friction is not defined by clear physical boundaries. In fact, it can be defined as a " dry " one only by convention, since it does not remain constant, and its value depends on the force acting on the elastic element. Judging from the damping vibrations process, figure 4, it can be assumed that the type of friction in an elastomeric elastic element is a viscous one by its physical nature. Calculating the damping ratio at the present stage of research we assume that the nonlinear nature of the amplitude envelope of a damped nonlinear process is stipulated by nonlinear internal friction, and the elastomer damping coefficient corresponds conventionally to the viscous friction type.

The conditional damping coefficient is calculated in the following order. The damping decrement is:

$$d = \frac{A_i}{A_{i+1}} \quad (4)$$

where: d is the damping decrement; A_i, A_{i+1} are the two consecutive values of the oscillation amplitudes.

The vibration damping coefficient is:

$$h = \frac{lnd}{T} \quad (5)$$

where: h is the vibration damping coefficient, [sec^{-1}]; T is the oscillation period, [sec]; lnd is the logarithmic decrement of the attenuation.

The conditional damping coefficient of the elastomeric element is:

$$k_э = 2 * h * M \quad (6)$$

where: M is the weight of the load reduced to the elastic element axis.

The results of the calculations for determining the damping coefficient of an elastomeric elastic element are summarized in Table 1.

Table 1. Test results of an elastomeric elastic element on a pendulum copra

Parameter	Measure units	Values
Load mass corresponding to the reduced load	M, kg	988
Average value of logarithmic decrement of the attenuation according to the results of all tests	*lnd*	0,588
The average value of the period of damped oscillations according to the results of all tests	T, s	0,73
The conditional damping coefficient	$k_э$, N*s/m	1592

In cases when acting friction forces [2] need to be taken into account while studying the suspension dynamic properties, they can be defined as the average value of the friction force $F_{тр}$ at the point of the elastic characteristic or for its limited area. Since the elastic characteristic is nonlinear, the friction force can be taken into account as the elastomeric element deformation function $f_э$, Figure 2. In some cases, in order to simplify the analysis of the suspension dynamic properties, it is possible to average the friction force within the most probable value of the elastic element deformation observed in the certain driving modes.

3. Determination of suspension parameters

For most design maps of cushioning systems the stiffness (and the suspension damping coefficient) reduced to the "wheel" differ significantly from the stiffness of elastic elements and the resistance coefficients of damping devices [3].

To determine the suspension parameters reduced "to the wheel", special calculations need to be done. Graphic-analytical techniques are mostly used ones. Figure 5. shows necessary graphical layout for a suspension with a guiding device on wishbones. Figure 5a presents the kinematic scheme of the suspension considered. Figures 5b and 5b show the forces acting on the steering knuckle with the wheel, figure 5g shows forces acting on the lower suspension arm.

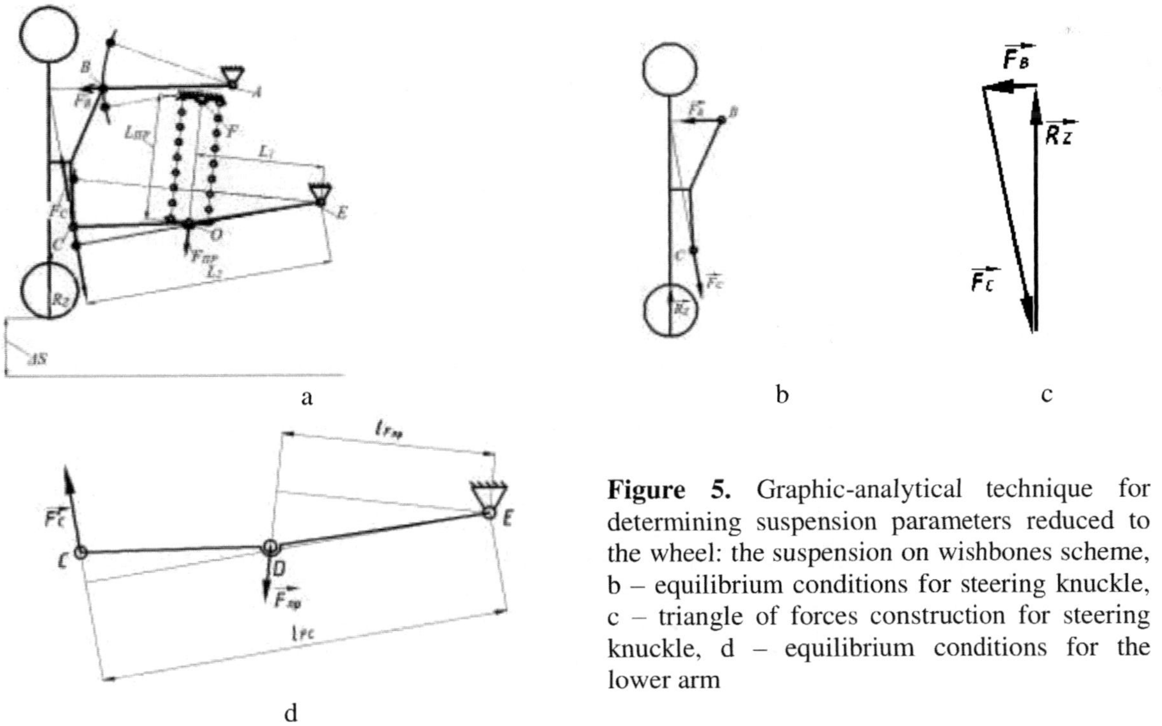

Figure 5. Graphic-analytical technique for determining suspension parameters reduced to the wheel: the suspension on wishbones scheme, b – equilibrium conditions for steering knuckle, c – triangle of forces construction for steering knuckle, d – equilibrium conditions for the lower arm

The forces acting on the suspension parts ensure their static balance. The values of the operating forces are determined for a fully loaded vehicle.

To make upper suspension arm equilibrious, the force $\overrightarrow{F_B'}$ arising in the hinge *B* must be directed along this lever. The upper lever is connected to the car frame and the steering knuckle with the help of hinges. Therefore, the force acting on the upper lever and the reaction acting on the steering knuckle are directed along the axis of the lever. The forces acting on the lower lever are shown in Fig. 5d. Since the elastic element is hinged at both ends, the force $\overrightarrow{F_{\text{пр}}}$ will be directed along the axis of the elastic element. The normal reaction *Rz* is applied at the point of contact of the wheel with the support surface and is directed vertically upwards. Since no moments are applied to the wheel and the steering knuckle, the steering knuckle equilibrium is ensured by the following condition: the sum of the vectors acting on the steering knuckle must be zero:

$$\overrightarrow{F_{\text{в}}} + \overrightarrow{F_{\text{C}}} + \overrightarrow{R_Z} = 0 \qquad (7)$$

The suspension elastic characteristic is the dependence of the normal reaction value *Rz* from the suspension " deflection " *fn*, which is actually defined as the wheel (its axis) shift with regard to the car frame, when the elastic element is deformed.

To construct the suspension elastic characteristic it is necessary to define the dependence *Rz* = *F(fn)*, that is, to determine the values of the normal force *Rz* at different values of the elastic element deformation. The graphic approach is considered to be the simplest engineering solution. The method involves the use of the suspension kinematic scheme. In the scheme, several (6 - 8) positions of the suspension levers are drawn with different elastic element deformations , larger and smaller static loads. The elastic force value $\overrightarrow{F_{\text{пр}}}$ is determined for each position of the levers.

The value of the normal reaction *Rz* is determined for each position of the levers. To do this, one needs to draw a diagram of the forces acting on the steering knuckle. At the equilibrium state of the steering knuckle, the vector sum of all the forces acting on it must be zero, which is graphically displayed as a closed polygon of forces. The directions of the force applied to the upper lever and the normal reaction are known. The direction of the force from the lower lever is determined by a well-known rule. The line that continues the force vector F_C must pass through the intersection of the action lines of the forces F_B and *Rz*, Figure 5 a.

The value of the normal reaction for each position of the suspension arms is determined from the force triangle. When constructing a force triangle, the directions of all forces acting on the steering knuckle are known. The line of action of the force F_C is drawn in the direction specified by the suspension kinematic scheme. The force vector F_C must be protracted in accordance with the scale of forces selected. From the end of the vector, a vertical action line of the normal reaction *Rz* is drawn, and from the beginning of the vector, the line of action for the force F_B is drawn. The intersection of the constructed lines determines the value of the normal reaction *Rz* for the considered position of the suspension arms, taking the selected scale into account. This is the way to determine the normal reaction value for each of the levers constructed positions and to construct a suspension elastic characteristic which determines the stiffness values reduced to the wheel.

The damping coefficient value reduced to the wheel is determined from the following conditions, figure 6.

The moment of the damping force in relation to the lower lever axis of rotation, created by the elastomer is equal to:

$$T_{k1} = k_1 * V_1 * l_1 \qquad (8)$$

where: $k_1$is the damping coefficient of the elastomer; V_1 is the elastomer deformation rate (displacement of p. O_1); $l_1$is the action moment arm of the damping resistance force.

The moment of the damping force reduced to the wheel is equal to:

$$T_{k2} = k_2 * V_2 * l_2 \qquad (9)$$

where: $k_1$is the damping coefficient reduced to the "wheel"; V_1 is travel speed of p. O_2; l_2 the action moment arm of the damping resistance force.

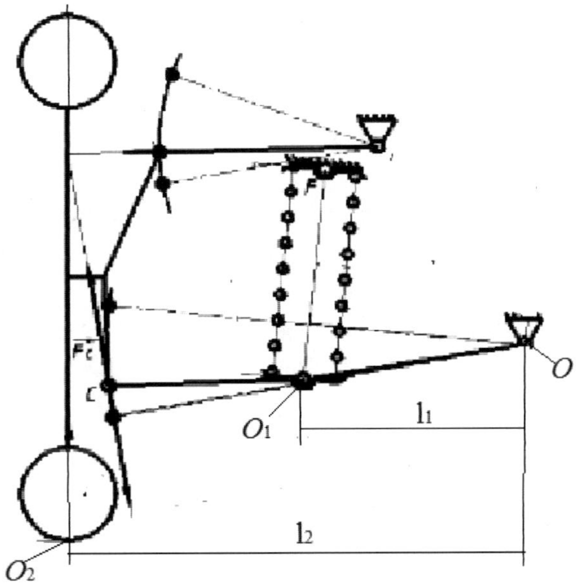

Figure 6. Determining of the damping ratio

It is obvious that $V_2 = \frac{V_2 l_2}{l_1}$. Reasoning from the equality of moments and the velocity ratio we get the formula of the damping ratio reduced to the "wheel":

$$k_2 = \frac{k_1 l_1^2}{l_2^2} \quad (10)$$

At the preliminary stage when choosing main design parameters, the natural oscillation frequency of the vehicle sprung mass can be set as the initial value, so the resulting value of the reduced suspension stiffness can be assessed taking the recommended oscillation frequency of the sprung masses into account.

The natural oscillation frequency of the car sprung masses ω_0 with design static load is used in customary calculations for preselecting suspension stiffness.

Static (conditional) deflection and oscillation frequency are related by the dependence[4]:

$$\omega_0^2 = \frac{g}{f_{CT}} \quad (11)$$

where: ω_0 is the natural oscillation frequency of the suspension sprung masses, [s^{-1}]; f_{CT} is the suspension static deflection of the suspension, [m].

Depending on the type of car, natural oscillation frequency can vary from 1 to 2.5 Hz [4-6]. A lower frequency value provides higher quality of the ride. At the stage of preliminary calculations for different kinds of vehicles the natural oscillation frequency at rated static load can be taken equal to: for passenger cars ω_0 = (6…8,5) s^{-1}; for buses ω_0 = (8…11) s^{-1}; for trucks ω_0 = (9…15) s^{-1}. According to the results of the experiment the natural oscillation frequency of the sprung masses for the considered suspension structure with elastomeric elements made ω_0 = 9,5 c^{-1}.

Choosing the value of dynamic deflection, it is necessary to consider that suspension must be energy-intensive enough to exclude "breakdowns", parts of the suspension guider or axles impacting on the structural elements of the body or frame. The energy intensity (capacity) of the suspension is understood as the deformation work of the suspension elastic element from the fully unloaded state to the maximum possible value corresponding to the suspension breakdown. In general, the deformation work is equal to [5]:

$$U = \int_0^{f_{max}} \Phi(f)df \quad (12)$$

where: U is the the suspension energy intensity; [kN*m]; $\Phi(f)$ is the function describing the curve of the the suspension elastic characteristic.

In numerical terms, the deformation work is determined by the area under the elastic characteristic curve (see Figure 2).

The suspension capacity is limited by the design and the maximum stresses allowed by the elastic element material. Bearing this in mind, one should choose the maximum possible suspension dynamic deflection, being guided by the suspension capacity of the prototype cars and the maximum value of stresses in the suspension elastic element.

In addition, the value of the dynamic deflection at the preliminary design stage can be estimated using the specific dynamic energy intensity $U_{уд}$ [3]:

$$U_{уд} = \frac{U_{д}}{F_{ст}} \tag{13}$$

where: $U_{д}$ is the suspension dynamic capacity, ($U_{д} = U - U_{ст}$); $U_{ст}$ is static energy intensity of the suspension; $F_{ст}$ – static load.

4. Results and conclusions

It is empirically supported that in order to minimize the probability of suspension breakdown when driving on all types of roads, the value of the specific dynamic energy intensity, which depends on the car type, should vary within: for passenger cars and buses $U_{уд}$ = (0,04…0,06) kNm/kN; for trucks $U_{уд}$ = (0,05…0,07) kNm/kN; for all-terrain vehicles $U_{уд}$ = (0,06…0,08) kNm/kN. For the considered suspension structure with elastomeric elements, the value of the specific dynamic energy intensity is $U_{уд}$ = 0,068 kNm/kN.

As a result of the study, the elastic characteristic of the elastomeric elements package was determined and the material internal friction was evaluated. To predict the ability of elastomers in the car suspension to absorb oscillation energy, we presented the technique describing the attenuation process using a "conditional" damping coefficient. We gave an outline of methods for reducing the suspension parameters to the "wheel", calculating the natural oscillation frequency of the sprung masses and determining the suspension energy intensity. Analysis of the natural oscillation frequency values and the specific energy intensity of the elastomeric suspension allows us to conclude that using the same sort of construction in the suspensions of trucks, buses and all-terrain vehicles is advisable.

Acknowledgment

The research was carried out with financial support from the Ministry of Education and Science as part of the project "Creating high-tech production range of vehicles GAZelle Next with the new electronic architecture of[electronic systems", under Agreement No. 075-11-2019-027 from 29.11.2019 (decree No. 218 of the Government of the Russian Federation dated April 09, 2010).

The experimental studies were carried out using the equipment of the NSTU Sharing Center "Transport systems".

References

[1] Kuleznev V N and Shershnev V A 2007 *Chemistry and Physics of Polymers* M. Colossus p 367
[2] Khachaturov A A 1976 *Dynamics of the road-tire-car-driver* Mechanical Engineering p 535
[3] Rotenberg R V 1972 *Car suspension* Mechanical Engineering p 392
[4] Kravets V N 2013 *Theory of the car* Nizhny Novgorod. state Tech. University n.a. R.E. Alekseeva p 413
[5] Yatsenko N N and Prutchikov O K 1968 *Smooth running of trucks* Mechanical Engineering p 220
[6] Wong D 1982 *Theory of land vehicles* Mechanical Engineering p 284

IOP Conf. Series: Materials Science and Engineering 1086 (2021) 012024 doi:10.1088/1757-899X/1086/1/012024

Features of macromodeling of small towns with a monostructured economy

K O Goncharov[1], M E Eliseev, K V Laschuk and A O Chikaev

Nizhny Novgorod State Technical University n.a. R.E. Alekseev, Minin str., 24, 603950, Nizhny Novgorod, Russia

E-mail: [1]goncharov@nntu.ru

Abstract. The article discusses the features of a development macromodels of small towns with a monostructural economy. For a more accurate setting of the model, which will correspond to reality, the additional parameter "check in for a colleague at work" was taken into account when development a general origin-destination matrix(matrix OD). The model with this method was tested on the 2019 models and has a high convergence according to GEH statistics.

1. Introduction

Macromodeling is currently one of the most important means of describing the transport systems of cities. The article discusses the features of development macromodels of small cities with a monostructural economy. By a small town with a monostructural economy, we mean a town with a population of less than 50,000 inhabitants, with a single town-forming enterprise or a group of enterprises in one sector of the economy. It is Implied that a most percentage of the working-age population works in this enterprise. It will also be implied that this institution is located at some distance from residential areas, i.e. or outside the town or within the town, but on the outskirts. The most interesting (characteristic) for this study will be the option of the location of the enterprise at a significant (more than 20 km) distance from the residential area of the town.

When modeling the transport systems of such cities, significant discrepancies arise with the existing theory. Consideration of the causes of their occurrence and the ways correcting the methods of modeling for these objects are the purpose of this research.

The research is based on models developed by the authors in 2019 for 4 cities in the Russian Federation: the town Polyarnye Zori and Kovdor and the towns of Borodino and Zelenogorsk in the. Town Zelenogorsk, strictly speaking, is somewhat beyond the scope of this research. its population is slightly higher - 61,915 inhabitants according to 2019 data, but, as will be shown, it has the same modeling features, therefore, it is also given in this work.

2. Materials and methods

The main research method is the method of macromodeling, in which transport demand and transport supply are considered, and the transport subarea is taken as the main element of modeling. At the initial stage of modeling, the initial data concerning the transport network and characteristics of subareas are collected, as well as the necessary measurements of the characteristics of traffic flows are made, which are necessary in the future for the calibration and validation of the model.

IOP Conf. Series: Materials Science and Engineering 1086 (2021) 012024 doi:10.1088/1757-899X/1086/1/012024

A large number of scientific works are devoted to transport modeling. An important role in further reasoning is played by displacement models depending on the purpose of travel [1-22]. Typically, all trips made by the respondents are subdivided into "home-work", "home-other", "work-home", "work-other point" and "other point-other point". The fundamentals of this approach are outlined in [2] and [3].

Models that consider longer chains of the type "home-work-home", "home-shop-home", etc. are called activity-based models. One of the first models of activity is the SAMS model developed by Kitamura et al. [14, 15].

In [4] and [5], the travel of employees (workers) during the day is considered, in [22] - the coordinated movement of family members, including the sharing of personal transport. This work is also devoted to the sharing of personal vehicles

3.Theory / calculation

The key factors for applying the approach described below are as follows:

- small linear size of the town;
- a monostructural town economy: the presence of a town-forming enterprise, which employs most of the able-bodied population;
- the remoteness of the enterprise from the town center should significantly exceed the linear dimensions of the town.

The most typical for the approach described below is the model of the town of Polyarnye Zori, therefore, on its example, the main theoretical provisions will be formulated. Its main features are as follows:

- the population is 16.5 thousand people, i.e. it belongs to small type cities
- the monostructural economy of the town: the main (town-forming) enterprise providing the vast majority of jobs (more than 2000);
- town-forming enterprise is significantly removed from residential areas of Polyarnye Zori (formally, it is one of the town's districts) - at a distance of about 13 km from the town center;
- The town has compact dimensions: the length of the route between its most distant parts is 2 - 2.5 km.
- grocery stores start working at 9:00.

Let's consider the features of development a model of the morning "rush hour". According to measured traffic flows, the highest traffic intensity falls on the time period from 6:30 am to 8:00 am. This is primarily due to the fact that the shift at the town-forming enterprise starts at 8:00. Due to the fact that shops start working at 9:00 am, there are no "home-to-shop" trips during the morning rush hour. The overwhelming majority of correspondence is made up of trips "home - work" and, much less, "work - home" (workers returning from the night shift). It is not difficult to determine the amount of traffic flow from the town to the gas station and vice versa, since a single road leaves the town in this direction (Figure 1). The colors correspond to the service levels, which represent the ratio of the average vehicle speed on the segment of street and road network to the free movement speed.

The respondents can get to the place of work either by personal or official transport; there is no public transport excluding for official transport and taxis. It is easy to determine the number of respondents using public transport by the occupancy of buses, which is recorded at the roundabout at the exit from the town (the "roundabout" icon in Fig. 1). At the same time, the flows inside the town should be directed exclusively towards the exit from the town, which is not observed. Further research showed that a significant portion of morning rush hour travel was "check-in for a work colleague". Due to the small linear size of the town (2-2.5 km) and the significant remoteness of the place of work (13 km from the town center), when driving for a colleague at work, the lengthening of the path is no more than 10% of the route, i.e. subjectively perceived as insignificant. Again, due to the small linear dimensions of the town, these trips do not depend on the distance, i.e. the respondent is equally likely to visit any of the town's districts.

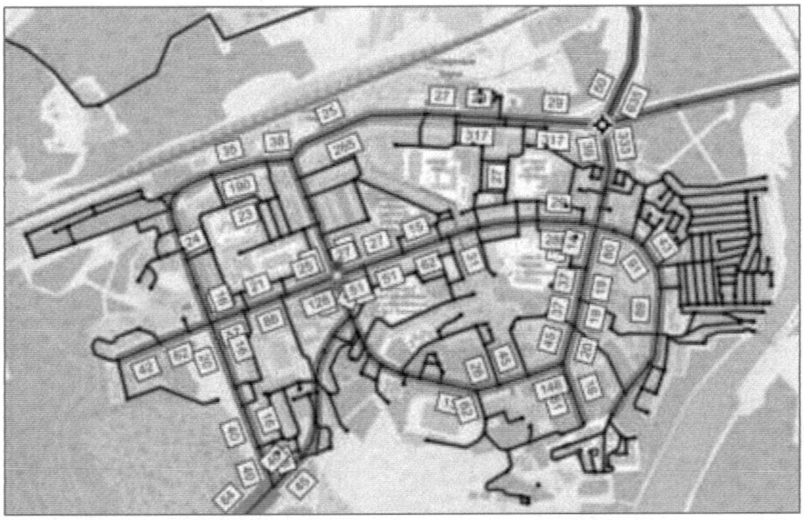

Figure 1. Model of traffic flows in Polyarnye Zori, morning "rush hour" (6:30 - 8:00)

The share of these trips can be calculated based on two measurements:
- calculation of the average number of respondents per one passenger car;
- measurements of traffic flows at key intersections.

The calculation formula for calculating the correspondence matrix OD "check in for a work colleague" is:

$$((a_{ij})) = \left(\left(KS \cdot \frac{W \cdot P_i \cdot P_j}{P^2}\right)\right), \qquad (1)$$

where KS-is the coefficient of a joint trip, W-is the number of employees of the town-forming enterprise, P_i is the number of residents of the i-th district, P_j is the number of residents of the j-th district, P is the number of residents of the town.

Verification of the model, supplemented with the matrix OD "check-in for a colleague at work", showed good correspondence of the model to real data: for the town of Polyarnye Zori, the value of GEH-statistics calculated by the formula:

$$GEH = \sqrt{\frac{2 \cdot (V - C)^2}{V + C}}$$

where V - simulated values; C - measured values, did not exceed 3 units for the sections under consideration, with an acceptable value of 5 (dimensionless units), a good value of 4. The values of the GEH statistics are given in Table 1.

Table 1. Assessment of the quality of calibration of the macromodel

Section of the road	Intensity (fact), units / h	Intensity (model), units / h	The quantity GEH statistics
To Kandalaksha	66	64	0,248
From Kandalaksha	43	45	0,302
St. Sivko d.16 (direction 1)	48	52	0,566
St. Sivko d.16 (direction 2)	80	99	2,008
To the Kola NPP	601	635	1,368
From the Kola NPP	48	50	0,286
Lomonosov 6 (direction 1)	80	89	0,980
Lomonosov 6 (direction 2)	147	129	1,532
Energetikov 33 (direction 1)	38	37	0,163
Energetikov 33 (direction 2)	170	210	2,902

The evening rush hour model was built on the assumption that the correspondence "check-in for a colleague at work" will be performed in the opposite direction due to the fact that the change at the town-forming enterprise lasts an equal number of hours and exit through the checkpoint is carried out in time. The hypothesis of the reversibility of the matrix OD " check-in for a colleague at work " in the evening rush hour was confirmed for the town of Polyarnye Zori. In the general case, the evening matrix OD requires correction.

In addition to the town of Polyarnye Zori, this approach was used to develop models for the cities of Kovdor towns and Borodino and Zelenogorsk. Several roads lead from the town to the checkpoint, but only one of them has an asphalt-concrete surface, so more than 98% of the traffic falls on it. The Borodino evening rush hour model is shown in Figure 2. The colors correspond to the service levels.

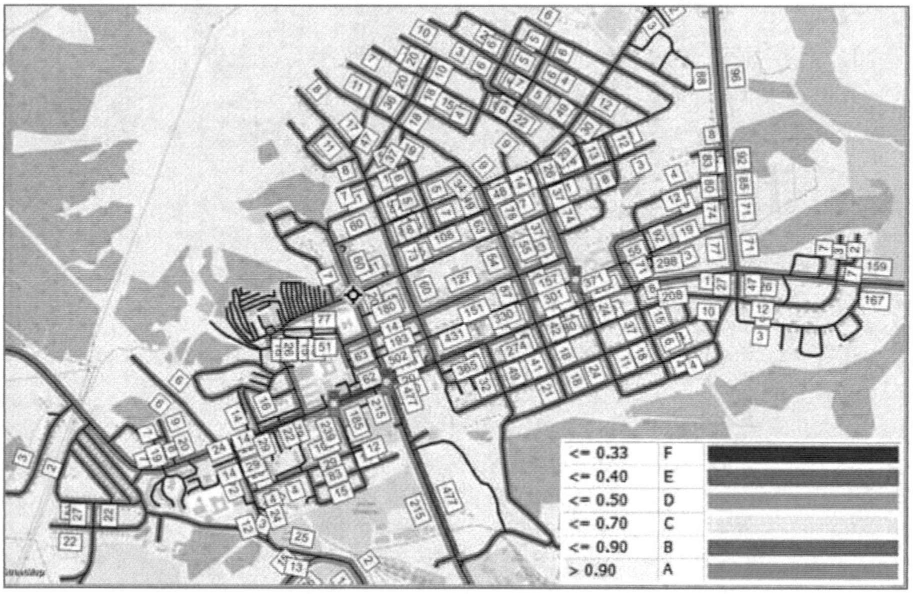

Figure 2. Model of the evening rush hour in Borodino

4. Results

The paper considers an approach that makes it possible to most accurately develop a macromodel of a small town with a monostructural economy. The determining factors influencing the accuracy of model development are revealed. The calculation formula (1) is given - the formula for calculating the matrix OD of the type "check in for a colleague".

5. Discussion

For the town of Kovdor the town-forming enterprise remoteness from the town center is much less than in the town of Polyarnye Zori, therefore the coefficient of a joint trip (KS) is much lower than for the cities of Polyarnye Zori and Borodino.

In Zelenogorsk feature of modeling is that the entire territory of the town and suburb is formally the territory of the town. At the same time, the town of Zelenogorsk and the settlements of Oktyabrsky, Ovrazhny, 1000 courtyards, Orlovka, which formally are part of Zelenogorsk and the enterprise are located on the territory of a town. All public transport on the territory of a town, with the exception of several intermunicipal routes, is therefore urban. Despite the developed public and official transport, the share of joint trips "to work" is significant here due to the significant remoteness of the town-forming enterprise from the town center.

IOP Conf. Series: Materials Science and Engineering 1086 (2021) 012024 doi:10.1088/1757-899X/1086/1/012024

6. Conclusions

The approach developed for calculating the matrix OD "to work" (matrix OD "check-in with a colleague") was tested when creating models of 4 cities in 2019.

The high degree of correspondence of the models to the data of measurements of traffic flows proves the validity of the formulas obtained.

References

[1] Arentze T, Timmermans H, Janssens D and Wets G 2006 Modeling Short-term Dynamics in Activity-Travel Patterns: From Aurora to Feathers *Presented at the Innovations in Travel Modeling Conference*

[2] Ben-Akiva M E and Lerman S R 1985 *Discrete Choice Analysis: Theory and Application to Travel Demand* MIT Press, Cambridge, MA

[3] Ben-Akiva M, Bowman J L and Gopinath D 1996 Travel demand model system for the information era *Transportation* **23** pp 241-266

[4] Bhat C 2001 A comprehensive and operational analysis framework for generating the daily activity-travel pattern of workers *Paper presented at the 78th Annual Meeting of the Transportation Research Board* Washington, D. C.

[5] Bhat C R and Singh S K 2000 A comprehensive daily activity-travel generation model system for workers *Transportation Research* **34(1)** pp 1-22

[6] Borgers A W J, Hofman F and Timmermans H J P 1997 *Activity-Based Modelling: prospects. In Activity-Based Approaches to Travel Analysis* Pergamon, Oxfoord, UK pp 339-351

[7]. Cullen I and Godson V 1975 Urban networks: The structure of activity patterns *Progress in Planning* **4(1)** pp 1-96

[8] Ettema D, K de Jong, Timmermans H and Bakema A 2006 *PUMA: Multi-agent modeling of urban systems* Transportation Research Board CD-ROM

[9]. Gliebe J P and Koppelman F S 2002 *A model of joint activity participation* Transportation **29** pp 49-72

[10] Goulias K G 2002 *Multilevel analysis of daily time use and time allocation to activity types accounting for complex covariance structures using correlated random effects* Transportation **29(1)** pp 31-48

[11] Goulias K G and Kim T 2005 An analysis of activity type classification and issues related to the with whom and for whom questions of an activity diary *Chapter 14 in Progress in Activity-Based Analysis* pp 309-334

[12] Jones P, Koppelman F and Orfeuil J 1990 Activity analysis: State-of-the-art and future directions. In Developments in Dynamic and Activity-Based Approaches to Travel Analysis *A compendium of papers from the 1989 Oxford Conference,* Avebury, UK. pp 34-55

[13] Jonnalagadda N, Freedman J, Davidson W A and Hunt J D 2001 Development of Microsimulation Activity-Based Model for San Francisco *Transportation Research Record* **1777** pp 25-35

[14] Kitamura R, Pas E I, Lula C V, Lawton T K and Benson P E 1996 The sequenced activity simulator (SAMS): an integrated approach to modeling transportation, land use and air quality *Transportation* **23** pp 267-291

[15] Kitamura R and Fujii S 1998 Two computational process models of activity-travel choice *In Theoretical Foundations of Travel Choice Modeling*, Pergamon pp 251-279

[16] Markovnina A I et al. 2019 Imitation modeling as a forecast of transport flows self-organization *IOP Conf. Series: Materials Science and Engineering* **632** 012001 doi:10.1088/1757-899X/632/1/012001

[17] Markovnina A I, Makarov and Belyakov V V 2019 Multicriteria optimization in the study of active vehicle safety *IOP Conf. Series: Materials Science and Engineering* **695** 012026 doi:10.1088/1757-899X/695/1/012026

[18] Markovnina A I, Makarov and Belyakov V V 2020 The employment of multi-criteria assessment methods in the study of road safety *IOP Conf. Series: Materials Science and Engineering* **819** 012003 doi:10.1088/1757-899X/819/1/012003
[19] Pendyala R M, Kitamura R, Kikuchi A, Yamamoto T and Fujii S 2005 The Florida Activity Mobility Simulator (FAMOS): An Overview and Preliminary Validation Results *Presented at the 84th Annual Transportation Research Board Conference*
[20] Pribyl O and Goulias K G 2005 Simulation of daily activity patterns *Progress in Activity-Based Analysis*, Elsevier Science, pp 43-65
[21] Vause M 1997 *A* Rule-Based Model of Activity Scheduling Behavior *Activity-Based Approaches to Travel Analysis,* Elsevier Science, Inc., New York, pp 73-88
[22] Zhang J, Timmermans H J P and Borgers A W J 2005 A model of household task allocation and time use *Transpn. Res. B* **39** pp 81-95

IOP Conf. Series: Materials Science and Engineering 1086 (2021) 012025 doi:10.1088/1757-899X/1086/1/012025

Formation of performance characteristics of a Formula SAE sports vehicle based on calculations and simulations of systemic elements

A Kulagin[1], K Goncharov, V Kuzmichev, M Golubev, M Gulin and V Khozerov

Nizhny Novgorod State Technical University n.a. R.E. Alekseev, Minin str., 24, 603950, Nizhny Novgorod, Russia

E-mail: [1]kulagin@nntu.ru

Abstract. The article presents a description of the research and development carried out by NNSTU employees as part of their participation in the international project Formula Student. The paper includes the calculated assessment of the operational properties of the vehicle being designed: the calculation of the center of mass coordinates, with the presentation of data based on the configuration features of the vehicle, the calculations of stability, the efficiency of the brake system, the input parameters of the steering simulation are given, on the basis of which the graphical dependences of the angles were obtained steering wheel rotation, as well as data on prototyping mechatronic transmission systems. The data obtained is compared with the technical regulations requirements of the international student competitions «Formula SAE».

1. Introduction

During developing a prototype of a vehicle, the performance characteristics of the vehicle are relevant indicators. These characteristics are evaluated on the basis of calculated dependencies, simulation processes, and a basis of experiments in the case of automation of existing systems. At the same time, it is necessary to take into account the features of the configuration and operational restrictions within which the system is being developed [1]. Researchers from Nizhny Novgorod State Technical University have many years of experience in the design of wheeled vehicles. Recently, there has been active research into the development of light commercial vehicles [2, 3] and off-road vehicles on ultra-low pressure tires [4-6]. These vehicles are produced at industrial enterprises in Nizhny Novgorod. Also at the university, students have developed different vehicles. This is a mini all-terrain vehicle "Korsak" [7-10] and a car developed within the framework of the Formula Student project [11, 13-15]. We will consider the modeling and design process in the article using a racing car as an example. Some of the calculations are fundamental and are necessary for the subsequent formation of vehicle systems calculations: safety cage, energy absorbing device, suspension, braking system, steering.

2. Calculation of the center of mass coordinates and sports car stability

The calculation of the center of mass coordinates is based on a detailed study of the vehicle configuration in the form of detailed 3D models of parts and assemblies. A number of 3D models are based on 3D scan data. In order to obtain data on the vehicle systems' center of gravity coordinates, the resulting virtual three-dimensional model is located at the origin of coordinates [11]. An additional parameter which involved in the calculations is the mass of elements. A systematized set of parameters for the mass and coordinates of the center of gravity of the sports car systems is presented in Table 1.

The value of the coordinate of the center of mass of the sports car is calculated by the ratios:

$$\boldsymbol{x_c} = \frac{\sum \boldsymbol{m_i x_i}}{\sum \boldsymbol{m_i}}, \text{м и } \boldsymbol{y_c} = \frac{\sum \boldsymbol{m_i y_i}}{\sum \boldsymbol{m_i}}, \qquad (1)$$

where x_c is the coordinate of the center of mass, plotted along the abscissa, m; y_c is the coordinate of the center of mass, plotted along the ordinate, m; m is the mass of the element, kg; i is the ordinal number of the element.

Table 1. Systematized set of parameters

System name	Weight mi (kg)	Coordinates	
		Xi (mm)	Yi (mm)
Frame	39.2	1140,505	361,136
Damping device	0.725	-99.1572	362.5
Partition wall	2,449	-1	362.5
Refractory walls	9.1	1495.3795	368.9754
Cockpit floor and front floor	2,269	852,146	161.671
Pilot	70	1041,108	450.8812
Armchair	2.2	1199.05	326.35
Engine	62,538	1780.05	417.2767
Fuel system	5,748	1664,793	531,5054
Exhaust system	five	1766,439	226.0459
Steering	5.807	664.5163	293.2873
Transmission	12,136	2150.09	282.5409
Pedal assembly	4.807	108.0867	238,9938
Front axle suspension	11,223	663,1569	272,1079
Rear axle suspension	10,705	2206,987	295,4099
Brake system. front axle	2.703	733,7443	263.9293
Brake system. rear axle	2.703	2283.83	270,1543
Wheels	52.56	1437,701	270,215

Calculation of the necessary products of the mass of an element by the corresponding coordinate along the abscissa axis $m_i x_i$ (kg · m) and ordinates $m_i y_i$ (kg · m), and determining the total values of the mass $m=\sum mi$ (the sum of the masses of the vehicle elements), on the basis of data table 1, produced by Excele.

Calculation result: The sum of the products over X = 412046.2 mm; The sum of products by Y = 89659.05 mm; Mass m = 301.8744 kg; wheelbase L = 1550 mm.

Determine the value of the coordinates of the center of mass of the car using the formula (1) we obtain: $x_c = 1{,}364$ м, $y_c = 0{,}297$ м

The determination of static loads on the axles of the chassis is based on the results of calculating the mass and calculating the coordinates of the center of gravity of a sports car. The parameter of the static load distributed along the vehicle axes used as input data for calculating the braking system, suspension elements, and the stability of a sports car as well.

Calculation layout (figure 1) of determine the static load is realized on the basis of the following algorithm:

1. Get a side image of the vehicle.

2. Enter the Cartesian coordinate system and the origin point into the resulting image. In this case, it is necessary to pre-select the direction of the coordinate system axes. The origin point must coincide with the origin entered during the finding the sports car center of gravity.

3. Apply the circuit location parameters of the center of mass coordinates, the distance from the coordinates of the center of mass to the front axis of the vehicle, the wheelbase and the force of gravity, which replaced by reactions based on the liberation axiom. It is necessary to control that the resulting system is statically determinate.

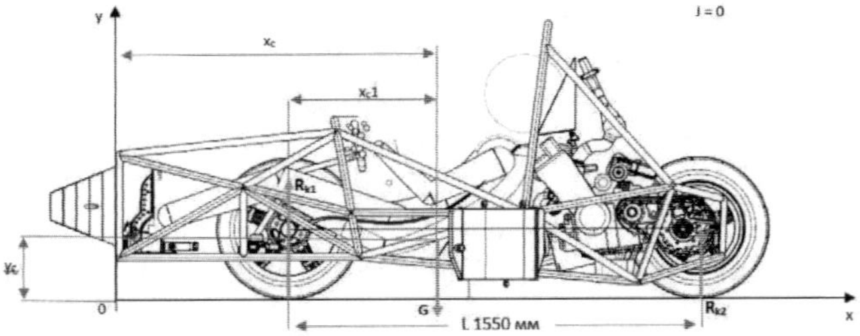

Figure 1. Calculation scheme for determining the normal load R_{k2} and R_{k1} of a stationary vehicle , where G – cars weight, H; x_{c_1}- distance from the front axle of the car to the coordinates of the center of mass, m; L is the cars wheelbase, m.

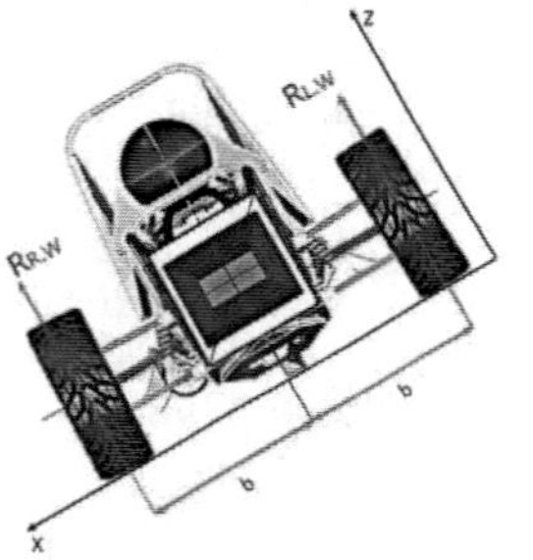

Figure 2. Calculation layout for determining normal reactions on the wheels of a sports car with a transverse tilt.

After building calculation layout, it is necessary to consider the equilibrium condition of the car, thereby to write the equations of statics, which are equivalent to zero the amount of projection on the axis of the Cartesian coordinates of all the forces acting on the body and the sum of the moments of these forces about the arbitrarily selected points or axes.

From the equation of the moments of forces acting relative to the axis of the front wheels, we determine the normal load acting on the rear axle of the vehicle:

$$R_{k_2} = \frac{G \cdot x_c}{L} = \frac{m \cdot g \cdot x_{c_1}}{L}, \tag{2}$$

Based on formula (2), we obtain the value of the load acting on the rear axle of a sports car $R_{k_2} = 1293{,}4578$ H.

We determine the normal load acting on the front axle of the car from the equation of the moments of forces acting relative to the axis of the rear wheels.

$$R_{k_1} = \frac{G \cdot (L - x_{c_1})}{L} = \frac{m \cdot g \cdot (L - x_{c_1})}{L}, \tag{3}$$

Based on the formula (3), we get the value of the load acting on the front axle of a sports car: $R_{k_1} = 1667{,}9300$ H.

Based on the results of calculations of the coordinates of the center of mass and static loads, the operational stability indicator of the car is formed. This indicator is characterized by a critical angle of deviation in the transverse direction. In calculating the critical angle of inclination, the calculation scheme is used (Figure 2). this scheme determines the normal reactions on the wheels of the car based on the loading condition of the car. In a static position, the car is affected by gravity.

Each of the corresponding forces from the rolling surface will cause normal reactions on the front and rear wheels $R_{A,B} = R_{K_1}$ and $R_{C,D} = R_{K_2}$, side on the left and right wheels $S_{A,B}$ and $S_{C,D}$. At the static position of the car, the bond reactions are equal in value and opposite to the corresponding active forces.

Normal reactions of rolling surface on the wheels of the front axle R_A and R_B and the rear axle R_C and R_D:

$$R_A = \frac{1}{2} \cdot G \cdot \zeta_A; \; R_B = \frac{1}{2} \cdot G \cdot \zeta_B; \; R_C = \frac{1}{2} \cdot G \cdot \zeta_C; \; R_D = \frac{1}{2} \cdot G \cdot \zeta_D, \tag{4}$$

where ζ_i is the coefficient of redistribution of normal reactions.

The coefficient of redistribution of normal reactions on the wheels of a car is determined by the expressions:

$$\zeta_{A,B} = \frac{l_2}{L} \cdot \cos\alpha_0 \cdot \cos\alpha_S \pm \frac{h_c}{L} \cdot \sin\alpha_0 + \frac{l_2}{L} \cdot \frac{h_c}{b} \cdot \cos\alpha_0 \cdot \sin\alpha_S,$$
$$\zeta_{C,D} = \frac{l_1}{L} \cdot \cos\alpha_0 \cdot \cos\alpha_S \mp \frac{h_c}{L} \cdot \sin\alpha_0 + \frac{l_1}{L} \cdot \frac{h_c}{b} \cdot \cos\alpha_0 \cdot \sin\alpha_S, \qquad (5)$$

where L is the track of the car; h_c is the distance from the reference surface to the coordinates of the center of mass; l_1 is the distance from the front axle to the coordinates of the center of mass of the car; l_2 is the distance from the rear axle to the coordinates of the center of mass of the car.

The sum of the normal reactions on the front and rear wheels R_{K_1} and R_{K_2} are equal:

$$R_{K_1} = R_A + R_B = \frac{1}{2} \cdot G \cdot \left(\zeta_A + \zeta_B\right) = \frac{1}{2} \cdot G \cdot \zeta_1,$$
$$R_{K_2} = R_C + R_D = \frac{1}{2} \cdot G \cdot (\zeta_C + \zeta_D) = \frac{1}{2} \cdot G \cdot \zeta_2, \qquad (6)$$

The total value of normal reactions on the left and right wheels of the car $R_{L.W}$ and $R_{R.W}$ are equal to:

$$R_{L.W} = R_A + R_C = \frac{1}{2} \cdot G \cdot \left(\zeta_A + \zeta_C\right) = \frac{1}{2} \cdot G \cdot \zeta_L,$$
$$R_{R.W} = R_B + R_D = \frac{1}{2} \cdot G \cdot (\zeta_B + \zeta_D) = \frac{1}{2} \cdot G \cdot \zeta_R, \qquad (7)$$

Equality to zero of the normal reaction on the front or rear wheels means their complete unloading, which is equivalent to overturning the car, then the condition for overturning the car is $R_{K_1} = 0$ *or* $\zeta_1 = 0$, then

$$tg\alpha_0 = tg\alpha_{0_{max}} = \frac{l_c}{h_c} \cdot \cos\alpha_S, \qquad (8)$$

Losing of lateral stability will occur when the right or left wheels located on the upper part of the slope are completely unloaded, in case of $R_{R.W} = 0$*or* $\zeta_R = 0$, then

$$tg\alpha_{S_{max}} = \frac{b_c}{h_c},$$
$$\alpha_{S_{max}} = arctg\left(tg\alpha_{S_{max}}\right), \qquad (9)$$

Angles of stability factor are the difference between the stability angles and the actual tilt angles of the car.

Based on the design layouts and dependencies, an indicator of vehicle stability was obtained, $\alpha_{S_{max}} = 63°39'44''$, while this indicator exceeds the established value of 60 degrees and makes it possible to pass competitions stability tests.

2. Calculation of indicators of the brake system

The braking performance parameter is determined by the design features of the braking system components. These features, specified by the requirements of the Formula SAE technical regulation [1], mean the presence of two independent brake circuits of the front and rear axle of the vehicle. The brake circuits are interconnected with the balance bar. That's the element that distributes the load on the rods of the brake master cylinders, set by the pilot during interacting with the brake pedal. Preparation of optimal braking performance index calculation is implemented by the brake actuator (figure 3).

The main parameters, determining the index of braking performance: the estimated brake fluid pressure in brake circuit P_{hyd}(10) and the magnitude of the brake fluid pressure, required to lock the front wheels of the car P_1(11);

$$P_{hyd} = \frac{F_{pp}}{A_{pi,h}}, \qquad (10)$$

where F_{pp} is the value of the force acting on the rods of the brake master cylinders, taking into account the balance bar, N; $A_{pi,h}$ is the area of the one brake master cylinder piston, mm^2;

$$\boldsymbol{P_1} = \frac{\boldsymbol{Rz_1 \cdot r_d}}{\boldsymbol{A_1}}, \tag{11}$$

where Rz_1 is the load on the front axle of the vehicle, N; A_1 is the ratio of the shoulder r_1 to the shoulder r_2, m; r_d is the dynamic radius of the wheel, m;

This type of vehicle implies movement on asphalt or cement surface, the dynamic radius r_d of the wheel is taken approximately equal to the static radius of the wheel r_{st} and is calculated using the following formula:

$$\mathbf{r_{st}} = \mathbf{0,5 \cdot d + \lambda_z \cdot \Delta_{tr} \cdot B_{tr}}, \tag{12}$$

where λ_z is the coefficient of normal tire deformation (λ_z= 0.8... 0.85); B_{tr} is the tire profile width, m; H_{tr} is the tire profile height, m; $\Delta_{tr}= \frac{H_{tr}}{B_{tr}}$ is the coefficient of the shape of the tire profile; d is the rim landing diameter, m; [12].

An important stage in the calculation is the selection of the optimal geometric parameters of the brake pedal. The shoulders, formed depending on the attachment points of the brake pedal axis and the balance bar axis, where: r_1 is the distance from the attachment point of the balance bar to the point of interaction between the pilot and the brake pedal, m; r_2 is the distance from the attachment point of the brake pedal axis to the attachment point of the balance bar axis, m;

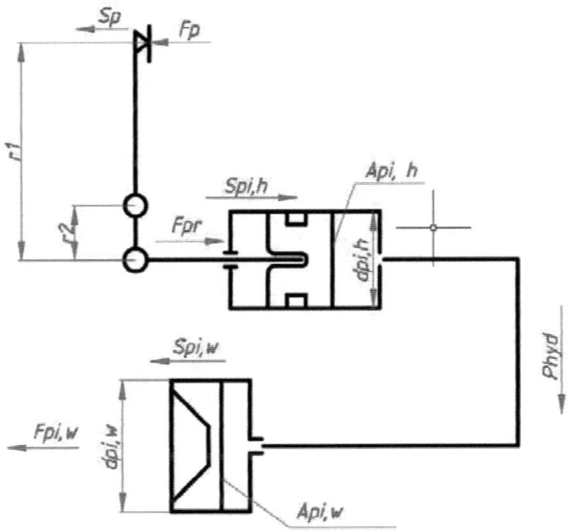

Figure 3. Brake drive circuit

Geometrical parameters (r_1, r_2) directly influence on the brake pedal S_p (13):

$$\boldsymbol{S_p} = \boldsymbol{S_{pi,h}} \cdot \frac{\boldsymbol{r_1}}{\boldsymbol{r_2}}, \tag{13}$$

where $S_{pi,h}$ is the piston movement in the brake master cylinder, mm;

$$\boldsymbol{S_{pi,h}} = \boldsymbol{j_b} \cdot \boldsymbol{j_{pi,w}} \cdot \frac{\boldsymbol{A_{pi,w}}}{\boldsymbol{A_{pi,h}}} \cdot \boldsymbol{S_{pi,w}}, \tag{14}$$

where j_b is the number of supports connected to the brake master cylinder; $j_{pi,w}$ is the number of pistons in the brake calipers; $A_{pi,w}$ *is the* caliper piston area, mm $^{2;}$ $S_{pi,w}$ *is the* movement of the piston in the support, mm;

Since the design of the brake drive implies the presence of a balance bar, the value of the force acting on the brake master cylinders' rods F_{pp} (15) is determined as follows:

$$\boldsymbol{F_{pp}} = \boldsymbol{F_{pr}} \frac{\boldsymbol{b_r}}{\boldsymbol{b_f + b_r}}, \tag{15}$$

where F_{pr} is the value of the total force acting on the rods of the brake master cylinders;

b_{r} *is the* distance from the plane of symmetry of the balance bar to the longitudinal axis of the brake master cylinders' rod of the rear axle, m; b_{f} *is the* distance from the plane of symmetry of the balance bar to the longitudinal axis of the brake master cylinders' rod of the front axle, m;

The variables included in the design formulas geometrically depend on the component base of the designed system.

$$F_{pr} = \frac{F_p \cdot r_1}{r_2}, \quad (16)$$

where F_p is the force acting on the brake pedal, N [5];

The parameters required for calculating the brake drive are presented in Table 2.

The results of the dependencies (10-16) of the performance indicators of the braking system of a sports car are shown in Table 3.

The calculated values given should correlate with the requirements of the Formula SAE technical regulation for car wheel locking. The value of the pressure reached in the brake circuit P_{hyd} should exceed the value of the pressure required to bring the front brakes to the full blocking $P_{1,}$ as a result of which the performance of the vehicle's braking system will be ensured and the vehicle's braking efficiency is optimal.

Table 2. Parameters required for calculating the brake drive.

	Parameters name	Parameters value
1	f_p is the force acting on the brake pedal, N	490
2	b_f *is the* distance from the plane of symmetry of the balance bar to the longitudinal axis of the brake master cylinder rod of the front axle, m	0.024
3	b_r is the distance from the plane of symmetry of the balance bar to the longitudinal axis of the brake master cylinder rod of the rear axle, m	0.0 15
4	$a_{pi,\,h}$ is the brake master cylinder piston area, mm ^{2}	283, 4
5	$a_{pi,\,w}$ is the caliper piston area, mm ^{2}	506.45
6	m_a is the total mass of the vehicle, kg	30 1, 87
7	r_{z1} is the load on the front axle of the vehicle, N	1667
8	$s_{pi,\,w}$ is the piston movement in the support, mm	1.4

Table 3. Brake system exploitation parameters results

	Vehicle calculation parameters	Estimated values of vehicle performance
1	r_1 is the distance from the attachment point of the balance bar to the point of interaction of the pilot with the brake pedal, m	0.25
2	r_2 is the distance from the attachment point of the brake pedal axis to the attachment point of the axis of the balance bar, m	0.03
3	$F_{pr,}$ is the value of the total force acting on the brake master cylinder rods, N	4083.3
4	F_{pp} is the value of the force acting on the brake master cylinder rod of the front axle, N	2512.8
5	P_{hyd} is the calculated value of the brake fluid pressure in the brake circuit, bar	88.6
6	r_d is the dynamic radius of the wheel, m	0.253
7	A_1 is the ratio of the arm r_1 to the arm $r_{2,}$ m	8.3
8	P_1 is the amount of pressure of the brake fluid in the brake circuit required to block the front wheels of the car, bar	50.08
9	Sp is the brake pedal travel, mm	83.4
10	$S_{pi,\,h}$ is the piston movement in the brake master cylinder, mm	ten

3. Calculation of steering performance

The key parameters for calculating steering are the wheelbase (L = 1550 mm), the front axle track width (B = 1200 mm), the scrub radius (r_f = 0 mm) and the distance between the points of intersection of the pivot axles of wheels with the supporting surface (l_0 = 1200 мм), they were obtained during the development of the suspension. The value of the minimum radius of a vehicle by rotation of the outer wheel should not exceed 4,500 mm (requirement FSAE Rules [1]).

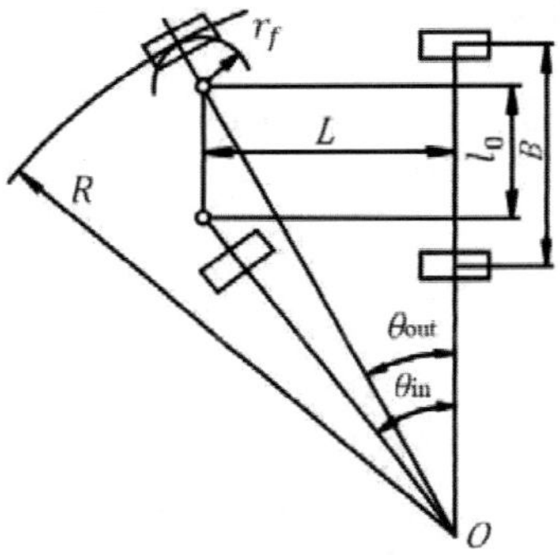

Figure 4. Vehicle turning layout.

To exclude the contact of the inner surface of the wheel disk and the suspension arms, as well as the vehicle entering the turn, the turning radius should be close to the value of 3700 mm.

In order for the wheels to roll without side slip, their axes must intersect at one point (Figure 4). This point (O) is called the center of rotation [12].

The steering angles of the steered wheels are related by the ratio:

$$\mathbf{ctg}\boldsymbol{\theta}_{out} - \mathbf{ctg}\boldsymbol{\theta}_{in} = \frac{l_0}{L}, \quad (17)$$

where θ_{out} is the angle of rotation of the outer wheel; θ_{in} is the angle of rotation of the inner wheel.

The swing angle of the outer wheel is calculated by the formula:

$$\boldsymbol{\theta}_{out} = \mathbf{arctg}\left(\frac{L}{\sqrt{(R-r_f)^2-L^2}}\right) = 24^\circ 46',, \quad (18)$$

Knowing the angle of rotation of the outer wheel, we calculate the angle of rotation of the inner wheel by transforming formula (17):

$$\boldsymbol{\theta}_{in} = \mathbf{arcctg}\left(\mathbf{ctg}\boldsymbol{\theta}_{out} - \frac{l_0}{L}\right) = 35^\circ 40', \quad (19)$$

Formulas (17,18,19) are valid only for a car with rigid wheels. In our case, the car moves on elastic tires, as a result of which there is a drift, since lateral forces act on the car during a cornering. Due to this, the slip angle must be taken into account when calculating. For sports cars, the slip angle of the outer wheel is close to the value $\alpha_н = 3^\circ$ [15], and for the inner wheel, the slip angle is very small, because the main part of the load in the turn falls on the outer wheel.

The actual steering angle of the outer wheel for a car with elastic tires will be calculated using the formula:

$$\boldsymbol{\theta}_{\mathbf{out}}^{actual} = \boldsymbol{\theta}_{\mathbf{out}} + \boldsymbol{\alpha}_{\mathbf{out}} = 27^\circ 46', \quad (20)$$

The obtained calculations must be put into the kinematic diagram of the suspension and steering system, the simulation model of which was formed in the Adams Car software by MSC Software (Figure 5), using the iterative selection method and taking into account such requirements as:

1. Coordination of steering kinematics and suspension kinematics;
2. Configuration requirements of individual units and parts of the car;
3. Preservation of the required angles of rotation of the steered wheels (18, 20).

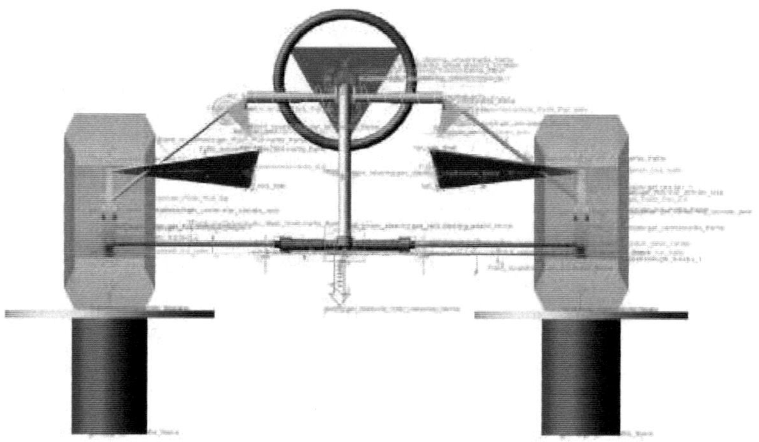

Figure 5. Steering system and suspension elements in Adams Car software

On the basis of simulation, a graph of the dependence of the steering angles on the steering rack travel was obtained (Figure 6).

To minimize the decrease in the indicators of points 1 and 2, the values of the angles of rotation of the steered wheels were changed in relation to the required ones, and for the inner and outer wheels were 34°48′ and 29° respectively [13]. Comparison of required and obtained values is shown in the graph (Figure 7). The upper curve is the required dependence, the lower is obtained from the simulation.

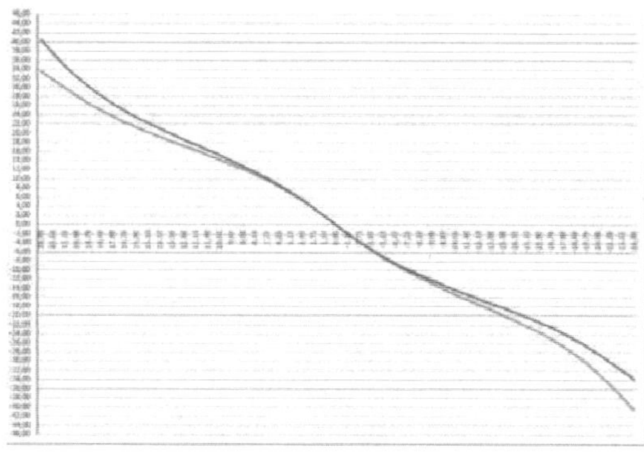

Figure 6. Angles of rotation of the steered wheels depending on the travel of the steering rack

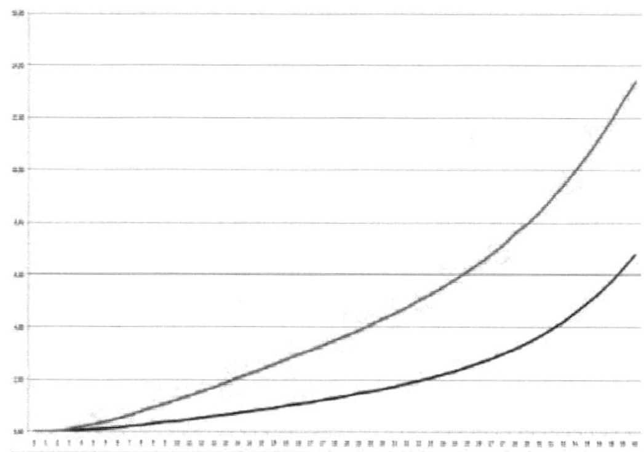

Figure 7. Dependence of the difference between the angles of rotation of the wheels on the angle of rotation of the inner wheel

4. Performance indicators of mechatronic systems

During developing the mechatronic system, experiments were carried out to measure the characteristics of the DC electric motor and the forces on the shift shaft and the clutch lever. Experimental data are shown in Table 4.

Table 4. Experimental data

	Parameter	Value
	Shift shaft	
1.	Maximum shaft rotation angle required to change gear, °	25
2.	Maximum torque required for gear shifting, N / m	7,056
	Clutch lever	
3.	Lever stroke, mm	27
4.	Lever arm, mm	36.5
5.	Maximum torque required to disengage the clutch, N / m	10.55215
6.	Lever rotation angle, °	42.4
	DC motor	
7.	Torque, N / m	3
8.	Shaft rotation frequency, rpm	65
9.	Starting torque, N / m	14

Based on the data obtained, the development of mechanisms was made that converts and transmits the torque of electric motors for shifting gears and disengaging the clutch. After that, the development, prototyping, assembly and testing of fasteners were carried out: system of supports for attaching the clutch actuator, fastening the clutch lever position sensor, mounting plate for the gearshift actuator, supporting element of the gearshift actuator, fastening the gearshift sensor.

At this stage of development, the results of the work are the ability, using mechatronic systems, to disengage the clutch, change gear and engage the clutch in 0.94 seconds, which is comparable to the time of shifting gears in manual mode. The "quick shift" feature reduces gear change times to 0.36 seconds.

5. References.

[1] *Formula SAE Rules* 2019 SAE International p 134
[2] Zezyulin D et al. 2014 Modeling of roads impacts for life prediction of light commercial vehicles parts *FISITA 2014 World Automotive Congress*
[3] Makarov V et al. 2018 Development of driver's assistant system of additional visual information of blind areas for Gazelle Next *IOP Conference Series: Materials Science and Engineering* **315(1)** 012016 doi: 10.1088/1757-899X/315/1/012016
[4] Belyaev A et al. 2019 Development of 8×8 all-terrain vehicle with individual wheel drive *VEHITS 2019 - Proceedings of the 5th International Conference on Vehicle Technology and Intelligent Transport Systems* pp 556-561
[5] Belyakov V at al. 2015 Multifunctional vehicle for coastal areas *12th International Conference on the Mediterranean Coastal Environment, MEDCOAST 2015* pp 945-951
[6] Anikin A et al. 2017 Calculation of traction capabilities of wheeled vehicles on low-pressure tires on snow *19th International and 14th European-African Regional Conference of the ISTVS*
[7] Kapustin A G et al. 2019 Development of design project of mini cross-country vehicle KORSAK *IOP Conference Series: Materials Science and Engineering* **695(1)** 012026 doi: 10.1088/1757-899X/695/1/012026
[8] Papunin A V et al. 2018 A dynamic model of unsupported pit traversal by a vehicle with 6×6 wheel arrangement *IOP Conference Series: Materials Science and Engineering* **386(1)** 012001 doi: 10.1088/1757-899X/386/1/012001
[9] Papunin A V et al. 2018 Calculation of trench width to be overcome by multi-axis wheeled vehicle *IOP Conference Series: Earth and Environmental Science* **194(6)** 062027 doi: 10.1088/1755-1315/194/6/062027

IOP Conf. Series: Materials Science and Engineering 1086 (2021) 012025 doi:10.1088/1757-899X/1086/1/012025

[10] Papunin A V et al. 2019 Field research of profile trafficability of 6×6 wheel assembly KORSAK vehicle *Journal of Physics: Conference Series* **1177(1)** 012051 doi: 10.1088/1742-6596/1177/1/012051
[11] Kuzmichev V A, Kulagin A L and Goncharov K O 2020 Calculation of the cordinate of the center of mass of a FORMULA STUDENT sports car *Collection of articles and abstracts of the 19th all-Russian youth scientific and technical conference "the Future of technical science"* NSTU, Nizhny Novgorod pp 149-151
[12] Kravets V N 2013 *Car theory: Educational guide* Nizhny Novgorod State Technical University n.a. R. E. Alekseev, Nizhny Novgorod p 413
[13] Gulin M S, Kulagin A L, Butin D A and Goncharov K O 2020 Determining the angles of rotation of controlled wheels and the location of steering elements *Collection of articles and abstracts of the 19th all-Russian youth scientific and technical conference "the Future of technical science"* NSTU. is the Nizhny Novgorod pp 136-137
[14] Golubev M S, Kulagin A L and Goncharov K O 2020 Loading the pedals of the working brake system of a sports car of the "FORMULA STUDENT" class *Collection of articles and abstracts of the 19th all-Russian youth scientific and technical conference "the Future of technical science"* NSTU, Nizhny Novgorod pp 135-136
[15] Belyakov V V et al. 2019 Calculation steering geometry of the car taking into account the slip angle tire *Paper of NNSTU named of R E Alekseev* **2** pp 156-162

IOP Conf. Series: Materials Science and Engineering 1086 (2021) 012026 doi:10.1088/1757-899X/1086/1/012026

Modeling of the total disturbing effects on the transmission elements from movement on different support bases and the engine

K Y Leliovsky

Nizhny Novgorod State Technical University n. a. R.E. Alekseev, Minin str., 24, 603950, Nizhny Novgorod, Russia

e-mail: kleliovskiy@mail.ru

Abstract. For the purpose of determining and predicting their mobility in operating conditions, it is necessary to create a comprehensive dynamic model that allows you to determine the parameters of power unit vibrations and torsional vibrations in the transmission, taking into account the relationship between torsional vibrations of the transmission, vertical, longitudinal-angular vibrations of the inertial masses of the vehicle, power unit and transfer case, elastic suspension elements, tires and power unit supports, disturbing effects from road roughness and the running engine. It allows you to determine the values of vibration loading of the body and associated components and aggregates of the vehicle, depending on the physical and technical parameters of the support base and the frequency response (speed of the car).

Introduction

The modeling of disturbing effects on the transmission elements of the objects of research caused by traffic on various roads, as well as engine operation, is considered. An equivalent model of their transmission was developed, which allows calculating the parameters of dynamic oscillatory processes occurring in it during operation [1-5]. To compile it, we use the model shown in Figure 1. It allows you to determine the values of vibration loading of the body and related components and aggregates of the vehicle, depending on the physical and technical parameters of the support base and the frequency impact (speed of the car). The loading effect from the road is modeled by the dependence was shown that the vibration parameters of the island are related to the values of loading moments and forces acting in the transmission. In this model, the structural elements of automotive equipment are represented by a concentrated inertial sprung mass m_c. In the developed model, we will replace it with a multi-mass equivalent dynamic transmission system.

Equivalent complex dynamic model of a transport and technological vehicle with a 4x4 wheel formula

The scheme of geometric parameters of an equivalent complex dynamic model of transport and technological vehicles with a 4x4 wheel formula, which are the objects of research, shown in Figure 2, is necessary for determining the kinetic, potential energy, as well as the scattering function. Here m_0, m_1, m_2, m_{ca} – mass of the body front, rear axle assembly with wheels, the power unit, respectively; $z_0(t)$ $z_1(t)$ $z_2(t)$, $z_{ca}(t)$ is the linear displacement of the body, front, rear axle assembly with wheels, the power unit, respectively; $C_1^{\Pi M}, C_2^{\Pi M}, C_1^{3M}, C_2^{3M}, C_{31}, C_{32}$ - stiffness of front and rear wheels, front and rear suspension of the machine, supports the power unit; $k_1^{\Pi M}, k_2^{\Pi M}, k_1^{3M}, k_2^{3M}, k_{31}, k_{32}$, - damping coefficients of front and rear wheels, front and rear suspension of the machine, supports the power unit; $\varphi_0^x(t), \varphi_0^y(t), \varphi_{ca}^x(t), \varphi_{ca}^y(t)$ - the angular displacement (oscillation) of the housing (body) and the power unit; I_1, I_2, I_3, I_4, I_5, I_6, I_7, $I_{\Pi M}$, I_{3M} – moments of inertia of the power unit, front and rear cardan shafts, main transmission, front and rear wheels, front and rear axles and the reduced moment of inertia of the translationally moving mass of the car; $C_{12}, C_{23}, C_{24}, C_{46}, C_{57}, C_{67}$ - torsional stiffness of the power unit, front and rear cardan shafts, front and rear axles, front and rear axles, tangential stiffness of tires; $k_{12}, k_{23}, k_{24}, k_{46}, k_{57}, k_{67}$ - coefficients of inelastic resistance of the power unit, front and rear cardan shafts, half-axles of the front and rear axles, tangential tire compliance; $C_{ca}, C_{\Pi M}, C_{3M}$ - stiffness of the power unit supports, as well as elastic elements of the front and rear suspension for twisting; $\varepsilon_1(t)$, $\varepsilon_2(t)$ - disturbing forces acting from the support base; $F_a(t)$ - external inertial force acting on the power unit; $T_e(t)$ – torque(disturbing) moment acting from the engine side; $T_{f1}(t)$, $T_{f2}(t)$ - moments of rolling resistance of wheels (disturbing moments) acting from the road side; l_1, l_2, l_3, l_4, l_5, l_6, l_7 – geometric parameters of the equivalent dynamic model (object of study).

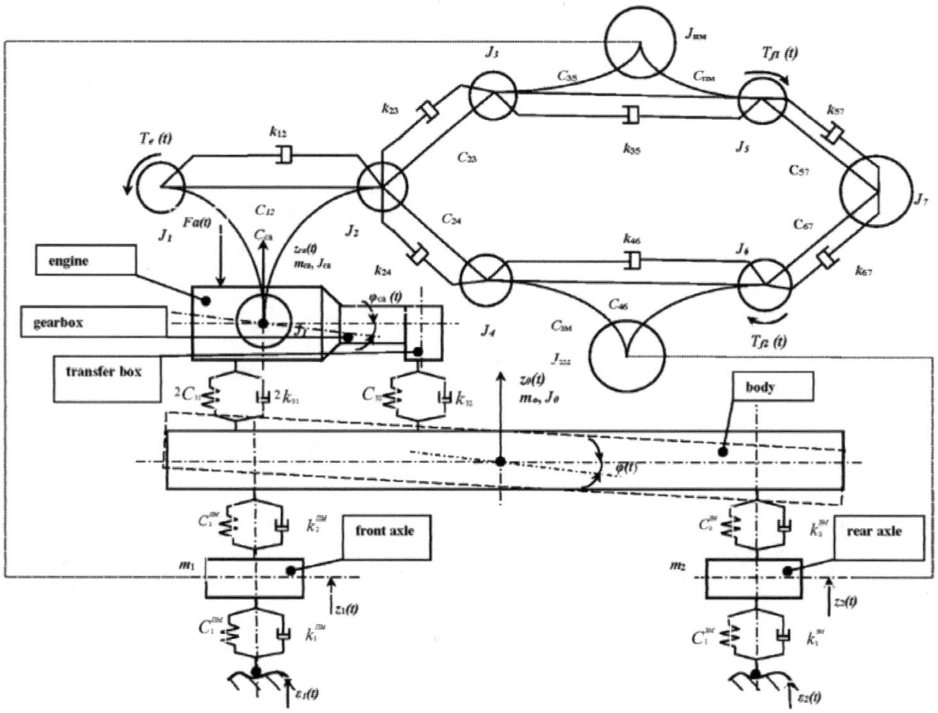

Figure 1. Equivalent complex dynamic model of a transport and technological vehicle with a 4x4 wheel formula for calculating vertical, longitudinal-angular, torsional vibrations of the transmission, as well as spatial vibrations of the power unit when moving on a different support base

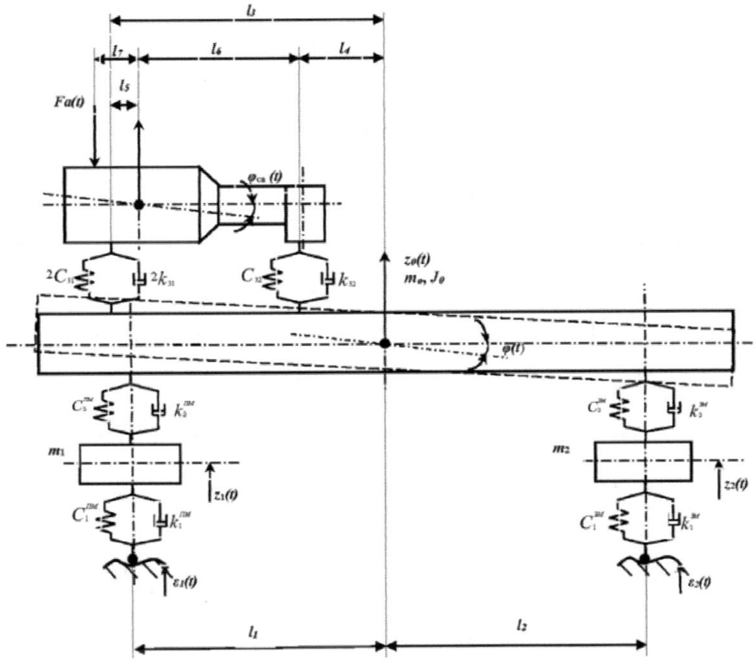

Figure 2. Diagram of geometric parameters of an equivalent complex dynamic model of a transport and technological vehicle with a 4x4 wheel formula

Let's write down the General equation of analytical mechanics for an equivalent complex dynamic model of a transport and technological vehicle with the 4x4 wheel formula shown in Figure 1. Based on the features of the construction of this system, as well as their compliance with the provisions of theoretical mechanics, there are sufficient grounds to compose it in the differential form of D'alembert – Euler. TTS 3007 and TTS 3910 were selected as research objects. Here are the technical parameters specified in the diagram that are typical for these machines.

Table 1. Geometric parameters of research objects for the model

Name	l_1,m	l_2,m	l_3,m	l_4,m	l_5,m	l_6,m	l_7,m
TTS 3007	1,725	1,725	2,150	0,250	0,155	1,525	0,050
TTS 3910	1,350	1,350	1,420	0,182	0,075	0,653	0,120

For a system whose equivalent model is shown in Figure 1, we use the Hamilton principle to write a generalized equation of analytical mechanics. Let's write down the General equation of analytical mechanics for an equivalent complex dynamic model of a transport and technological vehicle with the 4x4 wheel formula shown in Figure 1. Based on the features of the construction of this system, as well as their compliance with the provisions of theoretical mechanics, there are sufficient grounds to compose it in the differential form of D'alembert – Euler. Based on the assumptions made earlier, we assume that the system under study is quasi-conservative, i.e. it almost completely corresponds to the theorem on the conservation of total mechanical energy. In this case, we can write down its energy balance based on the Hamilton principle:

$$\delta\int_{t_0}^{t_1}(T-\Pi)dt=0, \quad (1)$$

where T is the kinetic energy of the dynamical system under consideration, Π is the potential energy of the dynamical system under consideration, (t_0, t_1) is the time interval for considering the dynamics of the system (integration).

Based on the energy expressions we have written for the dynamical system under study we will make a General equation of analytical mechanics in the form of a Lagrange of the second kind:

$$\frac{d}{dt}\left(\frac{\partial T}{\partial \dot{q}}\right)+\frac{\partial R}{\partial \dot{q}}-\frac{\partial (T-\Pi)}{\partial q}=F(t). \tag{2}$$

For the system under consideration, it is written as follows:

$$\begin{aligned}
& m_1\ddot{z}_1+2\left(k_1^{ПМ}+k_2^{ПМ}\right)\dot{z}_1+2\left(C_1^{ПМ}+C_2^{ПМ}\right)z_1-2k_2^{ПМ}\left(\dot{z}_0+l_1\dot{\varphi}_0\right)-2C_2^{ПМ}\left(z_0+l_1\varphi_0\right)=2k_1^{ПМ}\dot{\varepsilon}_1(t)+2C_1^{ПМ}\varepsilon_1(t)\\
& m_2\ddot{z}_2+2\left(k_1^{3M}+k_2^{3M}\right)\dot{z}_2+2\left(C_1^{3M}+C_2^{3M}\right)z_2-2k_2^{3M}\left(\dot{z}_0-l_1\dot{\varphi}_0\right)-2C_2^{3M}\left(z_0-l_2\varphi_0\right)=2k_1^{3M}\dot{\varepsilon}_2(t)+2C_1^{3M}\varepsilon_2(t)\\
& m_0\ddot{z}_0+2\left(k_2^{ПМ}+k_2^{3M}\right)\dot{z}_0+2\left(C_2^{ПМ}+C_2^{3M}\right)z_0+2\left(k_2^{ПМ}l_1-k_2^{3M}l_2\right)\dot{\varphi}_0+2\left(C_2^{ПМ}l_1-C_2^{3M}l_2\right)\varphi_0+2k_{31}\left(\dot{z}_0+l_3\dot{\varphi}_0\right)\oplus\\
& \oplus 2C_{31}\left(z_0+l_3\varphi_0\right)+k_{32}\left(\dot{z}_0+l_4\dot{\varphi}_0\right)+C_{32}\left(z_0+l_4\varphi_0\right)=2k_{31}\left(\dot{z}_{ca}+l_6\dot{\varphi}_{ca}^{y}\right)+2C_{31}\left(z_{ca}+l_6\varphi_{ca}^{y}\right)\oplus\\
& \oplus k_{32}\left(\dot{z}_{ca}-l_7\dot{\varphi}_{ca}^{y}\right)+C_{32}\left(z_{ca}-l_7\varphi_{ca}\right)+2k_2^{ПМ}\dot{z}_1+2C_2^{ПМ}z_1+2k_2^{3M}\dot{z}_2+2C_2^{3M}z_2\\
& I_0\ddot{\varphi}_0+2\left(k_2^{ПМ}l_1^2+k_2^{3M}l_2^2\right)\dot{\varphi}_0+2\left(C_2^{ПМ}l_1^2+C_2^{3M}l_2^2\right)\varphi_0+2\left(k_2^{ПМ}l_1-k_2^{3M}l_2\right)\dot{z}_0+2\left(C_2^{ПМ}l_1-C_2^{3M}l_2\right)z_0\oplus\\
& \oplus 2k_{31}l_3\left(\dot{z}_0+l_3\dot{\varphi}_0\right)+2C_{31}l_3\left(z_0+l_3\varphi_0\right)+k_{32}l_4\left(\dot{z}_0+l_4\dot{\varphi}_0\right)+C_{32}l_4\left(z_0+l_4\varphi_0\right)+C_{ПМ}\left(\varphi_0-\varphi_{ПМ}\right)\oplus\\
& \oplus C_{3M}\left(\varphi_0-\varphi_{3M}\right)=2k_{31}l_3\left(\dot{z}_{ca}+l_6\dot{\varphi}_{ca}^{y}\right)+2C_{31}l_3\left(z_{ca}+l_6\varphi_{ca}^{y}\right)+k_{32}l_4\left(\dot{z}_{ca}-l_7\dot{\varphi}_{ca}^{y}\right)+C_{32}l_4\left(z_{ca}-l_7\varphi_{ca}^{y}\right)\oplus\\
& \oplus 2l_1\left(k_2^{ПМ}\dot{z}_1+C_2^{ПМ}z_1\right)-2l_2\left(k_2^{3M}\dot{z}_2+C_2^{3M}z_2\right)\\
& m_{ca}\ddot{z}_{ca}+\left(2k_{31}+k_{32}\right)\dot{z}_{ca}+\left(2C_{31}+C_{32}\right)z_{ca}+\left(2k_{31}l_6-k_{32}l_7\right)\dot{\varphi}_{ca}^{y}+\left(2C_{31}l_6-C_{32}l_7\right)\dot{\varphi}_{ca}^{y}=2k_{31}\left(\dot{z}_0+l_3\dot{\varphi}_0\right)\oplus\\
& \oplus 2C_{31}\left(z_0+l_3\varphi_0\right)+k_{32}\left(\dot{z}_0+l_4\dot{\varphi}_0\right)+C_{32}\left(z_0+l_4\varphi_0\right)+F_a(t)\\
& I_{ca}^{y}\ddot{\varphi}_{ca}^{y}+\left(2k_{31}l_6^2+k_{32}l_7^2\right)\dot{\varphi}_{ca}^{y}+\left(2C_{31}l_6^2+C_{32}l_7^2\right)\varphi_{ca}^{y}+\left(2k_{31}l_6-k_{32}l_7\right)\dot{z}_{ca}+\left(2C_{31}l_6-C_{32}l_7\right)z_{ca}=\\
& =2k_{31}l_6\left(\dot{z}_0+l_3\dot{\varphi}_0\right)+2C_{31}l_6\left(z_0+l_3\varphi_0\right)-k_{32}l_7\left(\dot{z}_0+l_4\dot{\varphi}_0\right)-C_{32}l_7\left(z_0+l_4\varphi_0\right)-F_a(t)l_7\\
& I_{ca}^{x}\ddot{\varphi}_{ca}^{x}+k_{12}\left(\dot{\varphi}_1-\dot{\varphi}_2+\dot{\varphi}_{ca}^{x}\right)+C_{12}\left(\varphi_1-\varphi_2+\varphi_{ca}^{x}\right)+C_{ca}\left(\varphi_{ca}^{x}-\varphi_0\right)=0\\
& I_1\ddot{\varphi}_1+k_{12}\left(\dot{\varphi}_1-\dot{\varphi}_2+\dot{\varphi}_{ca}^{x}\right)+C_{12}\left(\varphi_1-\varphi_2+\varphi_{ca}^{x}\right)=T_e(t)\\
& I_2\ddot{\varphi}_2-k_{12}\left(\dot{\varphi}_1-\dot{\varphi}_2+\dot{\varphi}_{ca}^{x}\right)-C_{12}\left(\varphi_1-\varphi_2+\varphi_{ca}^{x}\right)+k_{23}\left(\dot{\varphi}_2-\dot{\varphi}_3\right)+C_{23}\left(\varphi_2-\varphi_3\right)+k_{24}\left(\dot{\varphi}_2-\dot{\varphi}_4\right)\oplus\\
& \oplus C_{24}\left(\varphi_2-\varphi_4\right)=0\\
& I_3\ddot{\varphi}_3-k_{23}\left(\dot{\varphi}_2-\dot{\varphi}_3\right)-C_{23}\left(\varphi_2-\varphi_3\right)+k_{35}\left(\dot{\varphi}_3-\dot{\varphi}_5+\dot{\varphi}_{ПМ}\right)+C_{35}\left(\varphi_3-\varphi_5+\varphi_{ПМ}\right)=0\\
& I_4\ddot{\varphi}_4-k_{24}\left(\dot{\varphi}_2-\dot{\varphi}_4\right)-C_{24}\left(\varphi_2-\varphi_4\right)+k_{46}\left(\dot{\varphi}_4-\dot{\varphi}_6+\dot{\varphi}_{3M}\right)+C_{46}\left(\varphi_4-\varphi_6+\varphi_{3M}\right)=0\\
& I_5\ddot{\varphi}_5-k_{35}\left(\dot{\varphi}_3-\dot{\varphi}_5+\dot{\varphi}_{ПМ}\right)-C_{35}\left(\varphi_3-\varphi_5+\varphi_{ПМ}\right)+k_{57}\left(\dot{\varphi}_5-\dot{\varphi}_7\right)+C_{57}\left(\varphi_5-\varphi_7\right)=-T_{f1}(t)\\
& I_6\ddot{\varphi}_6-k_{46}\left(\dot{\varphi}_4-\dot{\varphi}_6+\dot{\varphi}_{3M}\right)-C_{46}\left(\varphi_4-\varphi_6+\varphi_{3M}\right)+k_{67}\left(\dot{\varphi}_6-\dot{\varphi}_7\right)+C_{67}\left(\varphi_6-\varphi_7\right)=-T_{f2}(t)\\
& I_7\ddot{\varphi}_7-k_{57}\left(\dot{\varphi}_5-\dot{\varphi}_7\right)-C_{57}\left(\varphi_5-\varphi_7\right)-k_{67}\left(\dot{\varphi}_6-\dot{\varphi}_7\right)-C_{67}\left(\varphi_6-\varphi_7\right)=0\\
& I_{ПМ}\ddot{\varphi}_{ПМ}+k_{35}\left(\dot{\varphi}_3-\dot{\varphi}_5+\dot{\varphi}_{ПМ}\right)+C_{35}\left(\varphi_3-\varphi_5+\varphi_{ПМ}\right)-C_{ПМ}\left(\varphi_0-\varphi_{ПМ}\right)=0\\
& I_{3M}\ddot{\varphi}_{3M}+k_{46}\left(\dot{\varphi}_4-\dot{\varphi}_6+\dot{\varphi}_{3M}\right)+C_{46}\left(\varphi_4-\varphi_6+\varphi_{3M}\right)-C_{3M}\left(\varphi_0-\varphi_{3M}\right)=0
\end{aligned} \tag{3}$$

IOP Conf. Series: Materials Science and Engineering 1086 (2021) 012026 doi:10.1088/1757-899X/1086/1/012026

Since we are dealing with a dissipative dynamic oscillatory system with a finite number of degrees of freedom, it is necessary to write it in matrix form to simplify the solution of the system of differential equations.

$$M\ddot{x}_i + K\dot{x}_i + Cx_i = Q(t), \tag{4}$$

where x_i is a column vector of linear displacements and angular vibrations of inertial masses, M is a matrix of inertial coefficients of system elements, C is a matrix of linear and angular stiffness coefficients, K is a matrix of damping coefficients, and $Q(t)$ is a column vector of disturbing forces and moments.

The calculation of the root-mean-square value of the vibration acceleration of the power unit of a transport and technological machine with a 4x4 wheel formula under the disturbing influence of random irregularities of the micro profile of the support base during its movement is carried out in the MathLab software environment. Here is a block diagram of the algorithm for determining the RMS value of vibration accelerations of the power unit.

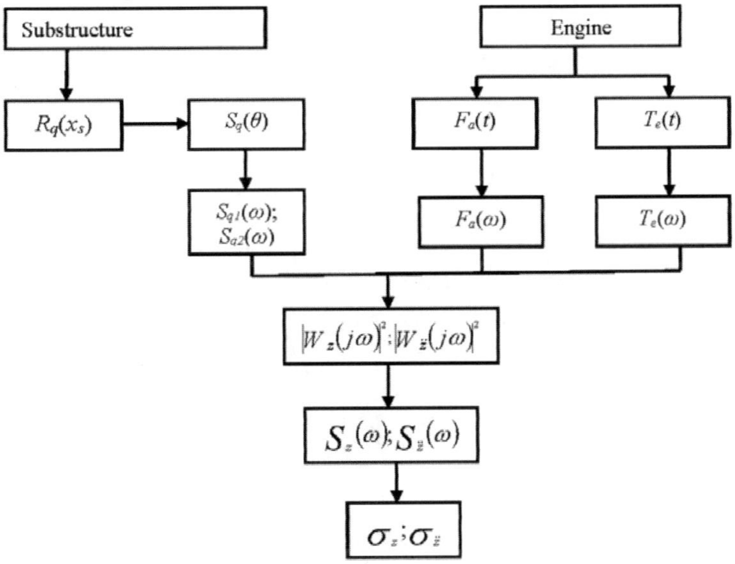

Figure 3. Block diagram for calculating the RMS value of vibration acceleration under random disturbing influence from the road and the engine

The spectral densities of the power unit vibrations when performing calculations by implementing the block diagram shown in Figure 3 are calculated using the following formulas:

$$S_z(\omega) = |W_z(j\omega)|^2 S_q(\omega), \tag{5}$$

$$S_{\ddot{z}}(\omega) = \omega^4 |W_z(j\omega)|^2 S_q(\omega). \tag{6}$$

The RMS value of vibration acceleration is calculated using the following formula:

$$\sigma_{\ddot{z}} = \sqrt{\int_{-\infty}^{+\infty} S_{\ddot{z}}(\omega) d\omega}. \tag{7}$$

Conclusion

In computational and theoretical experimental studies in the MathLab software environment of the spectral densities of vibration accelerations of the power unit of transport and technological machines that are the objects of research, based on the block diagram shown in Figure 3, the proper and partial vibration frequencies of dynamic links of the equivalent model shown in Figure 1 can be determined.

References

[1] Beresnev P et al. 2017 Simulation of road impacts for forecasting of durability of vehicle's suspension elements and supporting system *19th International and 14th European-African Regional Conference of the ISTVS*

[2] Latyshev G V 1973 Method for calculating the vibrations of the power unit of the car arising from the impact of road irregularities *Proceedings of NAMI* **145** pp 41-54

[3] Leliovsky K Y et al. 2019 Vibration load of transmission units at vehicle's motion over different roads. *Journal of Physics: Conference Series* **1177** 012053 doi:10.1088/1742-6596/1177/1/012053

[4] Tolsky V E 1988 *Vibroacoustics of the automobile* Moscow: Mashinostroenie p144

[5] Zezyulin D et al. 2014 Modeling of roads impacts for life prediction of light commercial vehicles parts *FISITA 2014 World Automotive Congress*

Modern methods of field tests of driver assist systems

Ye I Toropov[1], Yu P Trusov, A S Vashurin, P S Moshkov, V I Filatov and M E Gnenik

[1]Nizhny Novgorod State Technical University n.a. R.E. Alekseev, Minin st., 24, 603950, Nizhny Novgorod, Russia

E-mail: [1]evgeny.toropov@nntu.ru

In view of stringent vehicle safety requirements the automakers are urged to develop and implement electronic driver assist systems (ADAS). However, before starting the mass production, one has to make sure that these systems function correctly and efficiently. The current level of technologies, the simulation modelling allows debugging and testing of operation algorithms for electronic 'assistants' in laboratory conditions thereby reducing the costs of the prototype production. This approach is employed at intermediate development stages of the system. At the final stage the 'traditional' approach (field tests) is used by the automobile manufacturers, the reason being that the algorithm operation has to be tested in real conditions, and the certification of vehicles with ADAS functionality using methods of simulation modelling is not provided in current state standards. Keeping this in view, the problem arises of creating a measuring unit for geolocation of the vehicle with precision accuracy and synchronized video recording from multiple cameras simultaneously. This article presents the solution developed by employees from NNSTU n.a. R. E. Alekseev as regards the development of such a test system: a schematic diagram of the system is developed; the system is installed on a light commercial vehicle being tested; the system efficiency is checked. The system performance was checked when testing blind spot detectors, lane departure warning system and parking assistant.

Introduction

Over the recent decades, vehicle manufacturers are focused on the development of electronic active safety systems: anti-lock braking system (ABS), traction control system (TCS) and electronic stability control system (ESC). These functions allow the driver to maintain and exercise control over the vehicle in critical situations. However, given the advance of microelectronics and reduction in prices, access to technologies earlier unavailable to civil industries (radars, lidars, ultrasonic sensors, cameras, digital vision), the automakers are urged to implement the vehicle control automation. This research and development work resulted in the manufacture of mass production samples of auxiliary active safety systems, so-called assistants (DAS/ADAS). These systems can warn the driver of negative changes leading to a critical situation, and also interfere ahead in the vehicle control including the steering, braking, engine and transmission control systems [1, 2].

In the passenger and freight transport segment where light commercial, medium-duty and heavy-duty vehicles are used, the development and implementation of intelligent driver assist systems holds the highest potential; these systems are designed for monitoring 'blind' spots which is most critical for trucks, as well as preventing driving mistakes due to inattention, temporary distraction and physical fatigue of the driver [2-4]. The following systems meet these requirements:

– Lane Keep Assistance system (LKA). Its main function is to alert the driver to the vehicle departure from the lane, and adjust the vehicle motion path if the driver's actions are inadequate or none.

– Park Assist System (PAS) assists the driver in searching for a parking spot, and also controls the steering system [5].

– Blind Spot Monitoring (BSM) alerts the driver if the vehicle is a blind spot, and prevents collision with preceding traffic [6];

Implementation of these systems allows improvement of the traffic safety and brings down the costs of transportation companies related to vehicle idle time due to traffic accidents.

Today the technologies of software-based simulation *Model-in-the-loop* (MIL), *Software-in-the-loop* (SIL), and software and hardware-based simulation *Hardware-in-the-loop* (HIL) are most commonly used for testing of electronic systems thereby reducing the prototype production costs and helping to avoid mistakes at early stages of development. Despite all the benefits virtual tests cannot in full measure substitute 'full-scale' trials therefore for many years field tests have been the main development phase, and results of the tests a key factor for a decision whether or not the vehicle is ready for the market [7,8].

In this regard, the field testing methods of electronic driver assist systems as well as the necessary measuring equipment are still relevant.

Regulatory documents on functional characteristics of ADAS system

As of now, there are a number of documents in the Russian Federation regulating the requirements for electronic systems functions and describing the methods of their testing. Let us analyse each system separately:

1. Lane departure warning system. The system functioning is described in GOST R 58807-2020 whereby the following parameters are to be recorded:

– travel speed of the vehicle;
– speed of the vehicle leaving the traffic lane;
– time of the optical sensor actuation;
– position of the test object relative to the marking line;
– time of actuation of the malfunction indicator [9].

2. Assist the driver to decide on changing travel lanes. The system functioning is regulated by GOST R 58803 -2020 (the relevant part is taken from UN Regulation No. 130 Uniform provisions concerning the approval of motor vehicles with regard to the Lane Departure Warning System).

To evaluate the system efficiency, the following parameters are recorded:

– travel speed of the vehicle;
– steering effort;
– lateral acceleration;
– time intervals;
– time of actuation of the optical sensor activating the function, and the malfunction indication;
– video recording of the driver's actions (operating the turn signal lever);
– distance between the auxiliary (approaching) vehicle and the test object [10].

3. Lane keep assistance system (LKA) is regulated by GOST R 58804-2020. Measured parameters:

– travel speed of the vehicle;
– steering effort;
– lateral acceleration;
– time intervals;
– time of actuation of the optical and acoustic sensors activating the function, and the malfunction indication;
– video recording of the driver's actions [11].

4. Blind spot monitoring - GOST R 58808-2020. Measured parameters:

– travel speed of the vehicle;
– approach speed of the test and auxiliary vehicle vehicles;
– time intervals;

– longitudinal distance and lateral separation between the test vehicle and auxiliary vehicle;
– time of actuation of the optical sensor, and malfunction indication;
– video recording of the driver's actions [12].

5. Park Assist System is regulated by PNST 381-2019 (associated standard ISO 16787:2017 - Intelligent transport systems - Assisted parking system (APS) - Performance requirements and test procedures). Measured parameters:
– travel speed;
– lateral separation from parked vehicles;
– course angle distance between the test object and parked vehicles;
– time of actuation of the optical sensor, buzzer, and malfunction indication;
– test vehicle position given the parking spot geometry;
– time intervals;
– time of actuation of the optical sensor, and malfunction indication [13].

As can be concluded from the above requirements, the measuring unit installed on the test vehicle should have the following functionality:
– geolocation of the test object with precision accuracy including the auxiliary vehicle for a number of some tests;
– connectivity to on-board data bus CAN for recording time delays in human-machine interface operation;
– receiving video imagery with an option of synchronizing the recording with third-party equipment.

Development of measuring unit

Distances between moving test vehicles that is travelled distances, separation and intervals, cannot be measured with precision accuracy using 'traditional' equipment (e.g. optical travelled distance sensor, 'fifth wheel', linear laser distance sensor). The growing demand of the automakers for ADAS functions testing equipment urged the developers of measuring equipment to use the technologies earlier employed in the civilian sector only for geodetic measurements.

For maximum precision and convenience of ADAS-functions testing, advanced GNSS technologies are currently used by manufacturers of measurement systems. The vehicle positioning with 2 cm precision is achieved by using real-time kinematic technology (RTK). It is common knowledge that a single satellite device of any class cannot provide high precision positioning due to a large number of negative factors (atmospheric inhomogeneities, interference from stationary and mobile objects, indirect reflection or multipath propagation of signals), therefore, more accurate coordinates are determined by applying so-called 'RTK corrections' which are computed and transmitted by another GNSS-receiver (base station) the exact position of which is known (figure 1) [6, 14].

There are several ways to transmit RTK corrections from the base station to the receiver (rover):

1. GPRS channel – the downside of this method is that each receiver should have Internet connectivity which is not available everywhere;

2. GSM connection – it allows only 1 rover to be to connected, other drawbacks are similar to the first method;

3. A radio channel is the preferred option as the number of connected rovers is not limited, and the radio channel is not affected by the quality of cellular communication. The only downside of this option is that in the Russian Federation (as in many other countries) restrictions apply to the power rating of a radio frequency signal source for unlicensed use: for operating frequency 433 Mhz the maximum power rating shall not exceed 10 MW, for 2.4 Ghz frequency – 100 mW, given the maximum antenna suspension height of 10 meters. As evidenced from the experience, these power characteristics allow the coverage of a 500-600-meter radius area (subject to line-of-sight availability). A larger radio signal coverage area involving the application of a more powerful transmitter would require going through the procedure to obtain a license for the radio frequency and register the radio frequency signal source that can take as long as 6-8 months. It is practical if tests are always performed in the same location (vehicle test track); if the test track location is changed, all permits shall be renewed.

IOP Conf. Series: Materials Science and Engineering 1086 (2021) 012027 doi:10.1088/1757-899X/1086/1/012027

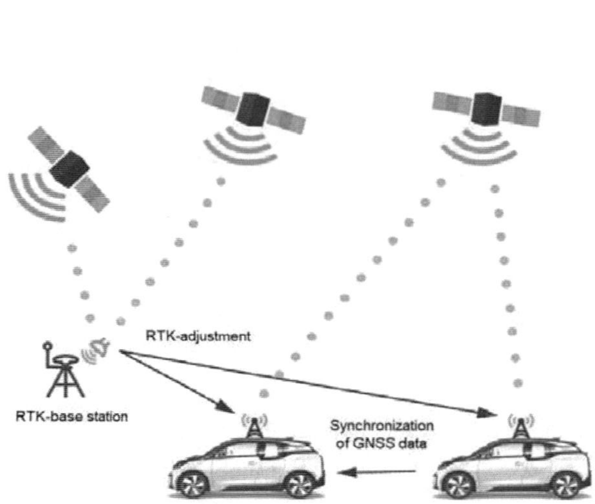

Figure 1. Schematic diagram of GNSS measurements for ADAS testing

Acquiring the exact coordinates from each of the rover receivers is not sufficient for testing and validating ADAS systems; it is necessary to calculate the test vehicle dynamic parameters (shortest, longitudinal distances and lateral separations, approach speed etc.) relative to auxiliary vehicles or the marking line. For this purpose, each vehicle is equipped with an additional set of telemetry equipment for synchronizing GNSS data on the target object (figure 1).

The specialists of the Nizhny Novgorod State Technical University were set a task to develop and implement a measuring unit for testing and validation of ADAS functions of a light commercial vehicle. The schematic diagram is shown in figure 2.

The main components of the measuring unit include:

1. High-precision measuring GNSS device Racelogic Vbox 3i 100 Hz (item 3, figure 2). It is installed on each vehicle for computation of the exact position, dynamic characteristics and ADAS functions evaluation parameters.
2. GNSS base station (item 12) is required for updating the position of the measuring instruments (with 2 cm accuracy). A local radio communication link is set up for transmission of RTL-corrections, that is the base station and all rover receivers are equipped with radio modules operating at 2.4 GHz frequency (item 6).
3. Additional 2.4 GHz telemetry (item 4) synchronizing GNSS data is required for calculation of ADAS parameters such as the relative distance between test vehicles, approach speed etc.
4. Racelogic CAN02 interface (item 1) is connected to the tested vehicle data bus to record the operation status of ADAS functions as well as a number of other parameters (engine rpm, operating brake and accelerator pedals, turn signal indication, engaged gear number etc.).
5. Speed display (item 8) showing the actual speed of the vehicle.
6. Multifunctional display MFD (item 9) allows displaying any of the measured parameters. For example, it is required for monitoring lateral separation between vehicles.
7. Racelogic Video Vbox Pro (item 10) is a GNSS device that can record a video streams from 4 video cameras simultaneously; it is required for recording the operation of light indicators, the vehicle crossing the marking line, video recording of the driver's actions etc. It is connected to Racelogic CAN bus (item 11) for synchronizing the recording with Vbox 3i.
8. Measuring steering wheel MSW (item 2). It is mounted right on a standard steering wheel and required for measuring the steering angle, rotation speed and torque. The measured data is transmitted to Vbox 3i via Racelogic CAN bus as well.

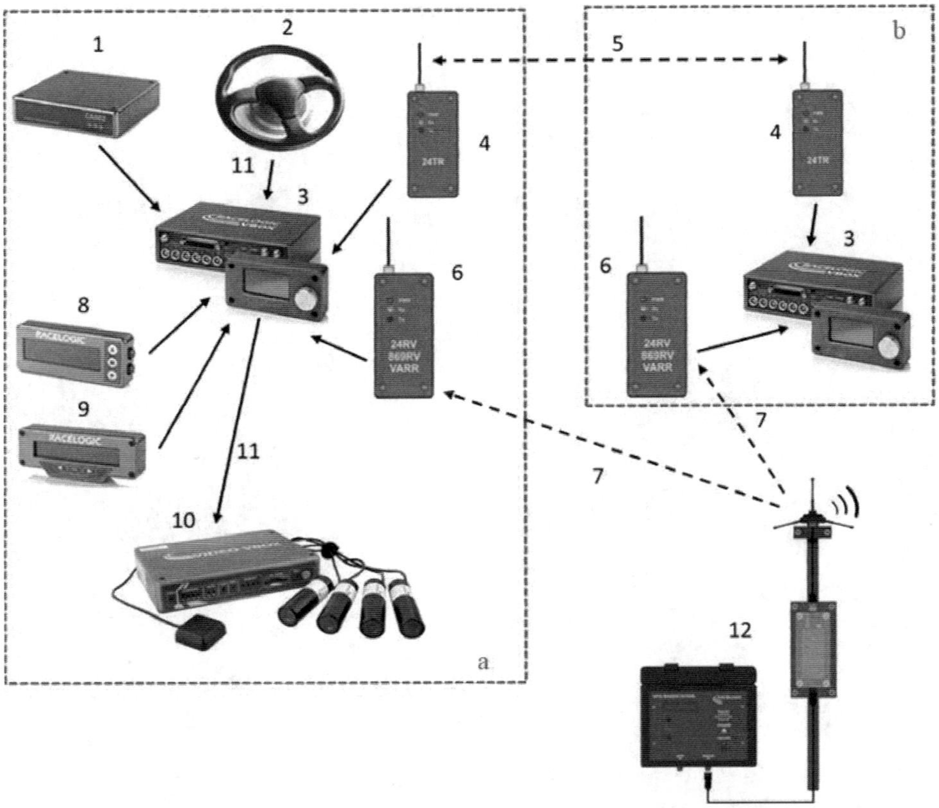

Figure 2. Schematic diagram of the measuring unit. *a* – component to be installed on the test vehicle; b – component to be installed on the auxiliary vehicle; 1 – CAN-interface to be connected to on-board data bus; *2* – measurement steering wheel (MSW); *3* – measuring GNSS device Racelogic Vbox 3i 100Hz RTK; *4* – telemetry 2.4 GHz for synchronizing GNSS data between the vehicles; *5* – radio communication link 2.4 Ghz for synchronizing GNSS data; *6* – radio receiver for RTK corrections; *7* – radio communication link 2.4 Ghz for transmission of RTK corrections; *8* – vehicle speed display; *9* – multifunctional display MFD; *10* – Video Vbox Pro; *11* – Racelogic CAN-bus; *12* – GNSS base station with 2.4 GHz telemetry for transmission of corrections.

The measuring equipment is mounted on the main and auxiliary vehicles as shown in figure 3 (a-e). Figure 3f. Location of GNSS base station with 2.4 GHz radio transmitter and the antenna mounted on on a telescopic rod. It is to be noted that given the maximum power 60 MW and the antenna suspension height 5 m, this radio frequency signal source can be used without any additional licenses.

Testing of ADAS lane detection functions require no auxiliary vehicles. In place of an auxiliary vehicle, lane marking lines are traced with geodetic accuracy using the 'lane marking system' (figure 4) and subsequently uploaded into measuring device Vbox 3i mounted on the test vehicle. This allows the calculation of driving parameters as required for assessment and validation the ADAS function: the distance from the wheels to the marking line and the speed of the vehicle leaving the lane.

IOP Conf. Series: Materials Science and Engineering 1086 (2021) 012027 doi:10.1088/1757-899X/1086/1/012027

a) b) c) d) e) f)

Figure 3. Mounted measuring equipment. a – Racelogic Vbox 3i 100Hz mounted in the main vehicle cabin; b – GPS/GLONASS equipment and telescopic antennae; c – Racelogic Vbox 3i 100Hz mounted in the auxiliary vehicle cabin; d – Video Vbox Pro cameras recording the right front wheel crossing the marking line; e – measurement steering wheel (MSW), Vbox display, Video Vbox cabin camera; f – GNSS base station with telemetry package.

Figure 4. Tracing of marking line

Testing of ADAS functions

The developed measuring system was tested during trials of the light commercial vehicle prototype equipped with driver assistance systems. The tests were conducted at the test ground of Berezovaya Poima Joint Research Center including the following functions:

– lane departure warning;
– blind spot monitoring;
– searching for a parking lot.

Some results of blind spot monitor function (BSM) testing are presented for the purpose of clarity. Test mode: the test vehicle travels at a speed of 30-35 km/h, the auxiliary vehicle accelerates to 40-45 km/h and overtakes on the blind side of the test vehicle. The objective of the test is to determine the time of actuation and operation of BSM assistant light and sound alarms. In addition to statistical data, video imagery was recorded using Racelogic Video Vbox Pro equipment. A video frame sample is shown in figure 5. As you can see, in addition to the option of synchronous video recording from 4 cameras at the same time, the software allows the application of virtual markings for evaluation of external objects' location. It is to be noted that Video Vbox features CAN interface with an option to record 32 additional channels received via the data bus: all numerical values in the video frame (figure 5) (BSM sound-optical indicator status, location of test vehicles relative to one other as well as contact points, i.e. the points around the vehicle outline corresponding to the minimum distance between test objects) are acquired using this method.

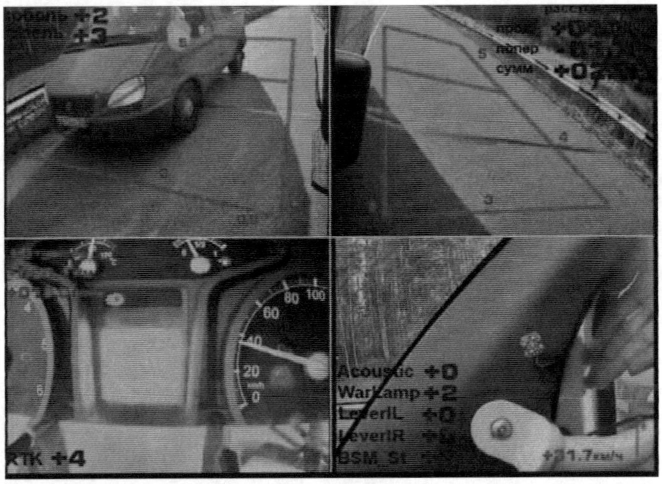

Figure 5. Vbox Pro video frame

Racelogic Vbox 3i 100 Hz measuring instrument is used for computation and recording of ADAS parameters including:

– travel speed of both test objects (figure 6a);
– distance between the vehicles: shortest and longitudinal distances and lateral separation (figure 6b);
– contact points of both vehicles (figure 6c). When setting up the measuring system, the contour of the test vehicles is marked (so-called contact points), i.e. the arrangement of markers around the outline of the vehicle relative to the GPS/GLONASS measuring antenna. In our case, these 4 points are arranged as follows: the first marker in the left front corner, with other markers numbered counter-clockwise;
– approach speed and accelerations of the objects relative to one another;
– angular distance between courses of the vehicles.

According to the test results, a set of statistical material is generated that allows ADAS system algorithm operation to be assessed (accuracy of functioning absence of false positives), validate the algorithm operation, check for compliance with high international and domestic requirements which is an integral part of development of any product.

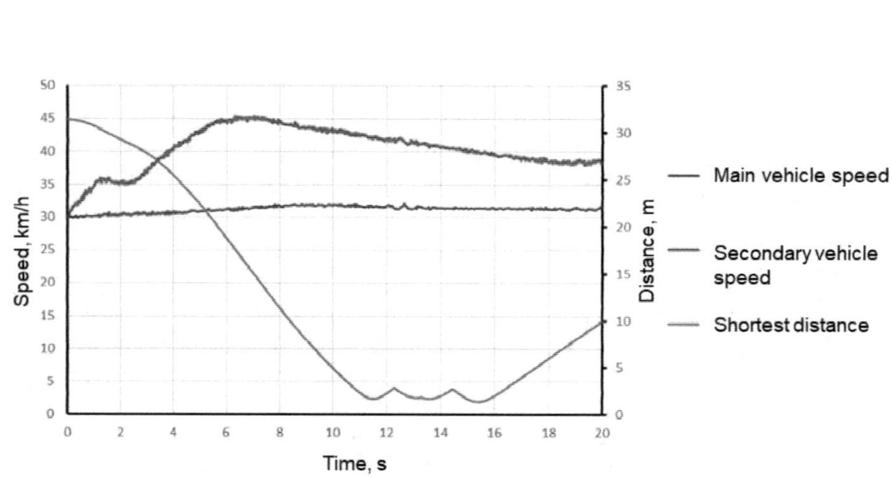

a)

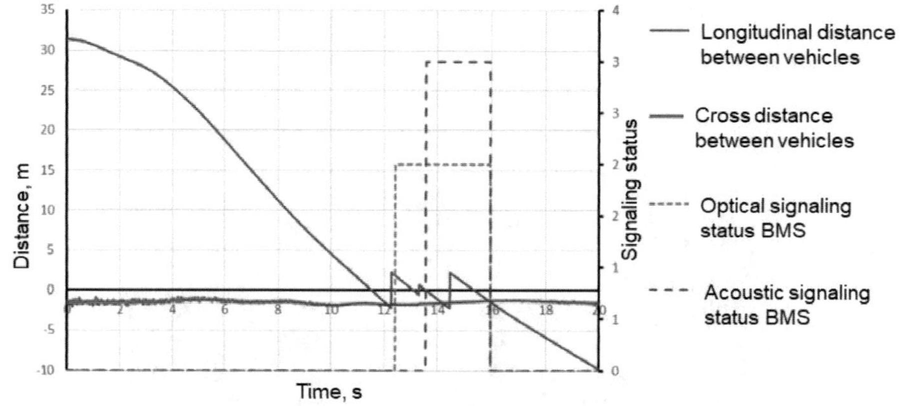

b)

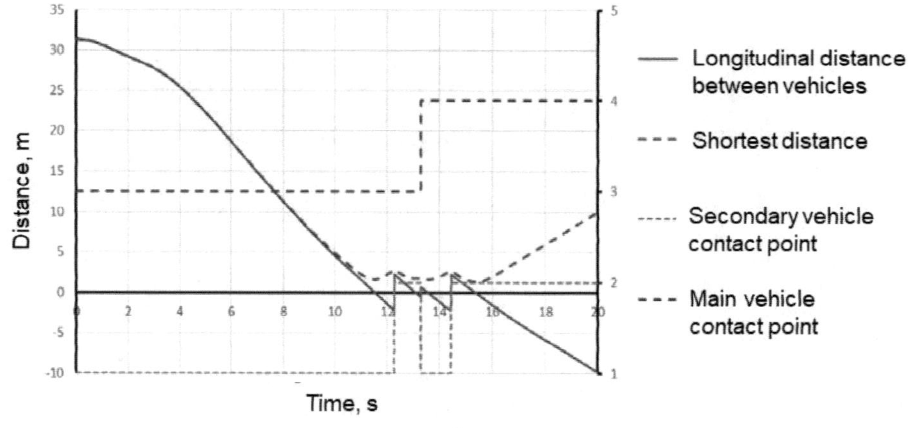

c)

Figure 6. BSM system test graphs. a – travel speed graph of the main and auxiliary vehicles; b – distance and lateral separation between the vehicles; BSM operation status; c – shortest distance between the vehicles; contact points

Summary

The developed measuring system was successfully test run when testing the driver assistance systems of the first series prototype. The results obtained in the course of the tests will be used for fine-tuning and debugging of the high-level algorithm of the electronic assistants. Also, these results will be again used for testing second- and third-series prototypes.

It is to be noted that this measuring system is compliant with international and domestic regulatory documents therefore the system can be used for certification tests of ADAS systems.

This measuring equipment potential is not limited to testing of the systems reviewed in this article; it can also be used for testing the Adaptive Cruise Control system (ACC), Autonomous Emergency Braking system (AEB), both with moving and stationary objects, Forward Collision Warning (FCW) and Traffic Sign Recognition (TSR).

These studies are conducted under the sponsorship of the Ministry of Education and Science of the Russian Federation as part of the project High-end production of the product range of GAZelle Next vehicle fitted with new electronic systems architecture under Agreement No. 075-11-2019-027 dated 29.11.2019 (Decree No.218 dated April 09, 2010 of the Government of the Russian Federation), using the measuring equipment of Transport systems, NNSTU shared resources centre.

References

[1] Prikhodko V M, Ivanov A M, Borisevich V B and Shadrin S S 2017 Methods of testing of the vehicle automated control systems *News bulletin of MADI* **14** pp 10-15

[2] Porubov D M et al. 2018 Development of the vehicle lane detection and keep system *Transactions of NNSTU n.a. R. E. Alekseev* **2** pp 25-29

[3] Tumasov A V et al. 2020 Development of the driver parking assist system for commercial vehicles *Transactions of NNSTU n.a. R. E. Alekseev* **3** p.p.132-140

[4] Makarov V et al. 2018 Development of driver's assistant system of additional visual information of blind areas for gazelle next *IOP Conference Series: Materials Science and Engineering* **315(1)** doi:10.1088/1757-899X/315/1/012016

[5] Shashkina K M et al. 2019 Development of the electronic steering unit to be integrated in driver assist systems of commerical vehicles *Science and business: development trends* **8** pp 15-22

[6] Hun-Jung Lim and Tai-Myoung Chung 2012 RP-ADAS: Relative Position-Advanced Drive Assistant System based on VANET (GNSS) *World Academy of Science, Engineering and Technology International Journal of Electrical and Computer Engineering* **6-3** pp 397-402

[7] Mardoyan G R et al. 2018 Modern approaches to testing ADAS systems at all stages of development *Transactions of NNSTU n.a. R. E. Alekseev* **4** pp 233-242

[8] Krejci P 2013 Testing of Automotive Park Assistant Control Unit by HIL Simulation *Mechatronics* pp 487-493 doi: 10.1007/978-3-319-02294-9_62

[9] GOST R 58807 2020 *Motor vehicles. Lane departure warning system. General technical requirements and test methods* M. Standardinform p 7

[10] GOST R 58803 2020 *Motor vehicles. Lane change driver assist systems. General technical requirements and test methods* M. Standardinform p 15

[11] GOST R 58804 2020 *Motor vehicles. Lane keep assistance systems. General technical requirements and test methods* M. Standardinform p 11

[12] GOST R 58808 2020 *Motor vehicles. Blind spot monitoring systems. General technical requirements and test methods* M. Standardinform p 15

[13] PNST 381 2019 *Intelligent vehicles. Park Assist Systems. Technical requirements and test methods* M. Standardinform p 22

[14] Kwanghee Lee, Jeong-Hyeon Bak and Chul-Hee Lee 2012 A Study on Vehicle Lateral Motion Control Using DGPS/RTK *IRAM 2012* CCIS 330 pp 159-16.

IOP Conf. Series: Materials Science and Engineering 1086 (2021) 012028 doi:10.1088/1757-899X/1086/1/012028

Natural nanomaterials as a basic for improving the environmental safety of transport systems

D O Chernyshev[1], L V Denezhko[2] and E E Bazhenov[2]

[1] Ural State Forest Engineering University, Siberian highway 37, Ekaterinburg, Russia
[2] Ural State Agrarian University, Karl Liebknecht street 42, Ekaterinburg, Russia

E-mail: [1]den_is-best@mail.ru

Abstract. Presented ways to improve the environmental safety of motor vehicles using liquefied petroleum gas and natural nanotubes - chrysotile asbestos. The basic components of petroleum gases, their physico-chemical properties, the percentage of fuel in the gas and the effect on engine operation, and the advantages and disadvantages are designated gaseous fuel. Presented defects in the operation of vehicles with gas-cylinder units associated with the violation of integrity of pipelines and systems. It specifies the requirements for the safe operation of the gas-cylinder cars. The aspects of the use of modern molecular techniques, the use of carbon nanotubes to solve the environmental problems created by cars.

1. Introduction

Perfection of the vehicle is determined not only using modern and advanced materials and technologies, not only equipment of automation and electronics systems, but also fuel-efficient and environmentally friendly, which ultimately characterize and technical and technological level of the country, which produces car.

There are several ways to increase efficiency and environmental vehicles: vehicle computerization, the use of cheap fuel with good performance properties, the use of modern technology for purification of exhaust gases from contaminants.

Affordable method to solve the problem of economic and efficient use of fuel resources while improving environmental performance is to create efficient designs LPG equipments and the widespread introduction of the gaseous fuel on trucks and passenger vehicles.

2. Experimental results

Gas fuel for internal combustion engines in Russia has been applied since 1974. This will save millions of tons of expensive liquid fuels. By creating and using LPG equipment operating characteristics of the vehicle should be different from the base (when operating on liquid fuel) is not more than 7%.

Liquefied petroleum gas - is lighter hydrocarbons, which is obtained by distillation, cracking or pyrolysis (high destructuring) petroleum products. Gases formed in the separation into fractions of associated and natural gases.

Petroleum gases at a relatively low excess pressure (1-2 MPa) and a wide temperature range are in a liquid state. The main components of petroleum gases are propane, butane and propylene.

The physicochemical properties of propane and butane differ significantly. The butane component is the most high-calorie and easily liquefied component, while propane and propylene ensure optimal vapor saturation in the gas cylinder. Even a slight change in the percentage of these gases in gas fuel affects the engine operating conditions. The composition of the gas fuel is regulated by GOST 57578-87. This standard provides for two grades of gas fuel: winter - PA (automobile propane), and summer - PBA (automobile propane-butane). PA contains 80 to 95% propane, and PBA 40-60%.

A feature of petroleum gases is that they are transported and stored in a liquid state, and are used as fuel in an engine - in a gaseous state.

Gas fuel must have good volatility and miscibility with air. This is necessary for the formation of a homogeneous and high-calorie combustible mixture with high anti-knock properties and a minimum content of resinous substances and mechanical impurities.

The main components of gas fuels (propane and butane) are colorless and odorless. To detect their leakage from the power supply system, odorants are added to the gas - substances with an unpleasant odor. Ethyl mercaptan is used as an odorant. This leads to the fact that the smell of gas is already felt at 0.5% by volume. Small amounts of odorants are completely harmless.

At atmospheric pressure, petroleum gases are non-toxic due to poor solubility in human blood. But, mixing with the air and getting into the human body, the components of the gas mixture reduce the oxygen content in the blood. Propane-butane mixtures, pouring out of the container in the liquid phase, evaporate, taking away heat, therefore, liquefied gas, getting on the human body, can cause frostbite.

An important advantage of gas fuel is its relatively low cost.

The operating time between overhauls of an engine running on liquefied gas is increased by 1.5 times compared to engines running on liquid fuel. Gas fuel does not wash off the oil film from the cylinder walls, which leads to improved lubrication conditions for the cylinder-piston group, does not give carbon deposits and varnish deposits in the engine and power system. When the engine is running on liquefied gas, the phenomenon of engine oil dilution and its contamination with coking products and other contaminants does not occur, which leads to an increase in the frequency of oil and oil filters change. The service life of spark plugs is significantly increased (up to 40%).

The octane number of gas fuel is in the range of 95 - 110 units. This makes it possible to use it on engines with high compression ratios (up to 12).

A huge advantage of gas fuel is its high environmental friendliness. Controlled toxic constituents in exhaust gases are significantly less than in gasoline engines: carbon monoxide (CO) by 3-4 times; hydrocarbons (CH) by 1.2-1.4 times; nitrogen oxides (NO) by 1.2-2 times. There are no harmful lead compounds in the gases. But these indicators are achieved due to the professional regulation of the gas supply equipment and its correct operation.

The combustion rate of the gas-air mixture is much lower than that of the fuel-air mixture. This leads to a noticeable decrease in engine noise (up to 10 dB), the engine runs smoother. This reduces the load in the crank mechanism, increasing its resource.

But not everything is so good. The use of gas fuel leads to a decrease in engine power by 5-7%. The reason is the lower heat of combustion of the fuel, a decrease in the filling factor and the speed of propagation of the flame front in the engine cylinders.

The starting qualities of a cold engine running on gas fuel at positive ambient temperatures and when the temperature drops to minus 5°C does not differ from starting the engine when operating on liquid fuel. At temperatures below minus 5°C, starting is difficult. It is usually recommended to start the engine with liquid fuel at low temperatures, and after warming up, turn on the power from the gas cylinder unit. A warm engine starts as if the ambient temperature is positive.

When operating vehicles with gas-cylinder installations, most of the defects are associated with leakage of pipelines and systems or with ruptured membranes. The appearance of internal depressurization in gas equipment, in contrast to gasoline equipment, causes a decrease in the operational properties of the vehicle and reduces the safety of operation. Efficient and safe operation of gas-cylinder vehicles requires an integrated approach, including, first of all, the development of gas-cylinder valves that are simple in design and reliable in operation. Secondly, the creation of a network of specialized enterprises for the installation of gas equipment and its technical support. But the most important thing is the training of highly qualified specialists of all levels, capable of ensuring competent operation and repair of equipment.

To purify exhaust gases from harmful impurities, it is advisable to use modern molecular technologies. Nanotechnology is little researched. The main discoveries predicted in this area have not yet been made. Nevertheless, the ongoing research is already showing good results. Nanotechnology is the next step in the development of high-tech industries; nanotechnology is the future in many areas of science and technology.

In developed countries, the results of work on nanotechnology are of great importance, which leads to the development of many programs for their development at the state level.

The technology for controlling the structure of matter from the atomic level makes it possible to obtain objects such as nanotubes and fullerenes of two types: organic (carbon and polymer) and inorganic. [1, 2]

Carbon nanotubes are a unique and promising material. Various products can be made from them. They are superior to traditional materials due to their developed surface and structural stability.

Carbon nanotubes are used in many fields. But they have not yet found wide application in ecology. And ecology plays one of the important roles in the modern world. The problem of air pollution in large cities and industrial centers is especially acute. The main source of pollution is road transport.

All over the world, they are trying to develop or find highly effective adsorbents that allow solving problems in the field of ecology (control of pollutants and methods of purifying water and air media).

Environmental problems can be combined in one direction based on the use of nanostructures - nanotubes. The use and use of carbon nanotubes will solve many environmental problems.

Nanotubes are peculiar cylindrical molecules with a diameter of about half a nanometer and a length of up to several micrometers. Nanotubes exhibit a whole spectrum of the most unexpected electrical, magnetic, and optical properties. Nanotubes can be conductors, semi-metals, and semiconductors. Superconductivity is also observed in them. Despite their apparent fragility, nanotubes turned out to be an extremely strong material, both in tension and bending. Moreover, under the action of mechanical stresses exceeding critical ones, nanotubes do not break, but rearrange their structure (Figure 1). [2, 4]

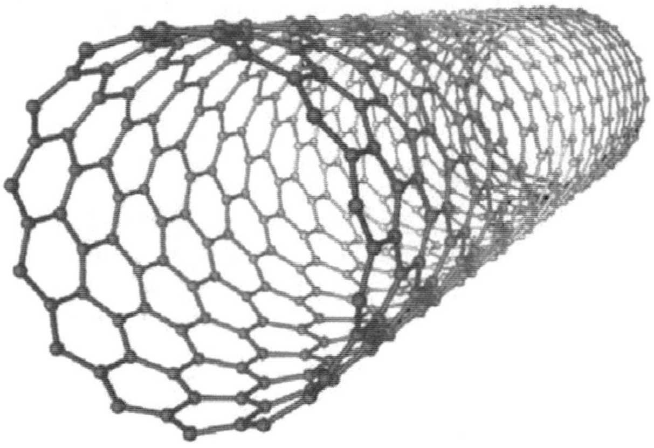

Figure 1. Nanotube structure

Inorganic nanotubes occur naturally in the form of minerals. One of these minerals is asbestos. There are two main types of asbestos - chrysotile asbestos and amphibole asbestos (Figure 2). [3]

Figure 2. Amphibole asbestos

Chrysotile asbestos is a fibrous mineral of the serpentine group, hydrous magnesium silicate $Mg_3Si_2O_5(OH)_4$. The chemical composition can vary depending on a particular deposit. Al, Fe, Ca, Ni, Mn, K, Na may be present in it as impurities. Chrysotile asbestos is the main type of asbestos used in industry today.

Chrysotile asbestos has a very interesting non-standard crystal structure (Figure 3).

Figure 3. Chrysotile asbestos

It consists of structural layers, which are bounded on the inside by a silicon-oxygen network, and on the outside - by a network corresponding to magnesium hydroxide. Since the size of the inner mesh is smaller than the outer one, the layers of chrysotile asbestos tend to roll into cylinders (tubes). [5]

In addition to fire resistance, resistance to acids and alkalis and other properties, the industrial value of asbestos is determined by the length of the fiber and its strength. So, according to the length of the fiber, chrysotile-asbestos is subdivided into 8 grades in Russia (from 0 to 7). For grade zero, the fiber length exceeds 13 mm, and for grade 7 - less than 1 mm. [3]

Chrysotile asbestos is used for the production of fire-resistant suits (for firemen), pipes and slates, and all kinds of cardboard and paper products. Iron-free chrysotile asbestos is an electrical insulator and is used in industry. The best long-fiber grades of chrysotile asbestos are used in the textile industry. The highest quality fiber is used to make filters.

None of the known asbestos substitute materials has the entire range of useful properties that chrysotile asbestos possesses [6, 7]:

- tensile strength over 3000 MPa (surpasses steel);
- density from 2.4 to 2.6 g/cm^3;

- melting point from 1450 to 1500 ^{0}C;
- coefficient of friction 0.8 units;
- alkali resistance from 9.1 to 10.3 pH;
- specific surface area 20 m^2/g.

Chrysotile asbestos is a naturally occurring nanosized tube with a spread in diameter, closely packed in a mineral and collected in bundles.

It should be noted that chrysotile asbestos is very similar in structure and properties to a carbon nanotube [9 - 11], and has many useful properties.

The main differences of chrysotile in comparison with carbon nanotubes [8, 12]:

- great rigidity;
- are initially multi-layered;
- tubes are formed open without fullerene hemispheres;
- the curvature (hence, the diameter of the nanotubes) is determined exclusively by the internal crystal structure;
- cheaper material.

3. Conclusion

Russia tops the world list of reserves and mining of natural and unique material chrysotile. Currently, in Russia and around the world, there is an acute problem of ecology, especially in motor transport. Scientists are trying to develop and find highly effective adsorbents that can solve problems in the field of ecology (control of pollutants and methods for purifying water and air media). The production and application of a new nanomaterial (chrysotile asbestos) can radically change the ecological situation.

References

[1] Eletskii A V 2002 Carbon nanotubes and their emission properties *Advances of Physical Sciences* (Moscow: UFN) p 401

[2] Eletskii A. V. 1997 Carbon nanotubes *Advances of Physical Sciences* (Moscow: UFN) p 954

[3] Lozovik Yu E and Popov A M 1997 Formation and growth of carbon nanostructures *Advances of Physical Sciences* (Moscow: UFN) p 263

[4] SRM 2483 Single-Wall Carbon Nanotubes 2013 (U.S. National Institute of Standards and Technology)

[5] ISO/TR 10929:2012 Nanotechnologies – Characterization of multiwall carbon nanotube (MWCNT) samples International Organization for Standardization

[6] Stando G, Lukawski D, Lisiecki F and Janas D 2019 *Intrinsic hydrophilic character of carbon nanotube networks* Applied Surface Science (Amsterdam: Elsevier) p 463

[7] Karousis N, Tagmatarchis N and Tasis D 2010 Current Progress on the Chemical Modification of Carbon Nanotubes *Chemical Reviews* (Vashington: American Chemical Society) p 5674

[8] WO16072959 2018 Method for Carbon Materials Surface Modification by the Fluorocarbons and Derivatives Patentscope.wipo.int.

[9] Haddon R, Zanello L, Zhao B and Hu H 2006 Bone Cell Proliferation on Carbon Nanotubes *Nano Letters* (Vashington: American Chemical Society) p 567

[10] Noyce S, Doherty J, Cheng Z, Han H, Bowen S and Franklin A 2019 Electronic Stability of Carbon Nanotube Transistors Under Long-Term Bias Stress *Nano Letters* (Vashington: American Chemical Society) p 1466

[11] Chin Wei T, Kok Hong T, Yit Thai O, Abdul Rahman M, Sharif Hussein Sharif Z and Soon Huat T 2012 Energy and environmental applications of carbon nanotubes *Environmental Chemistry Letters* (Switzerland: Springer Nature) p 1760

[12] Elishakoff I, Dujat K, Muscolino G, Bucas S, Natsuki T, Ming Wang C, Pentaras D, Versaci C, Storch J, Challamel N, Zhang Y and Ghyselinck G 2012 Carbon Nanotubes and Nanosensors: Vibration, Buckling and Balistic Impact (USA: WILEY) p 448

IOP Conf. Series: Materials Science and Engineering 1086 (2021) 012029 doi:10.1088/1757-899X/1086/1/012029

Optimization of traffic flow speed in urban conditions

E A Kindeyev

Department of Autotransport and Technosphere Safety, Vladimir State University n.a. Alexander and Nikolay Stoletovs, Vladimir, 87 Gorky Street, 600000, Russia

E-mail: kindeyev@yandex.ru

Abstract. The article is devoted to the problem of more complete use of the traffic network capacity in urban development. The aim of this article is to publish proposals for improving the efficiency of the existing traffic network to reduce the likelihood of traffic jams during high traffic hours. The results of the calculations showed that the maximum bandwidth of the traffic lane is achieved in a certain range of vehicle speeds and depends on the coefficient of adhesion.

1. Introduction

In urban conditions, the main task of the car driver is to get to the end point of his trip as quickly as possible, of course, ensuring the necessary road safety level. With an increasing number of private road users there is a situation in which the existing traffic network can no longer cope with the current traffic flow. To ensure the most complete usage of the existing traffic network efficiency, it is advisable to study the effect of the traffic flow speed on the lane capacity [1-3]. The study was conducted for conditions of steady traffic flow, without acceleration and braking.

2. Main text

As known [4], vehicle capacity is the maximum number of vehicles that can pass a given point during a specified period under prevailing roadway, traffic, and control conditions. Traffic lane vehicle capacity can be determined as [5]:

$$P = \frac{3600 \cdot V_a}{L_d}, \tag{1}$$

where

V_a is the speed of the traffic flow, m/s;

L_d is the safety clearance between successive vehicles, m.

The safety clearance can be defined as

$$L_d = l_a + S_0 + S_r, \tag{2}$$

where l_a is the length of the car, m; S_0 is the value of the stopping distance, m; S_r is the guaranteed reserve, m.

The guaranteed reserve S_r can be approximately considered equal to $0{,}1 \cdot V_a$ in meters, where V_a is the speed of the vehicle, measured in km/h. The stopping distance is determined for the vehicle without ABS emergency braking system [6]:

$$S_0 = V_0 \cdot (\tau_r + \tau_u + 0{,}5\tau_i) + \frac{V_0^2}{2 \cdot g \cdot \varphi}, \tag{3}$$

where S_o is the value of the stopping distance, m; V_0 is the initial velocity of vehicle in m/s; τ_r is the reaction time of driver, s; τ_u is the braking mechanism reaction time, s; τ_i is the deceleration rise time, s; g is the acceleration of free fall, m/s^2; φ is the coefficient of adhesion of the wheels with the road surface.

To solve the task, it is necessary to determine the dependence of the traffic lane capacity on the traffic flow rate in different road conditions with the appropriate values of the road adhesion coefficient: φ = 0.75 (dry asphalt concrete), φ = 0.55 (wet asphalt concrete), φ = 0.27 (rolled-up snow), φ = 0.09 (ice at a temperature of -5° C). For simplicity of calculations, we assume the following assumptions: car length l_a = 5 m, driver reaction time τ_r = 0.2 s, braking mechanism reaction time τ_u = 0,12 s, deceleration rise time τ_i = 0.2 s.

It is necessary to determine at what value of traffic flow speed in different road conditions the traffic lane capacity will be the largest (with an accuracy of 5 km/h). The total stopping distance for the traffic flow speed in various road conditions was calculated using the formula (3), the calculation results are shown in table 1.

Table 1. Total stopping distance in various road conditions

Speed (km/h)	Speed (m/s)	φ = 0.75	φ = 0.55	φ = 0.27	φ = 0.09
		Total stopping distance (m)			
5	1.39	0.72	0.76	0.95	1.68
10	2.78	1.69	1.88	2.63	5.54
15	4.17	2.93	3.36	5.03	11.6
20	5.56	4.44	5.20	8.17	19.84
25	6.94	6.19	7.38	12.01	30.19
30	8.33	8.21	9.93	16.6	42.79
35	9.72	10.5	12.84	21.92	57.59
40	11.11	13.05	16.10	27.97	74.57
45	12.50	15.87	19.73	34.75	93.74
50	13.89	18.95	23.71	42.25	115.09
55	15.28	22.28	28.05	50.49	138.64
60	16.67	25.89	32.75	59.46	164.37
65	18.06	29.75	37.81	69.16	192.30
70	19.44	33.85	43.19	79.50	222.18
75	20.83	38.23	48.96	90.65	254.47
80	22.22	42.89	55.09	102.53	288.94
85	23.61	47.80	61.57	115.14	325.60
90	25.00	52.97	68.42	128.48	364.45
95	26.39	58.41	75.62	142.55	405.48
100	27.78	64.11	83.18	157.35	448.71

If there are no experimental data on vehicle parameters, such as the road adhesion φ or the coefficient of braking performance, this is selected according to standard or average recommended values depending on the type and condition of the roadway surface. Now it is possible to calculate the traffic flow of the lane using the formulas (1) and (2). The calculation results are shown in the table 2.

Calculations show that at first, as the speed of traffic increases, the traffic lane capacity increases rapidly. However, quite quickly the traffic lane capacity reaches its highest value, after that it begins to decrease almost linearly and quite smoothly. The maximum traffic lane capacity occurs at a fairly low speed of traffic flow. The same calculations of the dependence of lane capacity on traffic flow speed for larger trucks and city buses were made. The following values were used for large trucks and city buses: car length l_a = 10 m, driver reaction time τ_r = 0.2 s, braking mechanism reaction time τ_u = 0.3 s, deceleration rise time τ_i = 0.3 s. The results show that the increase in the overall dimensions of vehicles and changes in the characteristics of braking mechanisms slightly changed the previously established patterns. The traffic lane capacity has decreased by about a quarter, and the values of the traffic flow speed at which the maximum traffic lane capacity is observed have increased by a third, as shown in tables 3 and 4.

Table 2. Traffic flow under various road conditions

Speed (km/h)	Speed (m/s)	$\varphi = 0.75$	$\varphi = 0.55$	$\varphi = 0.27$	$\varphi = 0.09$
		Traffic Flow (v/h)			
5	1.39	800	800	780	700
10	2.78	1300	1270	1160	870
15	4.17	1590	1520	1300	830
20	5.56	1750	1640	1320	750
25	6.94	1830	1680	1280	660
30	8.33	1850	1670	1220	590
35	9.72	1840	1640	1150	530
40	11.11	1810	1590	1080	480
45	12.50	1770	1540	1020	440
50	13.89	1730	1480	960	400
55	15.28	1680	1430	900	370
60	16.67	1630	1370	850	340
65	18.06	1580	1320	810	320
70	19.44	1530	1270	760	300
75	20.83	1480	1220	730	280
80	22.22	1430	1170	690	260
85	23.61	1390	1130	660	250
90	25.00	1340	1090	630	240
95	26.39	1300	1050	600	230
100	27.78	1260	1020	580	220

The traffic flow in localities is dominated by passenger cars, so the results of calculations obtained for passenger cars can be extended to the entire traffic flow. Discrepancies in the value of traffic flow do not exceed 5 percent and do not change the calculated functional dependence. Based on the above, the influence of large trucks and buses on the dependence of lane capacity on traffic flow speed can be ignored. The dependence of the traffic lane capacity on the speed of traffic flow, defined for a mixed urban traffic flow, is shown in figure 1.

IOP Conf. Series: Materials Science and Engineering 1086 (2021) 012029 doi:10.1088/1757-899X/1086/1/012029

Table 3. Large trucks and city buses total stopping distance

Speed (km/h)	Speed (m/s)	$\varphi = 0.75$	$\varphi = 0.55$	$\varphi = 0.27$	$\varphi = 0.09$
		Total stopping distance (m)			
5	1.39	1.03	1.08	1.27	2.00
10	2.78	2.33	2.52	3.26	6.18
15	4.17	3.89	4.32	5.99	12.54
20	5.56	5.71	6.47	9.44	21.09
25	6.94	7.79	8.98	13.62	31.82
30	8.33	10.14	11.85	18.53	44.74
35	9.72	12.74	15.08	24.16	59.85
40	11.11	15.61	18.66	30.53	77.14
45	12.50	18.74	22.60	37.62	96.61
50	13.89	22.14	26.90	45.44	118.27
55	15.28	25.79	31.56	53.99	142.11
60	16.67	29.71	36.57	63.27	168.14
65	18.06	33.89	41.95	73.28	196.36
70	19.44	38.33	47.68	84.01	226.76
75	20.83	43.04	53.76	95.47	259.34
80	22.22	48.00	60.21	107.67	294.11
85	23.61	53.23	67.01	120.58	331.06
90	25.00	58.72	74.17	134.23	370.20
95	26.39	64.48	81.69	148.61	411.52
100	27.78	70.49	89.56	163.71	455.03

Table 4. Large trucks and city buses traffic flow

Speed (km/h)	Speed (m/s)	$\varphi = 0.75$	$\varphi = 0.55$	$\varphi = 0.27$	$\varphi = 0.09$
		Traffic Flow (v/h)			
5	1.39	433.51	431.72	424.92	400.15
10	2.78	750.19	739.61	701.16	582.23
15	4.17	974.78	948.34	857.85	623.95
20	5.56	1129.40	1082.76	932.95	604.41
25	6.94	1232.06	1163.72	957.21	564.02
30	8.33	1296.68	1207.14	951.60	519.53
35	9.72	1333.69	1224.69	929.31	477.17
40	11.11	1350.80	1224.63	898.32	438.90
45	12.50	1353.65	1212.79	863.38	405.00
50	13.89	1346.37	1193.21	827.24	375.18
55	15.28	1331.96	1168.70	791.46	348.95
60	16.67	1312.61	1141.23	756.91	325.83
65	18.06	1289.92	1112.12	724.02	305.37
70	19.44	1265.07	1082.32	692.99	287.17
75	20.83	1238.91	1052.44	663.87	270.92
80	22.22	1212.05	1022.92	636.61	256.32
85	23.61	1184.95	994.04	611.14	243.16
90	25.00	1157.95	965.99	587.34	231.25
95	26.39	1131.26	938.87	565.11	220.41
100	27.78	1105.07	912.74	544.33	210.51

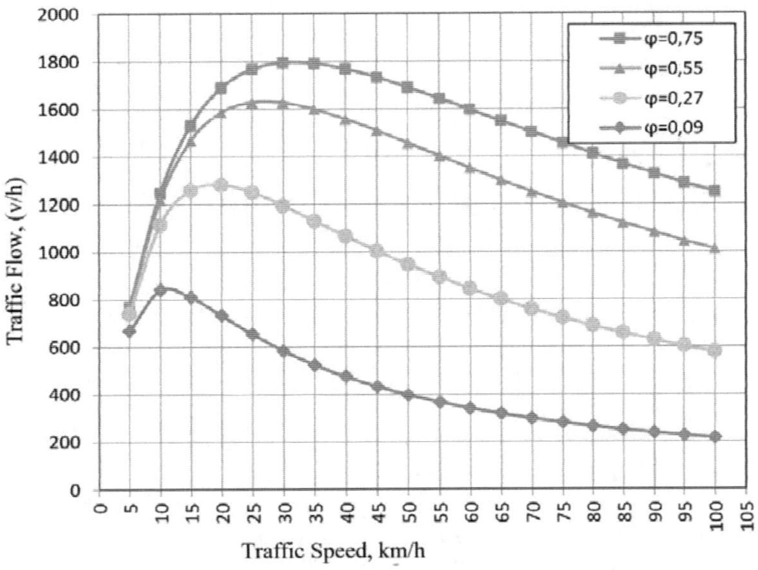

Figure 1. Traffic flow at different values of the road adhesion coefficient

For dry asphalt concrete (φ = 0.75), the largest traffic flow can be achieved at the traffic speed of 30 km/h. For wet asphalt concrete (φ = 0.55), the largest traffic flow can be achieved when the vehicle flow speed is 25 km/h. For rolled-up snow (φ = 0.27), the largest traffic flow can be achieved when the traffic flow speed is 20 km/h. For ice (φ = 0.09), the largest traffic flow can be reached at a speed of 10 km/h. Thus, in order to make better use of the capacity of the existing traffic network, it is advisable to reduce the maximum permitted speed in the city up to 30 km/h during rush hours.

References

[1] Markovnina A I et al. 2019 Imitation modeling as a forecast of transport flows self-organization *IOP Conference Series: Materials Science and Engineering* **632(1)** 012001 doi: 10.1088/1757-899X/632/1/012001

[2] Markovnina A I et al. 2019 Multicriteria optimization in the study of active vehicle safety *IOP Conference Series: Materials Science and Engineering* **695(1)** 012015 doi:10.1088/1757-899X/695/1/012015

[3] Markovnina A I et al. 2020 The employment of multi-criteria assessment methods in the study of road safety *IOP Conference Series: Materials Science and Engineering* **819(1)** 012003 doi:10.1088/1757-899X/819/1/012003

[4] Warne T R *et al* 2001 *A Policy on Geometric Design of Highways and Streets 2001* Washington: American Association of State Highway and Transportation Officials **2** p 74

[5] Zegeer J D et al. 2000 *Highway Capacity Manual 2000* Washington: National Academy of Sciences **7** pp 7-2–7-5

[6] Ginzburg G, Evtiukov S, Brylev I and Volkov S 2017 Reconstruction of Road Accidents Based on Braking Parameters of Category L3 Vehicles *Transportation Research Procedia* **20** pp 212–218

Possible solutions to the problem of recycling electric vehicle batteries on the example of global manufacturers

D A Kolesnichenko[1], D N Pechalinov, R V Vyatkin, A V Erofeeva

Nizhny Novgorod State Technical University n.a. R.E Alekseev, Minin str, 24, 603950, Nizhny Novgorod, Russia

Email: [1]D.kolesnichenko@nntu.ru

Abstract. The article discusses the methods of recycling and reusing of traction batteries of electric vehicles on the example of global automakers or companies specializing in this topic. Analyzed the legislative rules and regulations for the disposal of spent batteries in different countries. Warranties of electric vehicle's manufacturers were considered.

Introduction

In the global automotive market, the electric vehicle boom is growing. All leading automotive brands are investing actively in this area and have all-electric or hybrid vehicles in their product lines. Some governments are actively stimulating the sale and development of electric vehicles through subsidies or rather large tax breaks, and even point out the years when some countries will completely switch to electric cars. The wide usage of electric vehicles is necessary to protect the environment, primarily in large cities and megalopolises. Over time, this boom will come to Russia as well. All leading Russian vehicle manufacturers are developing modifications of vehicles on electric traction and organize the exploitation of electric prototypes. For example, in Moscow, there is an active introduction of electric buses.

The team of authors of this paper perform R&D in the field of developing of electric platforms for commercial vehicles. The team actively works with the GAZelle family of vehicles, recently a prototype of the electric platform for the GAZon Next was developed. In the process of development, the question of recycling and reuse of spent batteries is one of most difficult problems.

For all the obvious benefits of electric vehicles, their active development can lead to huge environmental problems. Batteries, which are the sources of energy in electric vehicles, have a certain resource, after which they must be properly disposed of, because they contain toxic substances. In Russia, the topic of recycling batteries is covered by extremely low scientific analytics.

Battery life can also be a big problem for electric vehicle consumers. Now the price for a battery is a significant part of the cost of an electric vehicle. For the customer, the possibility of an expensive replacement of the battery can become a serious problem and reason for refusing to purchase an electric car.

IOP Conf. Series: Materials Science and Engineering 1086 (2021) 012030 doi:10.1088/1757-899X/1086/1/012030

Methods of the analysis

The purpose of this work is to describe the methods of recycling batteries, reusing them on the example of few companies, including automobile manufacturers, as well as to study warranty offers for batteries from manufacturers of electric vehicles.

It is worth to underline that only a few countries have their own laws, rules and standards for the disposal and recycling of traction batteries. Clear rules are formulated only in the countries of the European Union and Japan, and they are also indirectly prescribed in Canada. China is only at the stage of developing such standards, including for the reuse of batteries in electric vehicles, while in the rest of the countries the rules are either completely absent or there is no car production at all (Figure 1).

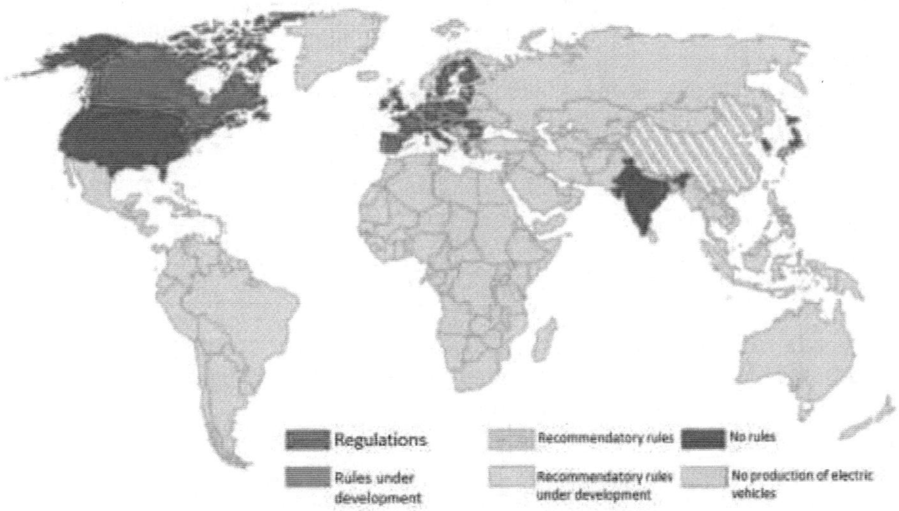

Figure 1. Requirements and regulations for recycling traction batteries in different countries

The fundamental documents in this area are the ISO 14000 series. In particular, the ISO 14040 - ISO 14043 standards provide for the assessment of the environmental performance of vehicles over the full life cycle, impose requirements to reduce the consumption of natural resources and energy, as well as the harmful impact on the environment at all stages life cycle of the vehicle. International standard ISO 22628 regulates the calculation of recycling rates and utilization of vehicles.

Results of the study

There are several large companies in Western Europe that are engaged in the disposal of traction batteries for electric vehicles. The technical process for recycling is reduced mainly to the recovery of metals - copper, nickel and cobalt. At the same time, lithium compounds are not processed due to the too high cost of such production, which is not covered by the utilization duty and makes such processed lithium much more expensive than extracted from the subsoil. Table 1 lists companies that work in the field of traction batteries recycling.

Table 1. European battery recycling companies

Name	Umicore	Redux	Duesenfeld	Fortun
Recast	+	+	+	+
Metal Restore	+	+	+	+
Battery reuse		Maintaining the electrical network		
Recycled part of the battery,%	70	70	up 91	over 80
Own battery production				+

This table confirms that nowadays traction batteries of electric vehicles are not 100% recycled. In addition to the disposal and recycling of used traction batteries, many companies are making great strides in reusing them. Some examples are described below.

One of the largest Japanese automakers Nissan and its subsidiary 4R Energy Corporation are planning to equip the streets of the Japanese city Namie with new streetlamps, which will be powered by a combination of spent batteries from an electric Nissan Leaf and solar panels. The project is called "The Reborn Light" and is supposed to provide lighting for the city to support the Namie recovery project after the 2011 tsunami.

The BMW Group has opened a battery repair and storage facility in Leizpig that can accept up to 700 batteries per day. This production should open new potential for cost savings and energy efficiency, as well as for reducing harmful emissions into the atmosphere. The German government actively supports such initiatives and provides its funding.

EVgo plans to use used traction batteries in its fast charging systems. Such a system is already in operation in Union City, USA, where batteries from the BMW i3 are used to reduce the load during peak hours in fast charging systems that are connected to the city's grid. The EVgo Union City system includes two chargers with fast charging capacity of 50 kW, as well as a secondary system that combines two BMW i3 batteries in one housing, which stores energy in the system with a capacity of 30 kW / 44 kWh.

Waste batteries from the BMW i3 electric car were connected to the British national power grid, forming one of the largest national energy storage facilities. It consists of six containers, five of which hold 500 batteries. In Ukraine, at one of the ski resorts, are going to use spent batteries for a backup power system. There are elevators in Paris that are powered by Renault electric vehicles. In Amsterdam, the stadium was powered by Nissan Leaf batteries.

As it can be seen there are many solutions for the reusing of traction batteries of electric vehicles. It is a good solution for storing and supplying energy as an auxiliary or emergency source. At the same time, as it mentioned above, there are big doubts for potential buyers of electric vehicles in case of possible risks of huge expenses for traction batteries replacement. In this regard, it is important to consider solutions and proposals for their customers from the world's leading manufacturers of electric vehicles. In accordance with such problem, it is worth to explain the warranty on batteries for new vehicles sold by various car manufacturers (Table 2).

Table 2. Battery warranty from leading manufacturers

Manufacturer	Mileage, km	Guarantee period, years	The cost of a new battery, $
Chevrolet	160 000, 240 00 (depends on the US state)	8, 10	15 700
SEAT	160 000	8	
Volkswagen	130 000	8	
FIAT	320 000	8	28 000
Tesla		8	16 000
Mitsubishi		10	
BMW	160 000	8	16 000
Renault		8	10 000
Nissan	160 000	8	6 200-7 800

As it can be seen from the table, almost all manufacturers of electric vehicles offer a guarantee for their batteries for a period of 8 years, while the possible mileage for such vehicles varies, but most often it is about 160,000 km, or 100,000 miles. At the same time, the cost of new batteries for different car brands varies greatly.

Naturally, the average consumer is not able to purchase a new battery at half the cost of the entire electric vehicle, so some manufacturers offer various solutions to reduce the price.

In 2018, the Japanese company Nissan launched a program at home to replace traction batteries in old Nissan Leaf electric vehicles, offering car owners the choice of a "refurbished" battery or a new one at a discounted price, while the old battery module had to be returned during replacement. Because its cost has already been included in the price of the replacement battery. The profit for the purchase of a new module was approximately 20%. In the USA, a similar company took place in 2016. Replacement work is carried out by licensed car dealers and service stations, which issue an appropriate certificate for the possibility of further registration of the car. If the traction battery is replaced with a new one, the car owner receives the same warranty that the Nissan Leaf has for 8 years or 160 000 kilometers.

French automaker Renault has introduced a new business model whereby, when selling an electric vehicle, the traction battery is rented to the owners of the vehicle, while the actual ownership of it remains with the manufacturer. After the battery has used up its resource, the manufacturer replaces them with new modules at the lowest price. The used batteries are then either "recovered" as spare batteries or recycled for reuse. Thanks to this approach, the cost of an electric vehicle is reduced by about 30%.

Summary

Manufacturers understand the problem of the high cost and relatively short resource of traction batteries for electric vehicles, so they are trying to offer consumers more favorable conditions to make electric vehicles more competitive in the market.

Based on the data reviewed, it can be concluded that, despite the little coverage of the issue of recycling and reuse of traction batteries, the world's leading car manufacturers are actively involved in this issue, including with government support. This gives great prospects for the development of electric transport, including Russia.

References

[1] *Proposal for an Electric Vehicle Regulatory Reference Guide* 2014 Economic Commission for Europe. United Nations

[2] *Battery recycling Umicore* URL: https://www.umicore.com (date of the application: 02.08.2020)

[3] *Battery recycling Redux* URL: https://www.redux-recycling.com (date of the application: 02.08.2020)

[4] *Battery recycling Duesenfeld* URL: https://www.duesenfeld.com/index.html (date of the application: 03.08.2020)

[5] *Battery recycling Fortum* URL: https://www.fortum.com/ (date of the application: 03.08.2020)

[6] *Batteries for Renault Zoe will be sold on credit* URL: https://www.abw.by/novosti/in_world/144059 (date of the application: 10.08.2020).

[7] *BMW i3 electric car battery life will get a second life* URL: https://3dnews.ru/935026 (date of the application: 11.08.2020)

[8] *Tesla Energy: rechargeable batteries for households and corporations* URL: https://3dnews.ru/913577 (date of the application: 12.08.2020)

[9] *How much does it cost to replace a battery in an electric car Chevrolet Bolt EV* URL: https://hevcars.com.ua/skolko-stoit-zamena-akkumulyatora-v-chevrolet-bolt-ev/ (date of the application: 13.08.2020)

[10] *SEAT Mii Electric review: interior, specifications, equipment* URL: https://akbinfo.ru/e-transport/seat-mii-electric-kompaktnyj-gorodskoj-elektromobil.html (date of the application: 15.08.2020)

[11] *Mitsubishi i-MiEV 2009* URL: https://elektro-mobili.ru/obzor/mitsubishi-i-miev/ (date of the application: 17.08.2020)

Predicting modeling of the service properties of carrying systems in the articulated-vehicle-based self-propelled artillery systems

E E Bazhenov[1], A N Kustovsky[2] and M L Usupov[3]

[1] Ural State Agrarian University, Karl Liebknecht street 42, Ekaterinburg, Russia
[2] Bauman Moscow State Technical University, Baumanskaya second street, 5/1, Moscow, Russia
[3] Ural State Agrarian University, Karl Liebknecht street 42, Ekaterinburg, Russia

E-mail: [1] st194@yandex.ru

Abstract. The work is devoted to the possibility and prospects of the use of articulated-vehicle-based self-propelled artillery systems in order to increase their effective firing rate abilities. The reasons for creating such systems are presented. A mathematical model and its description are given. The high convergence of the analytical and the experimental data shows the adequacy of the established mathematical model. The use of the articulated-vehicle construction in self-propelled artillery systems makes it possible to reduce the time between two shots by about 2.5 seconds. Work conditions of the crew are also substantially improved.

1. Introduction

The economic situation in the country and the political situation in the world sets tasks for the designers and manufacturers of self-propelled artillery, the solution of which will, on the one hand, provide a high degree of tactical, operational-tactical and strategic mobility, increase the firepower of self-propelled artillery pieces, and on the other hand, create products competitive in the world market.

One of the competitive advantages is the maximum customization of the products to the conditions of the buying country. Potential buyers of self-propelled artillery pieces are the countries of the Middle East. The conditions in which the equipment will be operated imply the use of wheeled running gear, which will undoubtedly increase the mobility of the machines [1,2].

The ammunition load of a self-propelled artillery piece (SPAP) is limited. An increase in it can be provided by the creation of articulated systems as part of SPAP and a loader vehicle, the ammunition load of which is three times higher than that of a single vehicle.

The effective rate of fire of SPAP largely depends on the time that elapses from the moment of the shot until the stop of hull oscillations and aiming recovery [3]. To achieve it, the elastic suspension elements must, on the one hand, perceive and damp the recoil energy from the shot, and on the other hand, have a high degree of dissipation to damp oscillations when the system is moving over the terrain [4, 5].

Ural State Federal University in conjunction with the central design bureau (CDB) Transmash has developed a complex mathematical model of the articulated vehicle-based SPAP which allows considering the SPAP-body oscillation process after the shot both for the SPAP as a single unit and in tandem with loader vehicle. Mathematical model also allows to determine a suspension parameters and elastic characteristics of the articulation joint that are required for increasing the effective rate of fire. If the SPAP-layout includes a loader vehicle, the mathematical model helps to estimate to what degree the oscillation damping time is affected by the presence of the second section (loader vehicle).

2. Theoretical part

The construction under consideration is characterized by a number of elastic constraints, control and self-adjustment elements and also by an accidental character of external and its internal forces [6]. The suggested approach allows to change the step of integration for each differential equation depending on motion character of any object without increasing the resources used, thus improving the quality of studied model significantly [7, 8].

Motion equation of the system:

$$\Theta\ddot{\vartheta} = F + R, \tag{1}$$

where ϑ - canonical coordinates for inertia forces;

Θ - inertia coefficients in the direction of canonical coordinates;

F – external forces reduced to canonical coordinates;

R – reduced internal forces;

The equation (1) can be represented in form of the tripartite graph (figure 1).

The graph defines the structure of the program during the integration of the system. In accordance with the graph structure it is required to elaborate the connections and the tops of the middle lobes of the graph. The scanning of the connections of the graph should be performed primarily between the middle and the left lobe, then between the middle one and the right one.

The interaction between the middle lobe tops is carried out in the conditions described in the left lobe. The internal action forces R represent a general class [9-11]:

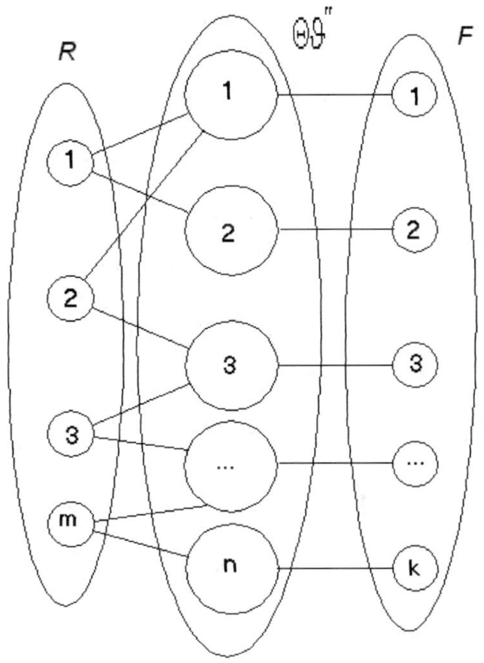

Figure 1. Tripartite graph for the motion equation

IOP Conf. Series: Materials Science and Engineering 1086 (2021) 012031 doi:10.1088/1757-899X/1086/1/012031

The figure 2 (a, b) represents the analytical model of the first and the second section of the articulated vehicle-based SPAP [12, 13]. On the figure 2 the force $P_ц$ represents the load on the trunnion pin from the gun during the shot. The force vector direction can change up to 360^0 about a Z_1 axis. During tactical operations a firing sector is considerably narrower and changes in a range of $\pm45^0$ about Z_1 [11].

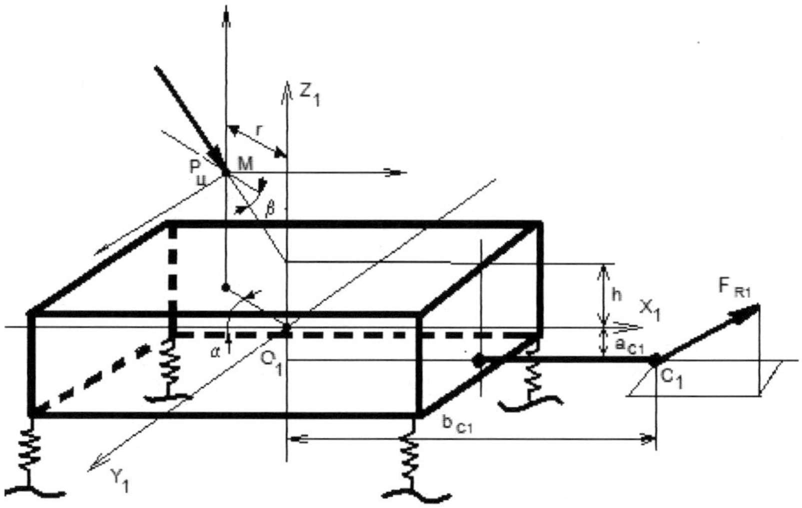

a

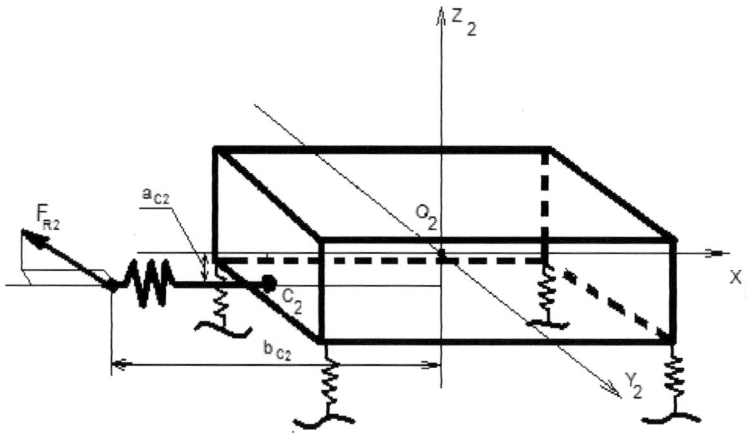

b

Figure 2. Analytical model of the articulated vehicle-based SPAP: a – first section (SPAP); b – second section (loader vehicle).

The interaction between the SPAP and the ground is carried out through the suspension elastic elements [14, 15]. As mentioned above, the running gear used in these type of vehicles is a wheeled running gear [16, 17]. On the figure 2 four support points are shown at every section. In the mathematical model the number of such points is not limited though.

The number of degrees of freedom of the real articulated vehicle-based SPAP is infinite. In order to mathematically describe a motion process of such system one should analyse a motion of the first and the second section separately. The coupling between the two sections should be substituted by an interaction force. When in motion, the connection between the two sections can be considered as rigid and when firing – as elastic. On the figure 2b this connection is represented by the elastic element with a coefficient of rigidity «c» and dissipation coefficient «k».

3. Experimental part

On the figure 3 is representing the diagram that shows displacements of the center of mass (CM) of the SPAP-bodies after the shot. In ordinate axis direction the CM-displacements along the longitudinal axis of the SPAP are shown. The abscissa axis represents time [10, 11].

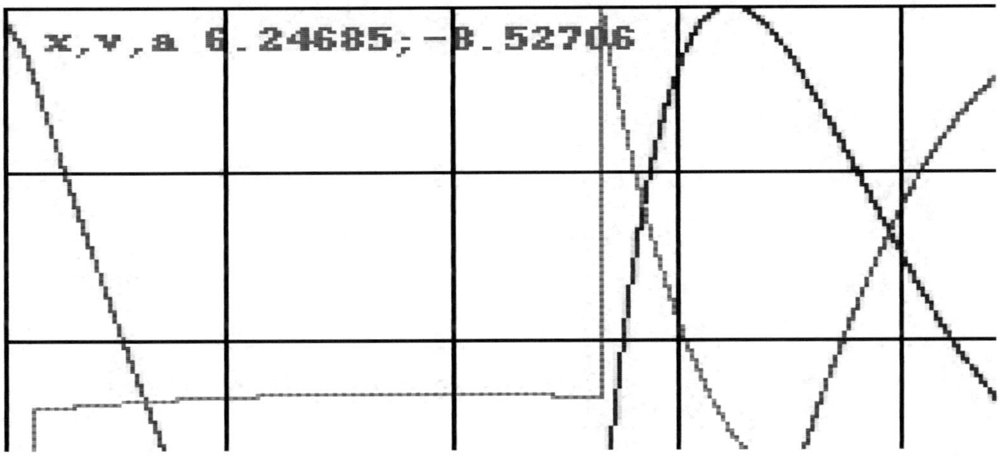

Figure 3. SPAP-bodies displacements after the shot and CM-acceleration of the first section: green – CM-displacement of the first section; red - CM-displacement of the second section; gray – CM-acceleration of the first section. (division value of the absciss axis – 0,5 s, the same for the ordinat axis – 0,075m)

4. Conclusion

The high convergence of the theoretical and experimental data allows for a conclusion that the established complex mathematical model can be used for an analysis of the SPAP-body oscillation process in case of its coupling with a loader vehicle, i.e. as an articulated vehicle-based SPAP.

The results of resolving of the mathematical model for an 2C19-SPAP in conjunction with the T3M334-unit have shown that the oscillation decay time after the shot has been significantly reduced: depending on a gun position, in average by 2,5 seconds. Acceleration magnitude at the commander's and the driver's sit is reduced by 20% in average.

As a result the use of the SPAP in tandem with the loader vehicle enables to create an articulated vehicle-based SPAP with enchanced firing performance.

Using the simulation modeling of the SPAP-bodies behavior allows to provide a parameters of the elastic and dissipative elements of the SPAP- and loader vehicle-suspension that enable on the design stage the prediction of the behavior of the SPAP-bodies after the shot and when in motion over the terrain, thus reducing the testing time required for creating a new units.

References

[1] Polungyan A A 2009 *Design of all-wheel drive vehicles* Moscow: MSTU them. N.E. Bauman p 488

[2] Bazhenov E E and Vlyukhin AV 2010 Application of articulated transport systems in the extractive industries of the national economy *Truck: Scientific and Technical. magazine* (Moscow) **4** p 34
[3] Dhir A and Sankar S 1997Analitical model for dynamic simulation of off-road Tracked Vehicles *Vehicle system dynamics* **27** p 37
[4] Bazhenov E E 2010 Modular principle of modeling of articulated transport-technological systems *Tractors and agricultural machinery* (Ekaterinburg) p 20
[5] Watanabe K and Kitano M 1986 Study on streerability of articulated tracked vehicles *Journal of Terramechanics* **23** p 69
[6] Bazhenov E E 2010 Modular principle of the synthesis of transport and technological systems *Izvestiya VUZov. The Volga region. Technical sciences* (Penza) **2** p 79
[7]. Veits V L 1969 *Dynamics of engine aggregates with internal combustion engines* Leinigrad, Mechanical Engineering, p 370
[8]. Tarasik V P 1998 Mathematical Modeling of Technical Systems, Moscow, Design-Pro, p 204-640
[9]. Chelomey V N 1980 *Vibration in Engineering: A Handbook* Moscow, Mechanical Engineering, p 544
[10] Buynachev S K and Bazhenov E E 2008 Modeling of the structure of mechanisms *Problems and achievements of the motor transport complex: a collection of materials of the 6th All-Russian Scientific and Technical Conference*, Ekaterinburg, p 27-29.
[11]. Bazhenov EE 2010 *Fundamentals of the theory of articulated transport systems* Ekaterinburg, UrFU, p 257
[12] Bazhenov E E 2009 *Articulated transport and technological systems* Ekaterinburg: USTU – UPI p 174
[13] Bazhenov E E, Buynachev S K and Kruchinin I N 2010 *Fundamentals of the theory of articulated transport systems* Yekaterinburg. UrFU p 257
[14] Bazhenov E E, Golomidov S I and Yushkov A I 1985 Determination of the drive power of the active axle of trailers *Exploitation of timber transport rolling stock: interuniversity*. Sat. sci. tr. (Sverdlovsk) p 72-75
[15] Platonov V F 1989 *All-wheel drive cars* Moscow: Mechanical Engineering p 312
[16] Belyaev A et al. 2019 Development of 8×8 all-terrain vehicle with individual wheel drive *VEHITS 2019 - Proceedings of the 5th International Conference on Vehicle Technology and Intelligent Transport Systems* pp 556-561
[17] Belyakov V at al. 2015 Multifunctional vehicle for coastal areas *12th International Conference on the Mediterranean Coastal Environment, MEDCOAST 2015* pp 945-951

IOP Conf. Series: Materials Science and Engineering 1086 (2021) 012032 doi:10.1088/1757-899X/1086/1/012032

Predicting the durability of piston engine crankshafts

A N Gots, Sh A Amirseyidov[1]

Vladimir State University n.a. Alexander and Nikolay Stoletovs, Vladimir, Gorky street 87, Russia, 600000

E-mail: [1] shihamir@mail.ru

Abstract. A method for testing crankshafts of automobile and tractor engines under the action of variable bending moment is proposed. The test machine consists of a frame on which the crankshaft is installed through elastic springs on the extreme root necks. On the extreme necks, pendulums are installed, one end of which is rigidly fixed, and at the opposite ends of the pendulums, two inertial (unbalanced) vibrators are installed, creating two equal forces in the horizontal plane, directed in opposite directions and changing their magnitude. This pair of forces creates a variable bending moment relative to the axis of the crankshaft, acting on all its elements, and causes fatigue failure in the weakest sections. Based on the test results, a graph is plotted in the coordinates of maximum stress – the number of loading cycles before failure, which allows you to determine the parameters of the crankshaft fatigue life model.
According to the corrected linear hypothesis of fatigue damage accumulation, shaft failure occurs when the sum of accumulated damage is equal to a certain value, which is determined by the results of fatigue tests.

1. Introduction

The main assumption, which is usually accepted in all theories for calculating the durability of a part under the action of a load spectrum, is that the effect of cyclic stresses of a given amplitude leads to fatigue damage, the value of which is determined by the number of cycles of stress of this amplitude, as well as the total number of such cycles before the destruction of the sample. As applied to crankshafts, it can be assumed that in the most loaded sections (usually this is the transition zone from the connecting rod neck to the cheek along the overlap line), the crystal lattice distortions occur, leading to the appearance of tears in some direction in which the greatest stresses act, which gradually pass under the influence of variable stresses into the crack. Under the influence of alternating stresses, the crack grows. When the total accumulated damage reaches a certain critical value, and the crack significantly weakens the cross-section, fatigue failure occurs.

2. Purpose of research

Development of a method for predicting the durability of piston engine crankshafts based on fatigue tests of full-scale samples

3. Materials and methods of research

In [1,3,4], the possibility of predicting the cyclic durability of crankshafts based on the results of fatigue tests is shown. Figure 1 shows an approximate graph of fatigue life for details [5,6]. Stress σ_{-1N_0} – the value of the endurance limit at the point of fracture A-for steel at $N = 2\cdot10^6...5\cdot10^7$ cycles. Its presence is associated with a change in the mechanism of predominant development of fatigue. At a high stress level, fatigue failure occurs at $N < 10^5$ cycles as a result of the accumulation of plastic deformations along the shear planes. This is low-cycle fatigue, where the patterns of static destruction are more pronounced. If the destruction occurs when the number of cycles $N = 2{\cdot}10^5...5{\cdot}10^6$, then in this case, the regularities of fatigue destruction appear [5,7,8].

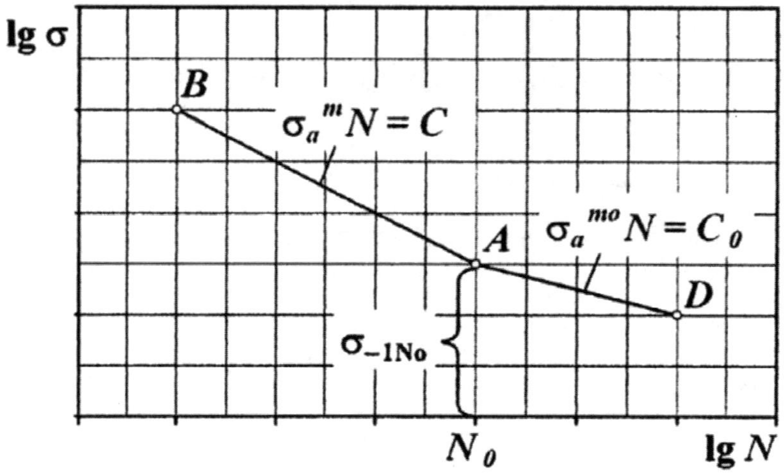

Figure 1. Approximate graph of fatigue life under regular loading

At a low stress level (along the straight line AD), diffusion processes of dislocation movement develop. However, to obtain a section of the fatigue curve, time-consuming and lengthy tests are required [7,8,9].

The diagram of the installation for fatigue testing of crankshafts, developed under the guidance of the author, is shown in Figure 2. the establishing consists of a frame 1, on which the crankshaft 3 is installed on the extreme root necks through elastic springs 2.

On the same necks, pendulums 4 are installed, one end of which is rigidly fixed to the necks, and at opposite ends of the pendulums, two inertial (unbalanced) vibrators 5 are installed to create cyclic alternating loads, rotating with the same angular velocity ω^2 and creating two equal forces $P = mr\omega^2$ in the horizontal plane.

On the same necks, pendulums 4 are installed, one end of which is rigidly fixed to the necks, and at opposite ends of the pendulums, two inertial (unbalanced) vibrators 5 are installed to create cyclic alternating loads, rotating with the same angular velocity ω and creating two equal forces $P = mr\omega^2$ in the horizontal plane [6].

Here m is the mass of the unbalanced balance weight of the vibrator located at the radius r; ω is the frequency of its rotation. The components of forces on the horizontal x-axis are equal to $P_g = mr\omega^2 \sin\varphi$, and on the vertical y – axis - $P_v = mr\omega^2 \cos\varphi$ ($\varphi = \omega t$ is the angle of rotation when the unbalanced mass m rotates around the vertical y axis, t is the current time).

In this case, the vertical component of the inertia force from the one balance weight load (Figure 2) $P_v = P{\cdot}\cos\varphi$ it is balanced by a similar component from the second balance weight, and the forces from the vibrators 5 on the left and right create equal and oppositely directed forces $2P{\cdot}\sin\varphi$. Thus, the crankshaft is loaded with a pure bend in a symmetrical cycle with a moment $M = 2Pl\sin\varphi$, where l is the length of the pendulum 4.for the same type of fatigue testing of crankshafts, executive document RD 23.3.62-89 "Methodology for accelerated fatigue testing of crankshafts of tractor and combine engines" was developed [5,6].

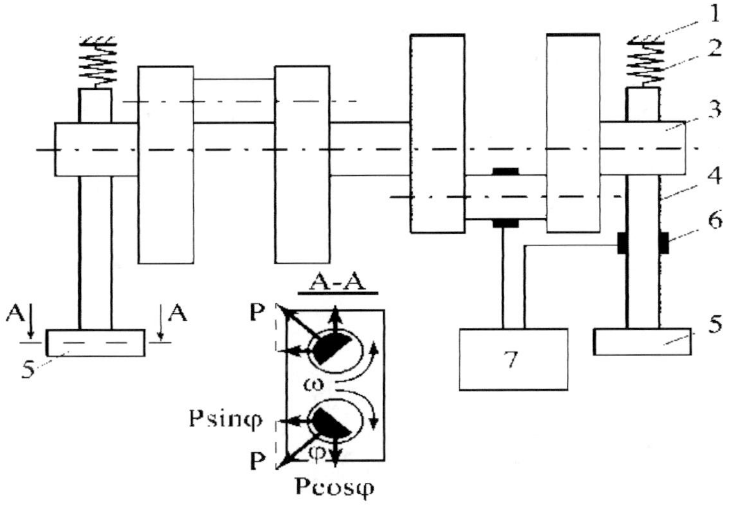

Figure 2. Installation diagram for testing crankshafts

To control the loading on the crankshaft rod journal in the section plane of the first crank pasted strain gauges 6, which are connected to the measuring instrument 7. Before the beginning of tests conducted calibration of the strain gages mounted on the crankshaft, which for both ends of the pendulum 4 (instead of system loading 5 – Figure 2) a static load was applied. The stand for calibration of strain gauges consisted of a rigid frame, supports in the form of prisms, a hydraulic Jack and a compression dynamometer DOSM-3 [4].

The appearance and growth of a fatigue crack in the test knee of the crankshaft is recorded by increasing the value of the signal received from the strain gages to the control and measuring equipment. The appearance of a crack in the cross section of the crankshaft reduces the frequency of its natural vibrations ω_s, and since the tests were carried out in the resonant mode, the vibration amplitude increased. Tests are terminated if the signal value increases by 15% from the original value [10, 11].

Figure 3. The crankshaft in the car for testing

The crankshaft can be installed in the test machine so that all the cranks are loaded simultaneously or each crank separately (Figure 3). the amount of bending moment acting in the *n-th* crank under test is monitored continuously throughout the entire test time. The control system consists of strain gages pasted on the connecting rod neck of the tested knee, a strain amplifier, and two oscilloscopes: electronic (for visual monitoring of the signal value from the strain gage) and loop (for periodic monitoring of the signal value after calibration under static loading). The drive of the vibrator 5 was carried out from a balancing machine.

The error in determining the controlled parameters during testing did not exceed the values indicated in table 1.

Table 1 Errors of controlled parameters

	Controlled parameters	Unit of measurement	Measurement errors, %
1	Bending moment, M_{ben}	Nm	2…3
2	Frequency of rotation of the drive shaft, n	min^{-1}	1
3	Operating time in load mode	h	0,1
4	Number of loading cycles, N	number	0,01

At loading of each crank pure bending in its plane with time Mesh destruction must occur in the cross section of the least stiffness from the fillets interfaces connecting rod neck of crankshaft with a cheek with access to a fillet of the pairing crank journal and cheeks [12-14].

An example of the destruction of the crankshaft in this section is shown in Figure 4.

Figure 4. Example of the destruction of the crankshaft of a diesel D144

When testing, it is necessary to obtain at least three values of cycles N_1, N_2, N_3 on the straight line *AB* (Figure 1), at which the destruction occurs by loading with a net moment M_1, M_2, M_3 . Since the value N_0 is selected in advance (usually taken N_0 = $2{\cdot}10^6 ... 5{\cdot}10^7$ cycles), such tests actually determine not only the maximum bending moment, but also the effect of the stress concentration in the fillet. In this case, it is quite simple to determine the effect of structural or technological measures on increasing the fatigue strength of the crankshaft [15-17].

4. The results of the study and their discussion

As deterministic models for the fatigue life of the crankshaft at the loading point (uniaxial stress state) can take it [7, 15]

$$\sigma_a^m \overline{N} = C , \tag{1}$$

where σ_a – the amplitude of the alternating voltage; $\overline{N}$ – the average number of cycles to failure; C and m – parameters of the test to the crankshaft, depending on the material, manufacturing technology and other factors.

It is known that fatigue tests with the constant amplitude of variable stresses are characterized by a large variation in the number of cycles before failure. Therefore, in equation (1), we always mean the average number of cycles $\overline{N}$ to failure. In addition, in tests, equation (1) will be valid if the stress σ_a is replaced by the moment at N_0 [15,16,17]

Since the crankshaft tests are performed at different values of the amplitude stresses σ_{ai}, it is possible to determine the value of the parameters C and m. Actually, using the model in the form of (1) and data on the cyclic durability of cranks for any two stress values σ_{a1}, σ_{a2} and the corresponding average values of cycles before failure $\overline{N}_1$ and $\overline{N}_2$, we obtain after logarithm two equations for determining C and m:

$$\begin{aligned} m\lg\sigma_{a1} + \lg\overline{N}_1 &= \lg C; \\ m\lg\sigma_{a2} + \lg\overline{N}_2 &= \lg C. \end{aligned} \tag{2}$$

The obtained values C and m allow us to find the endurance limit for the base number of cycles N_b = $2\cdot10^6$ or the value $\sigma_{-1\kappa} = K_\sigma\sigma_a/(\varepsilon_\sigma\beta_\sigma)$. Since the rated voltage can be calculated , the value of the parameter ratio $K_\sigma/(\varepsilon_\sigma\beta_\sigma)$ is determined based on experimental research data, which is taken into account when calculating at the design stage [18,19,20]

To determine the durability of the crankshaft in hours, assume that during the time ti of the engine in operation in one of the modes characterized by the value of the torque M_i and the angular speed of rotation of the shaft ω_i during the entire service life (until destruction) T, it will be produced n_i loading cycles, which for four-stroke engines is equal to

$$n_i = \frac{30\omega_i}{2\pi}60t_i = \frac{900\omega_i}{\pi}f_iT, \tag{3}$$

where $t_i = f_iT$; f_i is the probability density of operation in this mode.

From the equation of the actual endurance curves $\lg N - \lg M_{\max}$ of the crankshaft, we determine the number of loading cycles required for its destruction

$$N_i = \left(\frac{M_{\max 0}}{M_{\max i}}\right)^m N_0, \tag{4}$$

where m is an exponent numerically equal to the cotangent of the slope of the fatigue curve in logarithmic coordinates ($m = |ctg\alpha|$); α is the angle of inclination of the left branch of fatigue; $M_{\max i}$ – the amplitude of the maximum bending moment under the *i-th* loading mode.

Thus, taking into account (3) and (4), a certain amount of damage will be accumulated by the crankshaft over time t_i:

$$\frac{n_i}{N_i} = \frac{900\omega_i f_i T}{\pi\left(\frac{M_{\max 0}}{M_{\max i}}\right)^m \cdot N_0}. \tag{5}$$

According to the corrected linear hypothesis of fatigue damage accumulation, shaft failure occurs when, taking into account (5), the sum of accumulated damage is equal to:

$$\sum_{i=1}^{i=k}\frac{n_i}{N_i} = \frac{900T}{\pi N_0}\sum_{i=1}^{i=k}\omega_i f_i\left(\frac{M_{\max i}}{M_{\max 0}}\right)^m = a_p, \tag{6}$$

where k – the operation of the engine in operation.

Following [5] and considering that the stresses are replaced by limiting bending moments, the value a_p for bench accelerated tests [1] is also determined by the formula:

$$a_p = \frac{\varsigma M_{\max i} - 0{,}5M_{\max 0}}{M_{\max i} - 0{,}5M_{\max 0}}; \tag{7}$$

$$\varsigma = \sum\frac{M_{\max i}}{M_{\max 0}}\cdot\frac{\nu_{ai}}{\nu_m}\ (M_{\max i} > 0{,}5M_{\max 0}); \tag{8}$$

where $\nu_m = \sum_{i=1}^{i=k} \nu_{ai}$ – the total number of loading cycles. ν_{ai} – the number of cycles of repeating amplitudes in the loading block.

When using a limited endurance limit, you should assume:

$$a_p = \varsigma = \sum \frac{M_{\max i}}{M_{\max 0}} \cdot \frac{\nu_{ai}}{\nu_m}, \tag{9}$$

where the summation extends over all loading amplitudes without discarding small amplitudes.
According to the linear hypothesis of fatigue damage accumulation, crankshaft failure occurs when $a_p = 1$.

From (6), the crankshaft durability can be found by the fatigue strength resistance condition:

$$T = \frac{\pi N_0 a_p}{900 \sum_{i=1}^{i=k} \omega_i f_i \left(\frac{M_{\max i}}{M_{\max 0}} \right)^m}. \tag{10}$$

Therefore, to predict the durability of the crankshaft, it is necessary to have information about the fatigue strength limit $M_{\max 0}$, the probabilistic and statistical load f_i of engines in operation, as well as the loads $M_{\max i}$ on the crankshaft.

The relationship between the level of maximum bending moment $M_{\max i}$ and the number of loading cycles N_i before failure, even with the strictest observance of the identity of the crankshaft fatigue tests, due to the heterogeneity of the material itself, has a pronounced random character. To take this into account, we must assume that the parameters of the fatigue curve are random variables that obey some statistical distribution. In this regard, the durability of the crankshaft will be determined by the expression [6,21]:

$$T = \bar{T} \chi^m,$$

where $\vec{T}$ is the average life of the crankshaft, calculated by the formula (8); χ is a random parameter whose distribution density has the form:

$$f(\chi) = \frac{1}{\nu(1-\varepsilon)} \left(\frac{\chi - \varepsilon}{1-\varepsilon} \right)^{1/\nu} \exp\left[-\left(\frac{\chi - \varepsilon}{1-\varepsilon} \right)^{1/\nu} \right]$$

where v is the coefficient of variation; ε =0.6 is the fraction of the average fatigue limit $M_{\max 0}$ below which no fatigue damage accumulates.

Taking into account these factors, the probability of failure of the crankshaft due to fatigue failure during operation T is determined from the expression

$$F(T) = 1 - \exp\left\{ -\left[\frac{(T/\bar{T}) - \varepsilon}{1-\varepsilon} \right] \right\}. \tag{11}$$

5. Conclusions

A method for predicting the durability of the crankshaft comprises the following stages: the experimentally determined endurance limit of the crankshaft $M_{\max 0}$; and according to the loading conditions of the crankshaft load in the use – values of $M_{\max}$ and the number of blocks of loading, the formula (8) calculated the average durability of $\vec{T}$; in the formula (9) for a given resource T is the probability of failure of crankshaft due to fatigue fracture.

Referance

[1] Gots A N 2004 Scientific bases of calculation and accelerated testing of parts of the crank mechanism of tractor diesels at the design stage *Autoabstract. dissertations for the degree of doctor of technical Sciences* (*Vladimir*) p. 32

[2] Burok S 1982 *Reasons for failure of internal combustion engine crankshafts* J Automotive **37-2** pp 51–55
[3] Gots A N 1989 *Method for testing the crankshaft of an internal combustion engine* Collection of reports of the V national scientific and technical conference "Diesel engines" **2** pp 580-585
[4] Gots A N, Kudelya I N, Nekrasov S I 2000 *Predicting the durability of internal combustion engine crankshafts* J Tractors and agricultural machines **3** pp 23-25
[5] Kogaev V P 1977 *Calculations for strength under stresses, variables in time* Moscow: Machine-building Press p 232
[6] Gotz A N and Efros V V 2005 *Collection of scientific papers on engine building problems, dedicated to the 175-th anniversary of Bauman Moscow state technical University ed N A Ivashchenko and L V Grekhov* Moscow: Bauman Moscow state technical University Press pp 180-184
[7] Birger I A 1985 *Problems of strength of aircraft ed I F Obraztsov and A S Volmir* Moscow Mechanical engineering Press pp 105-150
[8] Birger I A 1982 *Deterministic and statistical models of fatigue* J Problems of strength **4** pp 24-28
[9] Dulnev R A 1971 *Summation of damage and strength condition in thermocyclic damage* J Problems of strength **10** pp 100-103
[10] Kordonsky H B and Friedman Ya F 1976 *Some questions of probabilistic description of fatigue life* J Factory laboratory **7** pp 828-847
[11] Stepnov M N 1985 *Statistical processing of results of mechanical tests* Moscow Mechanical engineering Press p 232
[12] Mavlyutov R R 1980 *Stress concentration in elements of aircraft structures* Moscow Nauka p 140
[13] Naudorf H, Rothe V, Ziese J 1976 *Influence on the strength of steel curving shafts for passenger car engines* J MTZ **5** pp 205-208
[14] Arai J 1965 *Test for reliability of machines and their elements* Moscow Mechanical Engineering Press pp 181
[15] Weibull 1964 *Fatigue tests and analysis of their results* Moscow Engineering Press p 279
[16] Bolotin V V 1984 *Forecasting the resource of machines and structures* Moscow: Mechanical Engineering Press p 312
[17] Barzilovich E Yu and Gnedenko B V 1985 *Problems of aircraft reliability* Collection of articles.Moscow: Engineering Press p 280
[18] Velichkin I N 1999 *Methodology and ways to reduce the duration of tests* J Tractors and agricultural machines **6** pp 30-33
[19] Velichkin I N 1999 *Development of a complex of accelerated tests* J Tractors and agricultural machines **11** pp 34-36
[20] Collins J A 1984 *Failure of materials in mechanical design* Moscow Mir Press p 624
[21] McClintock F A and Argon A S 1970 *Mechanical behavior of material* Moscow Mir Press p 614

IOP Conf. Series: Materials Science and Engineering 1086 (2021) 012033 doi:10.1088/1757-899X/1086/1/012033

Scaling and dynamics of special machines

I G Kuklina

Nizhny Novgorod State University of Architecture and Civil Engineering, Ilyinskaya str., 65, 603950, Nizhny Novgorod, Russia.

Email: irinalg15@yandex.ru

Abstract. Rotary screw machines (RSM) are widely known as multipurpose vehicles. RSM easily overcome impassable swamps, snow-covered, off-road, ice surface. A significant problem for the creators of vehicles with rotors is the high vibration loads on the operator and machine units, which require multiple experimental studies. A scaled-down model of a vehicle is often used for experiments. The model of the RSM is included in the experiment, replacing not only the object of research, but also the conditions in which it is studied. Strengthening the role of the theoretical side of research, there is a connection between the formulation of the experiment and its results with the object of research. Calculation and calculation of the elements of proportionality is a hard necessity. The article proposes a complex for calculating scale factors when testing vibration loading of machines equipped with screw rotors.

1. Introduction

The research of the similarity of oscillatory processes means that the data on the flow of oscillatory phenomena obtained when measuring specific vibratory displacement, can be extended to all cases of measurements such as this one. Amplitude-frequency and phase-frequency characteristics of the phenomena of displacement of screw rotors in a group of similar phenomena can be calculated by a simple review of the characteristics and scaling.

2. Mathematical model

Dimensional analysis of physical quantities in the studied processes will be based on the theory of similarity. We can get physical quantity of displacements Ψ only if we compare it with the quantity of the same physical nature – ψ_1. Ration rating in this case - Ψ_1, and ψ_1 is the unit of measurement, i.e.:

$$\frac{\Psi}{\psi_1} = \Psi_1 \qquad (1)$$

If it is necessary to change the unit of measurement a certain number of times - N, the same factor will increase or decrease the number$\Psi_1 = \Psi_1 \cdot \mathrm{N}$.

In classical physics the units of measurement are basic, if they are with arbitrary choice of size and derivatives. The system of units is a set of basic units and derivatives formed by certain rules. Increase in the number of physical dimensional constants increases the number of basic units of measurement.

In this study, it is necessary to make the equation of proportions of vibratory displacement quantities that is the basic unit of measurement of the mechanical system - acceleration, which in turn depends on the mass, distance and time. Dependence of the derivative from the main unit is a formula of dimension of a quantity, which reproduces the product of the raised to the power dimensions of the basic components of the unit. The dimensions of the basic formulas of vibratory displacements are shown in Equation 2:

$$\begin{gathered} \text{acceleration } (\mathrm{m/c^2})\ a = \frac{dv}{dt};\ \lfloor a \rfloor = \lfloor L \rfloor^1 \lfloor T \rfloor^{-2} \\ \text{power } (\text{Newtonкg } \mathrm{m/c^2})\ F = \frac{d(mv)}{dt};\ \lfloor F \rfloor = \lfloor M \rfloor^1 \lfloor L \rfloor^1 \lfloor T \rfloor^{-2} \\ \text{frequency } (\text{Hz } \ 1/\mathrm{c})\ \upsilon = \frac{n}{t};\ \lfloor \upsilon \rfloor = \lfloor T \rfloor^{-1} \\ oscillationperiod(c)\ \mathrm{T} = \frac{1}{\upsilon};\ \lfloor \mathrm{T} \rfloor = \lfloor T \rfloor^1 \\ amplitudeofoscillation(m)\mathrm{A};\ \lfloor \mathrm{A} \rfloor = \lfloor A \rfloor^1 \end{gathered} \tag{2}$$

In the experiments, it is necessary to make recalculation of parameters and, therefore, to move to a different system of measurement in which the units are:

$$\begin{gathered} \lfloor M_1 \rfloor = X \lfloor M \rfloor \\ \lfloor L_1 \rfloor = Y \lfloor L \rfloor \\ \lfloor T_1 \rfloor = Z \lfloor T \rfloor \end{gathered} \tag{3}$$

The equations of the derived quantities based on equations 2 taking into account equations 3 take look like:

$$\begin{gathered} \text{acceleration}(\mathrm{m/c^2})\ a = \frac{dv}{dt};\ \lfloor a_1 \rfloor = YZ^{-2} \lfloor a \rfloor = N_1 \lfloor a \rfloor \\ \text{power } (\text{Newtonкg } \mathrm{m/c^2})\ F = \frac{d(mv)}{dt};\ \lfloor F_1 \rfloor = XYZ^{-2} \lfloor F \rfloor = N_2 \lfloor F \rfloor \\ \text{frequency } (\text{Hz } \ 1/\mathrm{c})\ \upsilon = \frac{n}{t};\ \lfloor \upsilon_1 \rfloor = Z^{-1} \lfloor \upsilon \rfloor = N_3 \lfloor \upsilon \rfloor \\ \text{oscillationperiod(c)}\ \mathrm{T} = \frac{1}{\upsilon};\ \lfloor \mathrm{T}_1 \rfloor = Z \lfloor \mathrm{T} \rfloor = N_4 \lfloor \mathrm{T} \rfloor \\ \text{amplitudeofoscillation(m)} \quad \mathrm{A};\ \lfloor \mathrm{A}_1 \rfloor = N_5 \lfloor A \rfloor \end{gathered} \tag{4}$$

Thus, measuring process parameters of vibratory displacement of the body of the studied rotary screw machine are: acceleration, frequency and oscillation period, and then the received and amplitude-frequency and phase-frequency characteristics.

2. Experimental research

In order to study the vibratory displacement of the rotary screw machine on a viscoelastic suspension (previously there were constructions on a rigid or semi-rigid suspension only) we built a scaled model of transport and technological means. The designed mechanical system (Figures 1 and 2) was a complete simulation of physical processes occurring in the natural machine. The similarity of vibratory displacement of the model and the full-scale sample was achieved by proportional correspondence of parameters of the two systems of rotary screw machines.

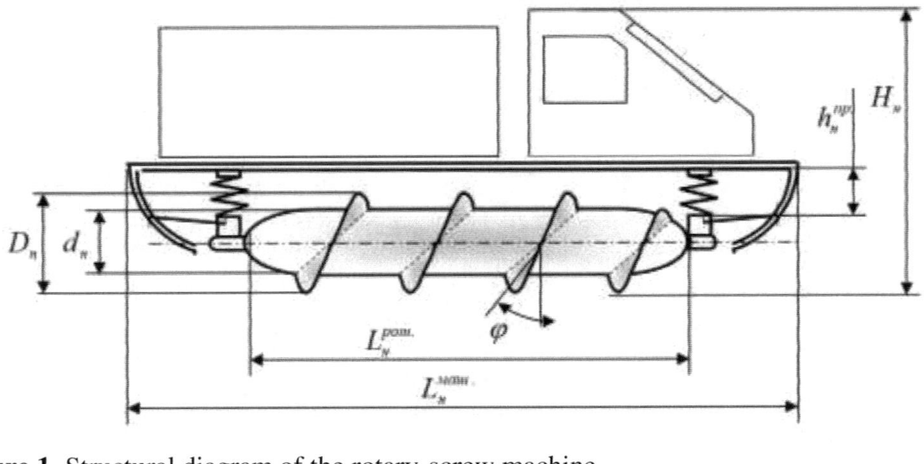

Figure 1. Structural diagram of the rotary-screw machine

IOP Conf. Series: Materials Science and Engineering 1086 (2021) 012033 doi:10.1088/1757-899X/1086/1/012033

$$Z_i/S_i = \mathfrak{m}_i \tag{5}$$

whereZ_iиS_i – are uniform parameters of the rotary screw machine individual points vibration process Z и S; $\mathfrak{m}_i$ – scale or similarity coefficient of the oscillating system.

Figure 2. The LFM-RVD-GPI-66, equipped with rotary screw drives and the model of rotary screw machine

Figure 2 shows photographs of ice-milling machine manufactured at the Gorky Polytechnic Institute and the model of rotary screw machine, made by the author (see photo) with the help of the staff and students of the Nizhny Novgorod Automotive College. The model is a likeness of the machine - the geometric (as opposed to an affine – $\mathfrak{m}_i = \mathfrak{m}_x = \mathfrak{m}_y = \mathfrak{m}_z$).

The difficulty in creating such an oscillatory research system is to meet the following requirements:

- compliance with the physical nature of similarity - all items are reduced both geometrically and in accordance with the characteristics of their functions;
- compliance with mathematical description of the oscillatory picture of the body placement - all previously obtained dynamic equations are completely valid for the scale model.

This research task also requires:

- kinematic similarity - similarity of velocity and acceleration
- dynamic similarity - similarity of forces.

We created similar conditions of full-scale systems and models using the basic postulates of the similarity of technical systems. We determined combination of parameters, called the similarity criteria, which have the same quantities.

Geometrical moving of the vehicle on the off-road is more often geometrically described by a harmonic function; this function is also used in a scale model studies, but the size of the amplitude and frequency are reduced accordingly.

I.e., two such processes are submitted:

$$\left.\begin{array}{c}\mathfrak{y}_1 + \mathfrak{y}_2 + \cdots + \mathfrak{y}_n = \sum_{i=1}^{n} \mathfrak{y}_i = 0 \\ \mathfrak{Y}_1 + \mathfrak{Y}_2 + \cdots + \mathfrak{Y}_n = \sum_{i=1}^{n} \mathfrak{Y}_i = 0\end{array}\right\} \tag{6}$$

$\mathfrak{Y}_n$и$\mathfrak{y}_n$are not equal to zero and equation 6 is rewritten as:

$$\left.\begin{array}{c}1 + \frac{\mathfrak{y}_1}{\mathfrak{y}_n} + \frac{\mathfrak{y}_2}{\mathfrak{y}_n} + \cdots + \frac{\mathfrak{y}_{n-1}}{\mathfrak{y}_n} = 0 \\ 1 + \frac{\mathfrak{Y}_1}{\mathfrak{Y}_n} + \frac{\mathfrak{Y}_2}{\mathfrak{Y}_n} + \cdots + \frac{\mathfrak{Y}_{n-1}}{\mathfrak{Y}_n} = 0\end{array}\right\} \tag{7}$$

We know, that function$\mathfrak{y}_i = \mathfrak{y}_i(Z_1, Z_2, \dots, Z_m), \mathfrak{Y}_i = \mathfrak{Y}_i(S_1, S_2, \dots, S_m)$have nonzero dimension, whereZ_iиS_i $(i = 1,2, \dots, m)$ – areidentical parameters of such processes of similarity with criterion of identity$\mathfrak{m}_i$ by formula5. So, the equations are true: $Z_i = \mathfrak{m}_i \cdot S_i$, а$\mathfrak{y}_i = \mathfrak{Y}_i \cdot N_i$, where$N_1 = N_2 = \cdots = N_n = N$. Taking into account these equations in systems 5.6 and 5.7, we get identical expressions:

$$\frac{\mathfrak{y}_1}{\mathfrak{y}_n} = \frac{\mathfrak{Y}_1}{\mathfrak{Y}_n}; \frac{\mathfrak{y}_2}{\mathfrak{y}_n} = \frac{\mathfrak{Y}_2}{\mathfrak{Y}_n}; \dots; \frac{\mathfrak{y}_{n-1}}{\mathfrak{y}_n} = \frac{\mathfrak{Y}_{n-1}}{\mathfrak{Y}_n}, \text{ and in generalized form: } \frac{\mathfrak{y}_i}{\mathfrak{y}_n} = idem \tag{8}$$

In a study of oscillations of the machine (in this case, describing the support surface we take$\mathfrak{y}_i = \sin \omega t$, then$\omega t = idem$. If we consider several types of similarity, we introduce similarity criterion - π_k, where$k$ – is a number of the criterion.

Any full equation of the physical process must be submitted according to the criteria of similarity of the members of the equation parameters.

Figure 3. Suspension elements of the rotary screw machine - hydraulic shock absorbers and a spring performed by physical scaling

When calculating the stiffness of the spring scale model the following equation is applied:

$$C = \frac{G \cdot d_D^4}{8 \cdot d_F^3 \cdot n} \tag{8}$$

where d_D – diameter of the wire, m; d_F –winding diameter, as measured from the axis of the wire, m; n –number of turns; G – shear modulus, Pa $(\frac{kg}{m \cdot s^2})$.

Equation 8 is a function of the parameters:

$f(d_D, d_F, n, G)$or in the general form:

$$f(Z_1, Z_2, \dots, Z_m) = 0 \tag{9}$$

$Z_{01}, Z_{02}, \dots, Z_{0m}$ - characteristic quantities that are related to the parameters of the general equation$Z_1, Z_2, \dots, Z_m$. Equality 9 takes the form:

$$f\left(\frac{Z_1}{Z_{01}}, \frac{Z_2}{Z_{02}}, \dots, \frac{Z_m}{Z_{0m}}\right) = 0 \tag{10}$$

In equation 8, the number of basic units is 4, and they are studied, and the rest will be their functions:

$$Z_{0s} = Z_{01}^{u_1} \cdot Z_{02}^{u_2} \cdot Z_{03}^{u_3} \cdot Z_{04}^{u_4} \quad (5 \le s \le m) \tag{11}$$

The ratio of the determinants of the system coefficients allows to obtain the exponents u_i. Independent variables are equal: $Z_{01} = Z_{01}$; $Z_{02} = Z_{02}$; $Z_{03} = Z_{03}$; $Z_{04} = Z_{04}$. Equality 10 is transformed as follows:

$$f\left(1, 1, 1, 1, \frac{Z_5}{Z_1^{x_1} \cdot Z_2^{x_2} \cdot Z_3^{x_3} \cdot Z_4^{x_4}}, \dots, \frac{Z_m}{Z_1^{s_1} \cdot Z_2^{s_2} \cdot Z_3^{s_3} \cdot Z_4^{s_4}}\right) = 0 \tag{12}$$

Dimension of the fractions elements are equal, then the equality is:

$$f(1, 1, 1, 1, \pi_1, \dots, \pi_{m-4}) = 0 \tag{13}$$

We have the following description of similarity: if there is $m - 5$ an independent criterion, under which the dependent criterion is satisfied, and the formula is:

$$\pi_1 = \Phi(\pi_2, \pi_3, \dots, \pi_{m-4}) = 0 \tag{14}$$

Proportionality of the congruent parameters in terms of uniqueness, and equality of similarity criteria of the oscillatory process is a necessary and sufficient condition for similarity.

Getting similarity criteria in the study of oscillatory system of the rotary screw machine.

Oscillatory system, which is rotary screw machine in all its details is unrealistic cumbersome and complicated for a possible solution of the equations, and the analysis of the results obtained is difficult. The equivalent system is simplified as far as statement of a problem permits. Its view in particular cases leads to the scheme for calculating the dynamics of rotary screw machine in the longitudinal and diametric planes.

Dynamic system equivalent to rotary screw machine consists of several masses: rotors, body, engine, cab et al., interconnected through elastic connections and dampers. A generalized equation of oscillatory process of the rotary screw machine on the viscoelastic suspension is as follows:

$$M_0\ddot{Z} + k_0\dot{Z} + cZ = H_x(t)\sin(\omega t) \quad (15)$$

That is function

$$f(M_0, H_x, \omega, Z, k, c, t) \quad (16)$$

whereM_0 –- Sprung mass, kg; H_x –the power of external influences kg m/s^2; ω –cyclic frequency, $1/c^1$; Z – displacements, m; k –coefficient of resistance of shock absorbers, Nc/m; c – spring rate, N/m; t – time, c.

Matrix components of the machine oscillation parameters in the system of unit, which takes the form:

	M	L	T
M_0	1	0	0
H_x	1	1	-2
ω	0	0	-1
Z	0	1	0
k	1	0	-1
c	1	0	-2
t	0	0	1

For this matrix it is easy to calculate the determinants of the third order. The quantities, of which determinants are different from zero, are independent combinations. Determinant of quantities M_0, H_x, Z equals 2, so they are an independent combination. Combinations: $\{\omega, k, c\}$; $\{M_0, c, t\}$; $\{M_0, k, t\}$ – calculation of determinants of which is zero are dependent of each other. Quantities Z, c, t are expressed through independent values of combinations M_0, H_x, ω. It is accepted: $Z = M_0^{x_1} H_x^{x_2} \omega^{x_3}$. We calculate the determinants derived by replacing of the lines. It is determined that:

$$Z = M_0^{-1} H_x^1 \omega^{-2}; c = M_0^1 H_x^0 \omega^2; k = M_0^1 H_x^0 \omega^1; t = M_0^0 H_x^0 \omega^{-1} \quad (17)$$

The result of the study is similarity criteria:

$$\pi_1 = \omega t;\ \pi_2 = \frac{Z}{M_0^{-1} H_x \omega^{-2}};\ \pi_3 = \frac{c}{M_0 \omega^2};\ \pi_3 = \frac{c}{M_0 \omega^1} \quad (18)$$

4. Conclusion

Thus, the effects are similar, if the determining criteria are invariant (the same). We set similarity of rotary screw machine the oscillatory system, undetectable criteria are considered to be same - due to their similarity. The similarity theory allows, without integrating the differential equations derived from them, to get similarity criteria and to set the criteria dependence displayed by equations 18 from of the experimental data.

References

[1] Gordeev B A and Kuklina I G 2012 *Adaptive hydraulic vibration support* Patent №127847 RU
[2] Kuklina I G 2013 *Hydraulic vibration support* Patent №132848 RU
[3] Kuklina I G and Kulyashov A P 2011 Estimation of oscillations inrotary screw machine taking into account the parameters of suspension and the line contact of its thrusters with the ground *Construction and Road Building Machinery* **7** pp 45–52
[4] Kuklina I G 2013 Kelvin–Voigt elements in studying dynamics of special machines *Nauchnoe Obozrenie*. **4** pp 59–67
[5] Kulyashov A P and Kuklina I G 2000 *Cross-country vehicle* Patent №15185 RU
[6] Kuklina I G 2013 *Cross-country vehicle* Patent №134896 RU
[7] Gordeev A B and Kuklina I G 2011 *Ultrasonic phase vibration transducer* Patent №2472109 RU

IOP Conf. Series: Materials Science and Engineering 1086 (2021) 012034 doi:10.1088/1757-899X/1086/1/012034

Simulation of conducted emissions in low voltage switching converters

A V Vasin [1], A A Novoselov, E E Samokhvalov and S V Lazarevich

[1]Nizhny Novgorod State University R.E. Alekseeva, Nizhny Novgorod, Russian Federation.

E-mail: [1]vasinlesha12345@mail.ru

Abstract. The article analyzes the computational method used to assess the electromagnetic compatibility of electronic devices in the automotive industry. The analysis is carried out using the example of a pulsed synchronous DC-DC boost converter. Calculations of the conducted interference of the converter, as well as methods of their reduction are presented. This calculation method is performed in the PSpice electrical circuit simulator.

1. Introduction

Nowadays, the problem of electromagnetic compatibility is becoming more and more important with the development of computer technology and electronics. The impulsive nature of the operation of automotive converters leads to significant emissions of electromagnetic energy in a wide frequency range. In this regard, the task arises to ensure reliable and uninterrupted operation of electronic systems.

Using this method allows you to find and correct errors at the design stage, the detection and correction of which on the finished product will bring significant costs. For the same reason, the application of this method in the design of devices can be accelerated, and the quality of the result achieved can be increased. The use of technologies for numerical analysis of the design of electronic devices allows with a high probability of passing certification and validation tests the first time, and also opens the way to optimizing the design, reducing cost and technically sound use of various types of components.

The importance of the problem of electromagnetic compatibility is growing not only with the increase in the number and complexity of electronic devices in production, but also with its underestimation, which is responsible for the normal functioning of automated equipment. This method, and its emergence, is associated with the large-scale deployment of electronic devices and their use in various types of environments.

2. Sources of EMF

To solve the EMC problem in electronic devices, it is necessary to understand the cause of their occurrence. As you know, EMF covers a wide range of frequencies, but now we are only interested in conducted interference, which ranges from 150 kHz to 108 MHz in accordance with the state automobile standard GOST R 51318.25-2012 (CISPR 25) [1]. Conducted noise, in turn, is divided into two components: differential and common-mode.

Differential interference generates differential currents that occur when signals pass through the communication lines of a two-wire transmission line, shown in Figure 1a.

Common-mode current flows in one direction from the switching power supply through the power input and returns back to the source at ground, as shown in Figure 1b [2].

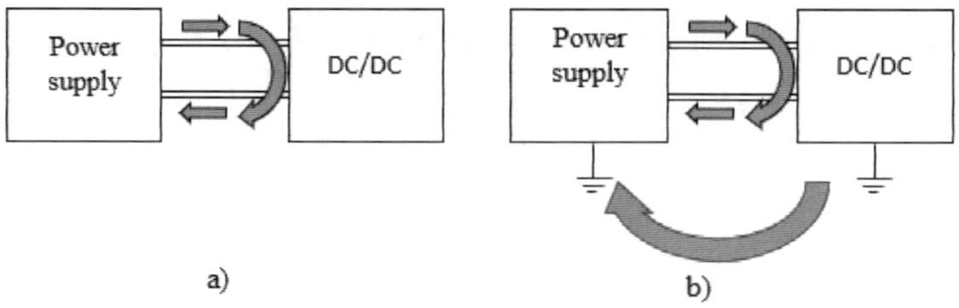

Figure 1. Diagram of voltage application of differential (a) common-mode (b) noise

3. Measurement of conducted emissions

To carry out measurements, a network equivalent is required, which serves both to supply the electronic device with mains voltage and to isolate the high-frequency components that need to be measured. In fact, the network equivalent consists of a high-pass filter and a low-pass filter (Figure 2). The measurement requires a dual delta network equivalent that can measure both common-mode and differential conducted emissions. The line impedance stabilization network connection diagram is shown in Figure 3.

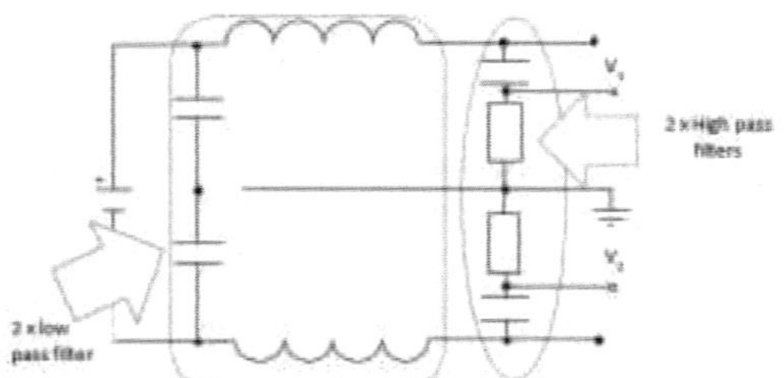

Figure 2. Schematic diagram of a line impedance stabilization network

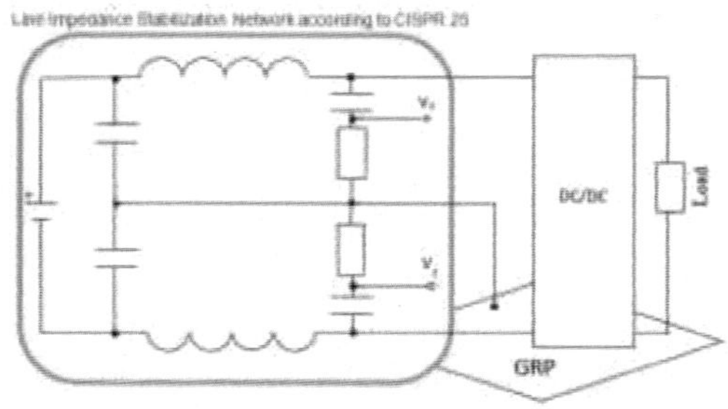

Figure 3. Connection diagram of the equivalent network for measurements of conductive mode

4. Object of research

The object of research is a synchronous pulse boost DC / DC converter with the following characteristics presented in Table 1. You can see the general view of the board in Figure 4.

Table 1. Characteristics of the step-up DC / DC converter.

Characteristic	Value	Units
Input voltage	12	V
Output voltage	24	V
Input current	18	A
Output current	nine	A
Working frequency	500	kHz
Power	216	W
Ripple current	20	%

The noise paths of a pulsed DC / DC converter are shown in Figure 5. Differential noise current occurs at the moment of switching the power switches of the converter. This current flows simultaneously in opposite directions on the power bus and on the power ground, as shown in the figure. The current flow of the differential component is closed and has a relatively small area [4].

Common mode currents occur in parasitic capacitances between the power bus lines, the power ground and the ground plane as shown in Figure 6. Their current is directly proportional to the rate of voltage change. Common-mode interference currents are much less than differential currents, but the area of the loop through which common-mode interference currents flow is larger. This circuit is an antenna, and therefore it is much more difficult to cope with these currents.

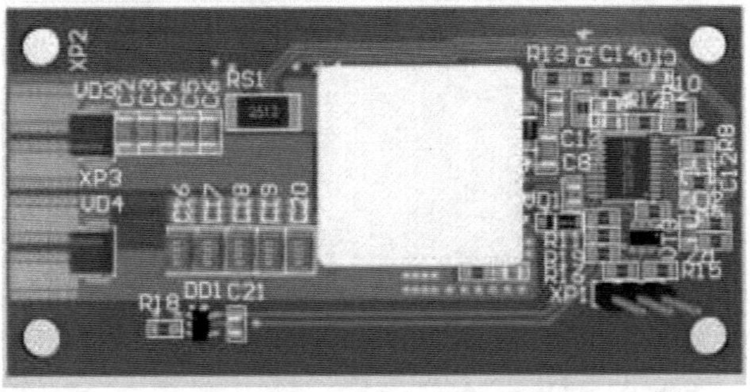

Figure 4. General view of the board without filters

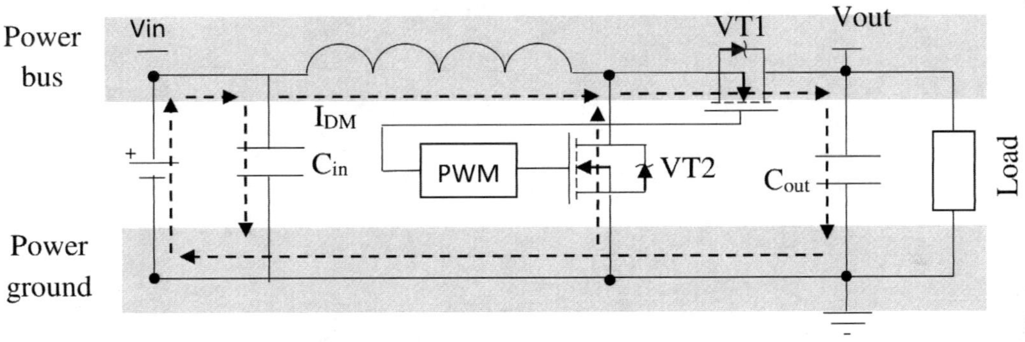

Figure 5. Differential noise paths for a DC / DC boost converter

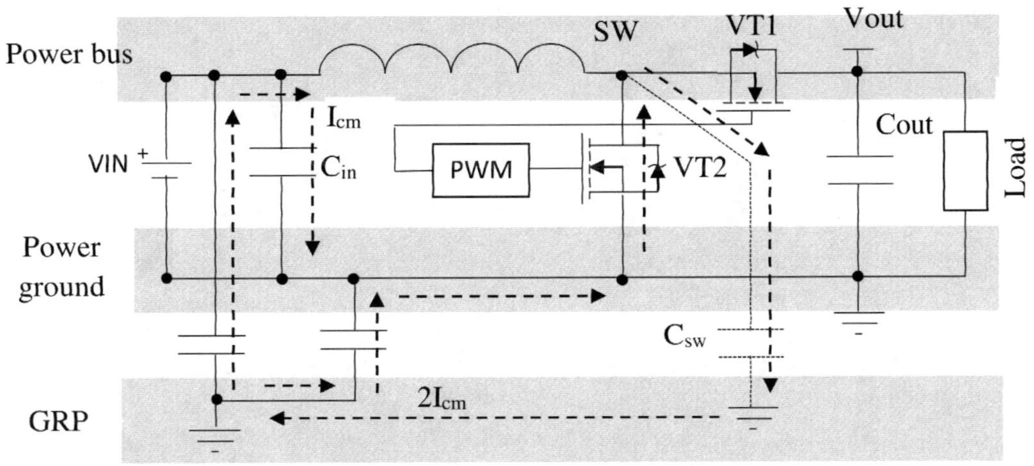

Figure 6. Common-mode noise paths of a DC / DC boost converter.

IOP Conf. Series: Materials Science and Engineering 1086 (2021) 012034 doi:10.1088/1757-899X/1086/1/012034

5. Simulation of conducted noise in the PSpice circuit simulator

To separate the common-mode and differential components of conducted disturbances, it is necessary to see the contribution of the common-mode and differential components in the network equivalent. When we measure the voltage between the two terminals V1 and V2, we get twice the differential noise voltage, which is shown in Figure 7. In common mode, the voltage flows in opposite directions, as shown in Figure 8, between the power supply and ground of the converter, in fact, it becomes almost zero. Figure 9 illustrates that the sum of the voltages V1 and the ground reference plane and V2 and the ground reference plane add up to almost zero for the differential component. If we measure the voltage between V1 and the ground reference plane and V2 and the ground reference plane, as shown in Figure 10, we get negative twice the common-mode voltage (Figure 11).

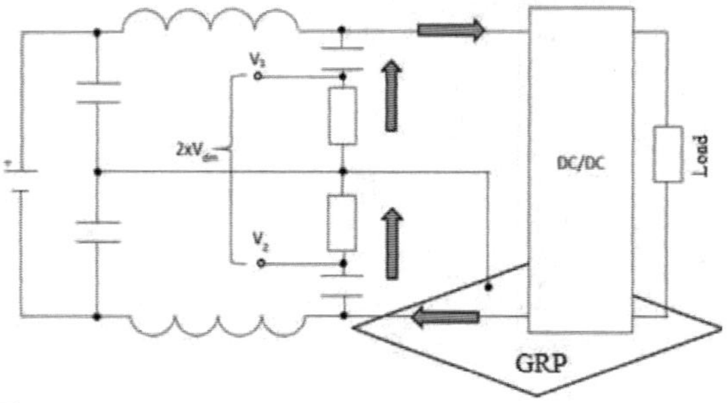

Figure 7. Voltage between V1 and V2 of the differential mode

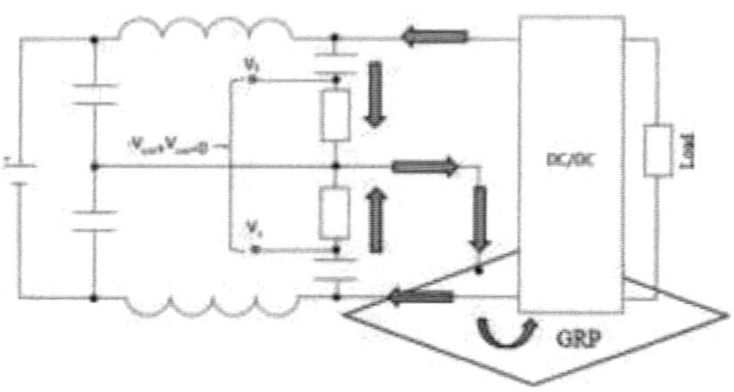

Figure 8. Voltage between pins V1 and V2 of the common-mode

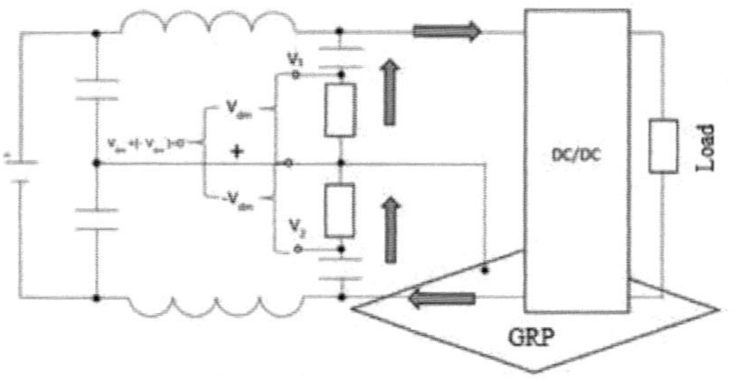

Figure 9. The sum of the voltages between V1 and the ground plane and V2 and the differential ground plane

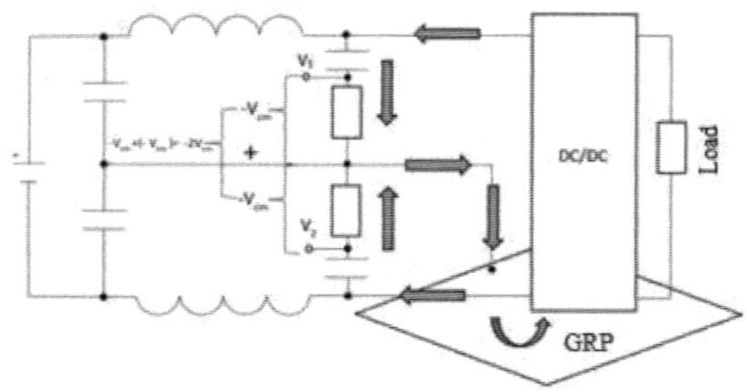

Figure 10. Sum of voltages between V1 and ground plane and V2 and common mode ground plane

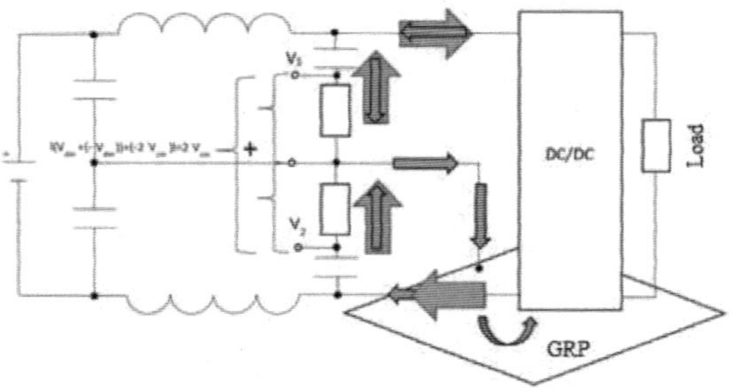

Figure 11. Getting twice the voltage for the differential mode

The result is the conducted noise spectrum of the DC / DC boost converter for the differential mode (Figure 12) and common-mode (Figure 13) noise components in PSpice.

As you can see from the interference pattern, the interference level exceeds the established limits, which means that the EMC certification tests will not be passed. It is necessary to reduce the conducted noise level.

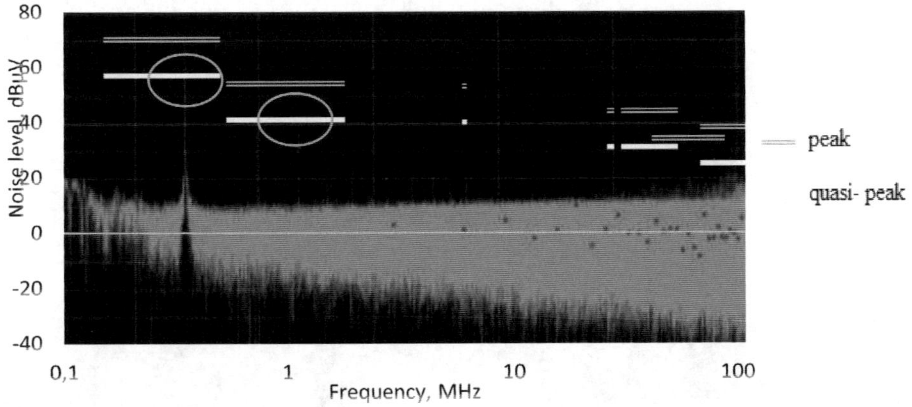

Figure 12. Differential noise spectrum of the converter

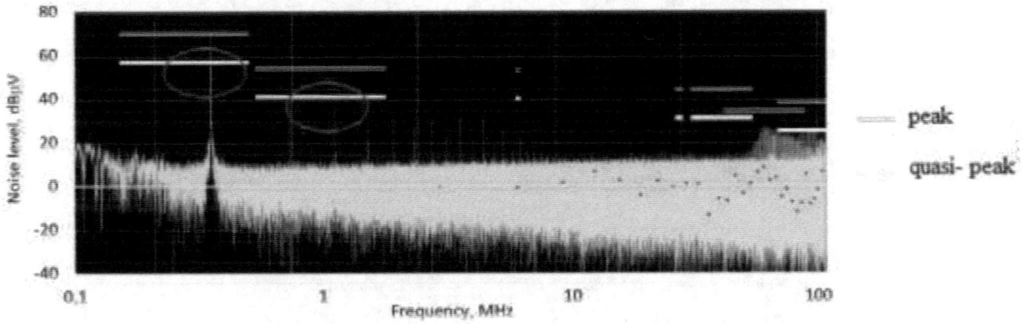

Figure 13. Common-mode noise spectrum of the converter

6 Ways to Reduce Conducted Noise

To successfully reduce the measured interference, both in the area of the fundamental operating frequency and its harmonics, it is most effective to use LC filters. They are able to reduce the level of interference due to their amplitude-frequency characteristic (AFC), which has a slope of up to 20 dB / Dec at a cutoff frequency of up to several kilohertz. It is necessary to take the following important rule: you need to set the filter cutoff frequency in the amount of about one tenth of the operating frequency of the inverter. The cutoff frequency can be calculated using formula 1. This approach will provide suppression of the EMC spectrum by about 40 dB µV [3].

$$f_c = \frac{1}{2\pi\sqrt{LC}} \tag{1}$$

In the case of an LC filter, we select an inductor with a rating of 0.625µH. as well as a 33µF electrolytic capacitor. We also convert the LC filter to a T filter for successful suppression of high frequency EMI. Let's choose a ferrite SMD element WE-MPSB. WE-MPSB is a multilayer ferrite SMD component, which is a special inductance [5].

Since the T filter is a resonant circuit, a damper is required. The snubber is a series RC circuit, which is shown in Figure 14. The snubber resistor value can be calculated using formula 2. The snubber capacitor depends on the input capacitance and should be about four times the nominal value of the standard input capacitor.

$$\mathrm{R}d = L_{filter}/\mathrm{C}_{filter} \tag{2}$$

where $L_{filter}\mathrm{C}_{filter}$ is the inductance of the LC filter and is the capacitance of the LC filter.

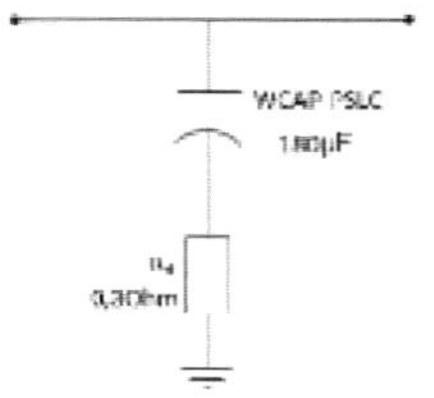

Figure 14. Damper

After installing the filter, we get the following picture for the common mode and differential components of the interference, shown in Figures 15 and 16, respectively.

As you can see from the obtained EMI spectrum, after installing the T filter, we do not exceed the peak and quasi-peak values of the standard, which means we are more likely to pass the certification tests. A general view of the board with filters is shown in Figure 17.

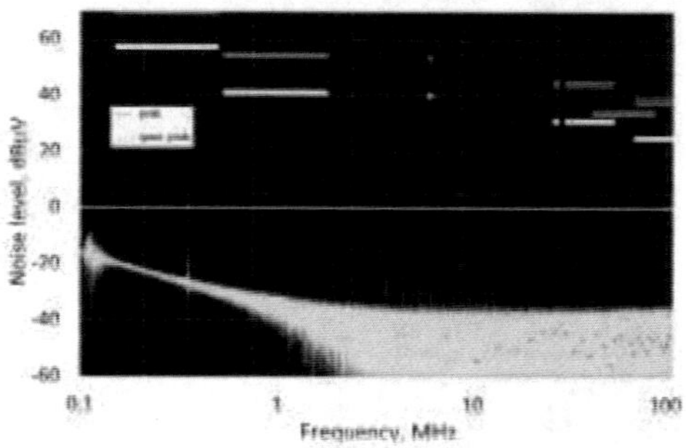

Figure 15. Differential noise spectrum of the converter

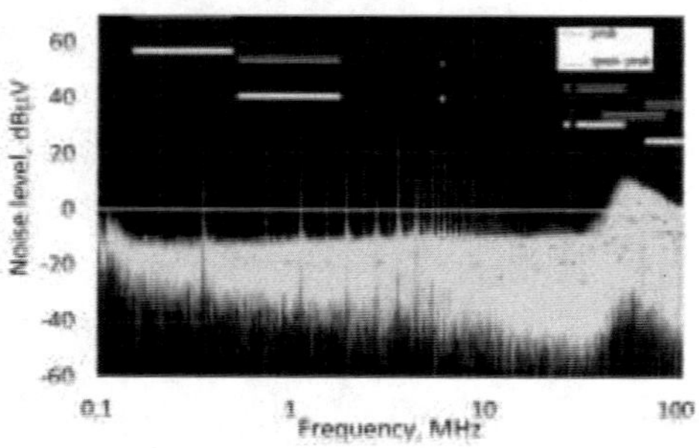

Figure 16. Common-mode noise spectrum of the converter

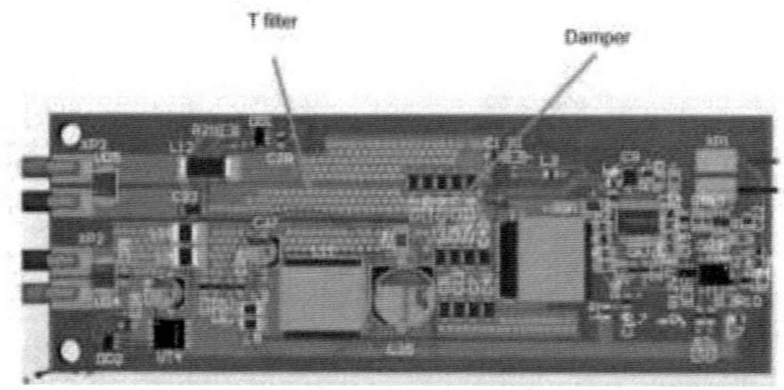

Figure 17. General view of the board with a filter.

7. Conclusion

This method made it possible to determine the level of conducted noise of the step-up converter already at the design stage of the electrical device, and also to correct its operation.

This research done with the financial support from Ministry of Science and Higher Education of the Russian Federation in the frame of the complex project "Design of a high-tech production of a range of GAZelle Next vehicles with a new electronic architecture of electronic systems" according to Agreement No. 075-11-2019-027 from November 29, 2019 (Governmental Regulation №218 from 09.04.2010).

References:

[1] Badalov A L and Mikhailov A S 1990 *Standards for the parameters of electromagnetic compatibility AREO. Directory* M. Radio and communication p 272

[2] Petrovsky V I and Sedelnikov Yu E 1986 *Electromagnetic compatibility of radio electronic means* M. Radio and communication p 216

[3] Sedelnikov Yu E 2006 *Electromagnetic compatibility of radio electronic means: a tutorial* Kazan. CJSC "New knowledge" p 304

[4] White Germantown J R 1977 *Electromagnetic compatibility of radio electronic equipment and unintentional interference* Maryland **1** p 352

[5] Kechiev L N 2007 *Design of printed circuit boards for digital high-speed equipment* p 616

IOP Conf. Series: Materials Science and Engineering 1086 (2021) 012035 doi:10.1088/1757-899X/1086/1/012035

Simulation of the passive part of an active car muffler

A Okunev[1], A Kocherov, S Mantserov, M Kolobov, D Bushmanov

Nizhny Novgorod State Technical University n.a. R.E. Alekseev, Nizhny Novgorod, Russian Federation

E-mail: [1]okunev14av17@mail.ru

Abstract. Active noise reduction is a new method for solving acoustic noise problems. Acoustic and thermal measurements of the passive system were performed to determine the design requirements for the future active system for the internal engine. Further, the study of this system by the finite element method was carried out. The article presents the results of measurements and analysis of the system during modelling.

1. Introduction
One of the most important global problems of modern civilization is noise. It has a negative impact on people's performance and health. The main sources of noise are power plants, namely internal combustion engines. The creation of machines with a reduced level of vibration and noise is an important scientific and technical problem.

Methods of dealing with noise has been divided into two types: active and passive. Passive methods are represented by reactive silencers made in the form of partitions and pipes, and resistive silencers, the channel of which is lined with sound-absorbing material. These types of silencers are valued for their high attenuation over a wide frequency range. However, they have a number of disadvantages, such as a large volume of construction and a low level of efficiency at low frequencies, as well as the creation of back pressure by the reactive part of the muffler, which often makes a passive approach to noise reduction impractical.

The second method in the fight against acoustic noise is active suppression systems. An active silencer contains an electroacoustic device that dampens unwanted noise, creating an audio signal with the same amplitude, but inverted in phase. Using the principles of active noise reduction, light and compact exhaust systems can be manufactured that effectively reduce the weight of the car. The main advantage of the active system is to reduce the hydraulic resistance of the exhaust system.

During the development of the active system prototype, acoustic and thermal tests of the passive exhaust system were performed to determine the requirements for the future design. Based on the test results, a simulation of a passive Bank was performed. This article discusses the simulation of the Gazelle Next silencer by the finite element method.

2. Sound tests
To determine the requirements for the future active noise reduction system, experimental measurements of the main operating parameters in the exhaust path of the internal combustion engine of the Gazelle NEXT car were carried out.

A microphone was used to measure the noise level in dB at various engine speeds. Previously, the car was warmed up to operating temperature. The measurements were performed according to the acoustic measurement standards.

Measurements were made in the idle mode of the engine (800 rpm), and then in steps of 1000 rpm. All measurements were performed without load. The measurement results are shown in figure 1.

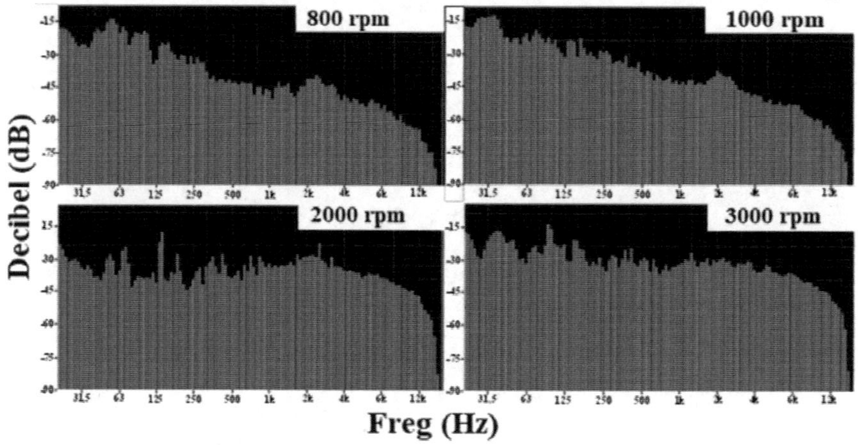

Figure 1. The results of the acoustic measurements at different values of engine speed

3. Thermal test

The temperature of the muffler was measured with a thermal imager. The measurement data were used to determine the boundary conditions for further modeling. The measurements were taken at various locations in the exhaust system. The results are shown in Figure 2.

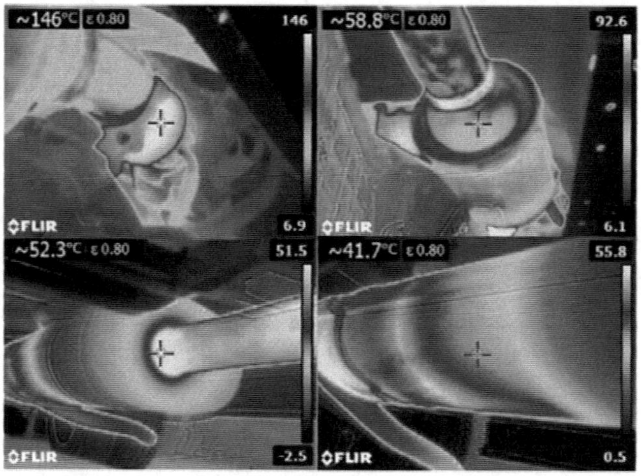

Figure 2. Thermal measurements of passive systems

4. Simulation of a passive silencer

A typical automobile exhaust system is a hybrid design consisting of a combination of reflective and diffusing silencers. The reflective parts are usually tuned to remove the dominant low-frequency harmonics of the motor, while the scattering parts are designed to eliminate high-frequency noise.

The silencer analyzed here is an example of a complex hybrid silencer in which the scattering element is created entirely by flow through perforated pipes and plates.

When developing the silencer model without fibrous materials, the following aspects were taken into account:

- Geometry. Very similar to commercially available car silencers, namely the Gazelle NEXT car silencer (figure 3);

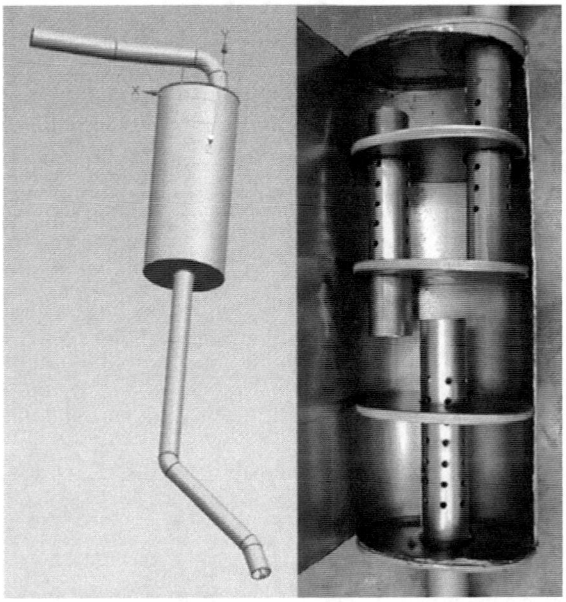

Figure 3. Gazelle Next Silencer

- Average flow distribution - the Mach number in the exhaust system is usually less than 0.3. This means that in silencers with flow expansion, the average Mach number is quite small (less than 0.1). In such cases, the effects of convective flow can be ignored, and the only important effect of the average flow is its effect on the resistance of perforated pipes/plates. This model applies to the case when there is no medium flow in the silencer;
- Temperature distribution - when the engine is running, the air temperature inside the muffler is usually in the range of 300-400° C. The silencer also has a temperature gradient. However, the acoustic effect of this gradient is small, and the average temperature is usually used to calculate the speed of sound. In this case, the experiments were performed at room temperature (20 ° C). Therefore, the model assumes a constant temperature in the silencer and uses the default values for air density and sound speed at 1 ATM and 20° C.

In General, a silencer can be represented as a kind of system for converting sound vibrations coming to its input. A grid is defined as the process of discretizing an infinite geometric domain into a finite number of elements and nodes. In this study, the use of tetrahedral mesh elements. The maximum element length is calculated based on the wave number and wavelength. For finite element analysis, 10 elements per wavelength were used for the maximum scan frequency, which is 1 kHz.

Calculating the wavelength:

$$\lambda = \frac{c}{f} \tag{1}$$

where λ - sound wavelength; c - Speed of sound (343 m / s), f - Maximum frequency (1000 Hz) $\lambda = 0{,}343$ м.

IOP Conf. Series: Materials Science and Engineering 1086 (2021) 012035 doi:10.1088/1757-899X/1086/1/012035

The maximum element length is 0.0343 m. This corresponds to one tenth of the shortest wavelength. The final grid of the calculation model is shown in figure 4.

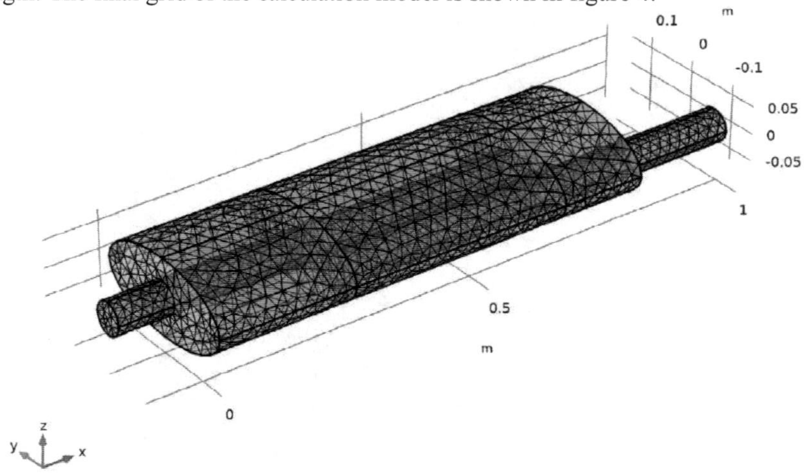

Figure 4. Mesh model of a car muffler

The Helmholtz equation is used for problems in the frequency domain, and the classical scalar wave equation is used for research in the time domain.

The model equation is a modified Helmholtz equation for acoustic pressure p:

$$\nabla\cdot\left(-\frac{\nabla p}{\rho}\right)-\frac{\omega^2}{c^2\rho}=0 \tag{2}$$

where p - density; c - Speed of sound; ω - Angular frequency.

Density should be included in the equation in cases where there are differences in density in different materials. The model assumes that reactive damping prevails in the low-frequency range.

There are three different types of boundary conditions. Rigid (wall-mounted) boundary conditions are used on all solid boundaries, which include the external walls of the silencer, the partitions between the resonator chambers, and the walls of pipes:

$$\left(-\frac{\nabla p}{\rho}\right)\cdot n=0 \tag{3}$$

A combination of incoming and outgoing plane waves is assumed at the input boundary:

$$\left(-\frac{\nabla p}{\rho}\right)\cdot n=\frac{i\omega}{\rho c}p-\frac{2i\omega}{\rho c}p_0 \tag{4}$$

In this equation, p_0 denotes the applied external pressure, and i is an imaginary unit. An outgoing plane wave is defined at the output boundary:

$$\left(-\frac{\nabla p}{\rho}\right)\cdot n=\frac{i\omega}{\rho c}p \tag{5}$$

The following equation defines the transmission loss in the silencer:

$$TL=10\log\left(\frac{P_{in}}{P_{out}}\right) \tag{6}$$

Here, P_{in} and P_{out} denote the acoustic effect at the input and output, respectively. The acoustic effect is calculated using the following equations:

$$P_{in}=\int_{\partial\Omega}\frac{p_0^2}{2\rho c}dA\,,\quad P_{out}=\int_{\partial\Omega}\frac{|p_0|^2}{2\rho c}dA \tag{7}$$

5. Simulation result

In this experiment, a CAD model of a car muffler was developed and imported into a computational program for numerical simulation. The problem was solved in the frequency domain using the pressure Acoustics, Frequency Domain interface. The model geometry consists of four separate resonator chambers separated by thin walls (figure 5).

The inlet and outlet correspond to the connection in the direction of the engine and free air, respectively.

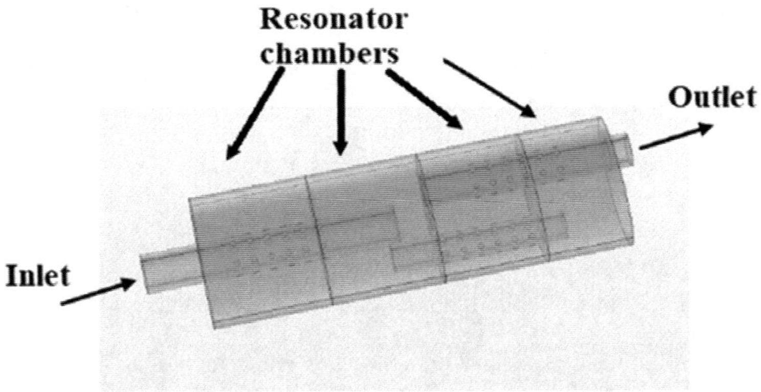

Figure 5. The geometry of the computational model

Figure 6 shows the result of studying the amount of transmission loss when changing the frequency. This graph shows that damping is better at higher frequencies, with the exception of a few deep dips across the entire frequency range. The simulation results will be used in the design of the active exhaust system of the Gazelle Next car.

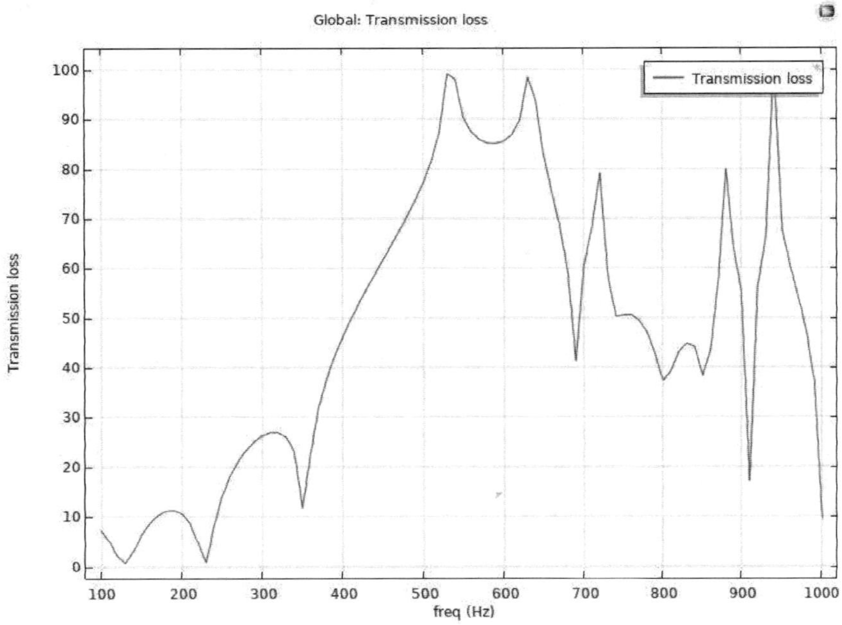

Figure 6. Damping (dB) in the silencer as a function of frequency (Hz)

IOP Conf. Series: Materials Science and Engineering 1086 (2021) 012035 doi:10.1088/1757-899X/1086/1/012035

6. Conclusion

Acoustic and thermal measurements of the Gazelle NEXT car were made. The measurement results were used to determine the technical characteristics of the active type system. The measurement results were also used for modelling the exhaust system. The simulation was performed using the finite element method, using a slightly modified Helmholtz equation for the calculation. A graph of transmission losses of a passive silencer in the frequency range from 100 to 1000 Hz was obtained. Measurements were made for the resistive part of the silencer.

This research done with the financial support from Ministry of Science and Higher Education of the Russian Federation in the frame of the complex project "Design of a high-tech production of a range of GAZelle Next vehicles with a new electronic architecture of electronic systems" according to Agreement No. 075-11-2019-027 from November 29, 2019 (Governmental Regulation №218 from 09.04.2010).

7. References

[1] Mantserov S A, Kocherov A V and Okunev A V 2019 Development of a model of active noise reduction in internal combustion engines *IEEE Xplore*

[2] Nelson P A, Elliott S J 1992 *Active Control of Sound* Academic Press. San Diego. CA p 432

[3] Kuznetsov A N and Polivaev O I 2010 Prospects for the use of active noise cancellation systems *VESTNIK* **1** p 46

[4] Wu J D and Bai M 2000 Digital signal processor implementation of active noise control systems for broadband noise cancellation in engine exhaust systems *Jpn. J. Appl. Phys* **39** pp 4982-4986

[5] Kuo S M and Morgan D R 1999 Active noise control: A tutorial review *Proc. IEEE* **87 - 6** pp 943–973

[6] Simranjit S 2013 *Implementation of Active Noise Cancellation in a Duct* p 82

[7] Shaw J 2002 Design and Control of Active Muffler In Engine Exhaust Systems *Taipei* p 6

[8] Boonen P S 1998 Development of an active exhaust silencer for internal combustion engines *Heverlee* p 8

[9] Gonzalez A et al. 2006 Multichannel active noise equalization of interior noise *IEEE Trans. Audio, Speech, Lang. Process.* **14 – 1** pp 110–122

Simulation of the process of interaction between the vehicle and snow in the environment of DEM and MBD-systems

D S Teslenko[1] and V V Belyakov[2]

[1]Moscow Polytechnic University, Bolshaya Semenowskaya str., 38, 107023, Moscow, Russia
[2]Nizhny Novgorod State Technical University n.a. R.E. Alekseev, Minina str., 24, 603950, Nizhny Novgorod, Russia.

[1]E-mail: mark-b52@inbox.ru

Abstract: The problem of increasing the mobility of tractors on snow in our country is very important. This is due to the long duration of the snow cover, primarily in those regions that are of great national and economic importance in terms of mining. There are many methods to increase the mobility of the vehicle, but all of them are associated with increased traction, as well as bearing characteristics of the wheel. The question of calculating the resistance to movement on the part of the body parts submerged in the snow, as well as transmission and engine for a long time did not receive sufficient attention, despite the fact that their contribution to the overall resistance to movement can reach 60%. In this article the existing methods of calculation and modeling of the value of resistance to movement on the side of housing parts are considered, their advantages and disadvantages are given. The design of a soil canal to study the movement of large-scale models of cars in the snow is considered, and the conclusion is made as to the feasibility of using this method at the present time. The paper also suggests a new method of drag calculation based on the joint use of discrete element modelling (DEM) and multibody dynamics (MBD) software packages. This approach allows both realistically simulate snow due to the discrete element modeling approach, and enter into the model the values of loads on the suspension, traction force, etc. through the use of MBD. The article deals with the technique of modeling the interaction of the body with snow in the EDEM package of DEM Solutions Ltd. The scheme of interaction between DEM and MBD-packages on the example of EDEM and MSC Adams software is described, as well as the technique of using Adams/Co-simulation Interface (ACSI).

1. Introduction

Modern economic development of Russia is connected with the increase of productivity of existing transport and technological machines. Recently customers, designers and researchers are interested not so much in passability of motor transport vehicles (MTV), as in their mobility - the integral index describing set of qualities [1]. It includes, except for cross-country ability, such parameters, as speed, fuel consumption, resource of motor transport, and also such properties, as reliability, maintainability, maintainability. There are many ways to increase mobility (Figure1), each of which has been thoroughly studied in many works, such as [2, 3].

At the same time, there are still a number of particularly difficult cases of machine mobility loss that need to be considered in more detail. One of these problems is mobility on weak soils. They traditionally

IOP Conf. Series: Materials Science and Engineering 1086 (2021) 012036 doi:10.1088/1757-899X/1086/1/012036

include marshy soils, water-saturated clay, as well as sand and snow. And if the mobility on the sands is of little relevance because of the almost complete absence of economic activity in the desert zone and is interesting rather in the defense aspect, the mobility on saturated soils is relevant in agriculture, on marshy soils - for the tasks of geology and mining.

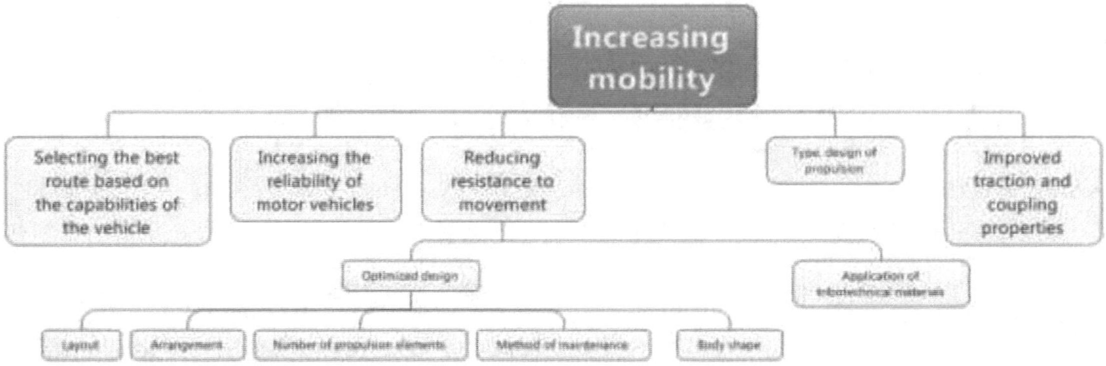

Figure 1. Ways to increase the mobility of the MTV

2. Relevance of the research

Of the weak soils, snow is the most common in the Russian Federation. The appearance of snow varies from year to year and quite often does not differ much from the average for many years of observation. However, in some regions there may be trends towards significant time shift. In addition, in recent decades the dates of the first snowfall on the territory of Russia have fluctuated within 2-4 decades for all regions [4-9]. According to [10], about 80% of the territory of the former Soviet Union is covered by it for a long period of time. The authors of the book [10] point out that in the Far North, on the coast of the Arctic Ocean the snow cover is 8-10 months a year, and in some areas of the North (Turukhansk region) the average monthly maximum height of snow cover reaches 1.2 m. Such duration, stability and very large depth of snow cover is a significant feature of our country's climate and has a great impact on its economy and lifestyle.

The range of wheeled machines in winter is extremely wide. This includes the transportation of goods, construction and maintenance of winter roads in normal conditions, and work to combat snow drifts. There is a large volume of transportation and snow retention work in agriculture, as well as a wide range of work in forestry.At the same time, the wheeled machines themselves are not adapted to moving on weak-bearing support bases, and the experience of operation shows that the efficiency of using wheeled machines in winter time is 2...3 times lower than crawler ones, and 5...6 times lower than in summer period [11]. In such conditions of operation the movement of wheeled machines is not only difficult, but often excluded altogether. Therefore, the use of wheeled machines, equipped with means of cross-country ability, is most often the only possibility of technological operations.

Movement on deep snow [12] wheeled cars is connected with dive of wheels and running gear in snow on considerable depth. The great depth of the track laid in the snow requires overcoming great resistance to movement. At the same time, movement on the snowy virgin land is almost always associated with increased wear and tear as a propulsor [13] (tires at reduced pressure dramatically lose their life), and the car as a whole. In addition, rutting requires quite high energy consumption, which leads to increased fuel consumption. Speed of movement does not usually exceed 25 km/hour.

Methods of calculation of wheeled cars cross-country ability on snow without taking into account excavating and bulldozing effects and interaction of the car bottom or protruding parts of the body do not reflect the real picture observed in operation [14]. When calculating by these methods the machine has a reserve of traction force when moving on any snow and, therefore, the cross-country ability is provided in any conditions of movement, which is in sharp contrast to practice. Taking into account the real character of interaction between the wheeled mover and the snow cover significantly clarifies the prediction of cross-country ability of the machine. In this case, for each type of snow there is a certain depth of occurrence, at which there is a loss of cross-country ability due to the fall of the reserve of traction force.

3. Analysis of publications

If the carrying capacity of the "snow-machine" system is such that the wheel is immersed in the snow by an amount exceeding the ground clearance, the probability of loss of cross-country ability of the machine sharply increases due to changes in both resistance and traction forces. On the one hand, there is a sharp increase in resistance to movement, and on the other hand, the traction force, which can be realized by the engine when interacting with snow, is significantly reduced [15]. Despite the presence of studies [11, 14, 16], which propose methods for calculating the value of resistance to movement from the bottom and body elements, this question is difficult to call solved. Available methods do not allow to predict the flow of snow masses along the machine, the influence of structural features of the body on the value of resistance.

Of particular importance is the problem of increasing mobility when moving on weak-bearing support bases is for small-size robotics, mainly operating in difficult conditions (Figure 2). Due to the scale effect, the requirements for their mobility are much higher than for vehicles of traditional size. The best known example of the importance of this problem is the Mars rover, immobilized due to hitting the sand dune. This was the reason why the Mars Exploration Rover mission worth several hundred million dollars ended in 2009.

The reason for this phenomenon is that the classical methods of terra-mechanics (engineering science studying the interaction between vehicles and deformable landscape) were developed mainly for large, heavy (more than a ton of weight) vehicles and were not originally intended for use on small, lightweight robots [17]. To date, there has been a boom in the development of small (less than 50 kg in weight and less than 1 m in length) robotic systems designed to address a wide range of both military and civilian tasks [18] . Traditional calculation methods based on the theory of Becker and other founders of terra-mechanics [19] are not always applicable to them, which actually leads to the creation of a new sciense – micro-terramechanics.

The proposed article sets a task to search for a method that allows you to predict the value of resistance when moving on the snow for small automatic, stand-alone automatic telephone exchanges. This research will allow to create more mobile, reliable and economical robotic systems.

Let's consider different methods used to determine the resistance force of movement in the snow. Their number is relatively small in comparison with the methods used for solid soils, the accuracy is relatively low. The reason for this is that the number of snow types and conditions is much higher than the variety of soils. Additional difficulties are added by the snow metamorphism, the prevailing effect of temperature, greater heterogeneity and porosity. The shear process takes place over time and, naturally, the physical state of snow changes [4-9]. The limit of snow compressive strength changes in time, and there is no long-term limit of tensile strength [22].

Figure 2. Test of a combat robotic complex in winter conditions [20]

It is most obvious to use methods similar to those used in snowplow calculations to determine resistance. Their traction and energetic calculations are made with the help of conditional snow resistance coefficient for cutting k_{res} . [23]. On the physical side, cutting resistance is the total resistance of elementary types of deformation, covering the resistance to compression, tearing, adhesion and friction (internal and external). Quantitatively, it expresses the force required to cut a layer of snow with a cross section equal to one 1 m^2.

According to the results of studies by A.L. Gorbunov, the values of the cutting resistance coefficient for different types of snow were obtained [23]. Tests of snow in its natural state (undisturbed, without compaction) were carried out on "frontal cutting", where the cutting edge was perpendicular to the direction of movement of the knife, delivered at a cutting angle of 37°. These data were obtained when cutting the layer with s=0.02 m^2 section 0.4 m wide and 0.05 m thick. According to the experiments of A.A. Kungurtsev and other researchers [244], the value of snow resistance to cutting is significantly influenced by cutting speed.

The analysis of literature sources shows that little attention was paid to the issue of snow cutting process earlier and that there are a number of contradictions in published works. Cutting forces calculated using well-known methods of V.M. Gusev, A.L. Gorbunov, V.F. Kulepov have values that differ significantly from each other [25].

The practice of designing the working equipment of snow plows shows that the available data on cutting forces are not enough [26, 27]. The cutting resistance coefficient does not take into account the influence of chip thickness, cutting width and cutting angles. The cutting resistance coefficient is determined experimentally and does not take into account the physical basis of the fracture process.

F.M. Zakirov [28] notes that the use of empirical dependencies in determining the cutting resistance of soil and snow becomes an obstacle to the improvement of dumps and other working bodies of earthmoving and snow-plow machines [29].

As a result of the review of studies in the field of fracture mechanics and theory of the limit state of bulk and bonded media, a qualitative commonality of the process of snow destruction by cutting with the process of destruction of ice, frozen soil and rocks was established [28, 30]. It is known that the mechanical characteristic of the material is the envelope of the limiting circles of the Sea, the shape of which depends only on the properties of the material. If the envelope of limiting circles for the material is set, then at any stress state it is possible to determine the safety margin, or to solve the problem of minimizing the energy intensity of the process of material destruction [31]. For a number of continuous media such as soils, rocks, snow and ice, there is a quality community of material destruction process. This qualitative generality consists in the fact that compression of a constricted volume of the material creates a compacted core, under the influence of which there are and spill cracks in inclined directions, there is a cyclic separation of the sliding bodies from the array [32].

According to A.N. Zelenin's method, the envelope of fracture is tangential to the tensile limit circle, and tilted to the α axis at φ of internal friction. Under this condition, the full identity of calculation and experimental data for a wide range of anisotropic materials is achieved. However, for the application of A.N. Zelenin's idea it is necessary to have the limits of compressive, tensile and internal friction strength of the material. Due to the variety of types and states of snow, obtaining these characteristics is associated not only with a large volume of laboratory tests, but also with the need to identify the type and current state of snow in real conditions.

In M.F. Zakirov's article [27] it is proposed to construct a line of snow destruction not by the main stresses but by the components of the medium resistance vector, which are associated with energy consumption for snow destruction. This allows to simulate the forces of resistance that occur when towing any snow plow. The proposed method is confirmed by experimental material, confirming the theory of the author.

4. Description of the model

Modeling the simulation of the interaction of automatic telephone exchange with the snow cover to determine the value of resistance to the movement of real automatic telephone exchange is complicated by the fact that its shape is very different from a plough snowplow, it is very complex, therefore, this method is clearly not applicable. The problem is solvable if we divide the surface of the MTV model interacting with snow into separate elementary surfaces, "facets" (Figure 3). In this approach, each of the facets having an area, the coefficient of friction, depending on the material, the angle of interaction with snow, can be calculated by the method of plough snowplow. The total resistance is the sum of the resistance of all the facets. This problem is insoluble manually, but quite in the power of modern personal computers.

To implement such a hybrid model, it is necessary to create brand new software, which makes the method expensive. The simplest way to determine the value of resistance to the MTV movement in snow is suggested in [16]. The method suggests reducing all submerged in snow or other weakly-bearing ground parts of automatic telephone exchange (suspension, transmission, engine crankcase, housing structures, etc.) to a generalized body (Figure 4).

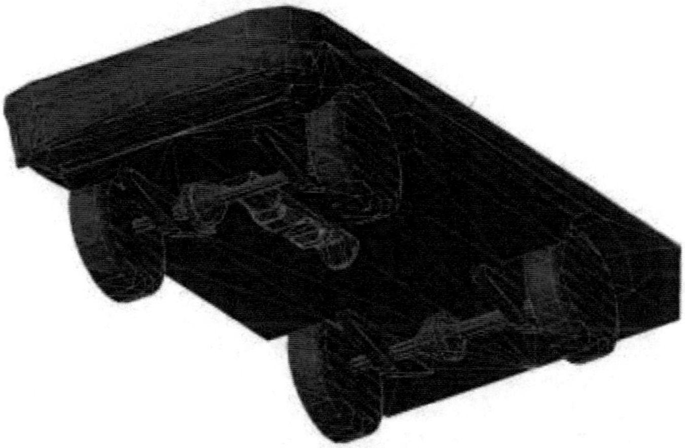

Figure 3. The faceted representation of the body

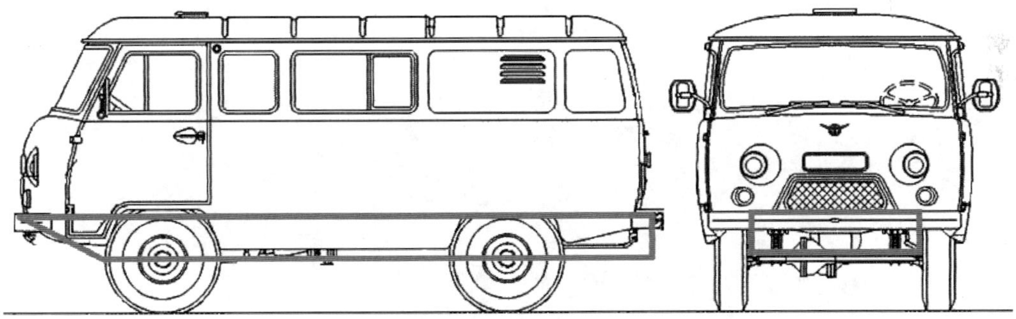

Figure 4. Side and frontal MTV projections (Generalized body is shown on the projections with red line) [16]

The proposed method allows to qualitatively compare the cross-country ability of cars on the snowy virgin land, but does not allow to assess the mobility. For this purpose it is necessary to calculate the value of resistance provided to the MTV movement by snow, which requires to take into account the shape of structural elements immersed in the snow. Obviously, cars with the same area of the front projection, but with a different shape of the bottom will have different mobility.

Historically, the first way to test the MTV were runs - long runs between different climatographic zones in its course. For all the proximity to the real road conditions this method has serious disadvantages - firstly, the lack of opportunity to conduct a serious repair or modification of the MTV during the passage of the route and secondly, significant time and money costs to organize the mileage [32].

Therefore, by the middle of XX century, the mileage was abandoned in favor of specially prepared polygons that have sections of roads with imitation of different surfaces, as well as simulations of obstacles. This made it possible to significantly reduce the testing time, facilitate the testing of new designs and commissioning of new machines.

But the test site is quite expensive in construction and operation. Often, to test individual elements of the design it is redundant, so use a highly specialized stands that reflect this or that aspect of the PBX. One of them is a soil canal, used mainly to determine the parameters of interaction between the wheel and the road [33]. The scheme of one of such stands is shown in Figure 5. Stand includes: test object; propulsor 1, propulsor drive 2; soil canal 3; load simulating normal load 4; cart 5; dynamometer to measure the traction force 6; drive 7; adjustable brake mechanism 8. In addition, can be used to measure torque, angular and linear movement of the wheel, tire deformation, ground deformation, normal contact pressures, specific tangential forces in contact.

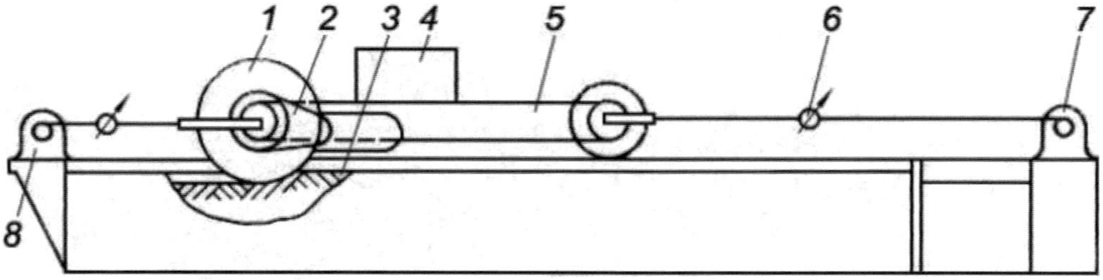

Figure 5. Stand to study the interaction of the wheel with deformable soil [33]

Basically, the ground channel is used for modeling the interaction between the wheel and the ground, but its application by M.G. Becker is known [34] for the task of finding the optimal propeller in the conditions of the Moon surface. Unlike traditional ground channels, which use a life-size wheel, he created reduced models, which allowed to simplify modeling by accelerating and reducing the cost of creating the models themselves. Unfortunately, it is currently unknown which soil simulator Becker used in his experiments half a century ago, but judging by the preserved photo (Figure 6), it is a loose material that resembles flour or starch, imitating a weakly-bearing soil in the form of moon dust.

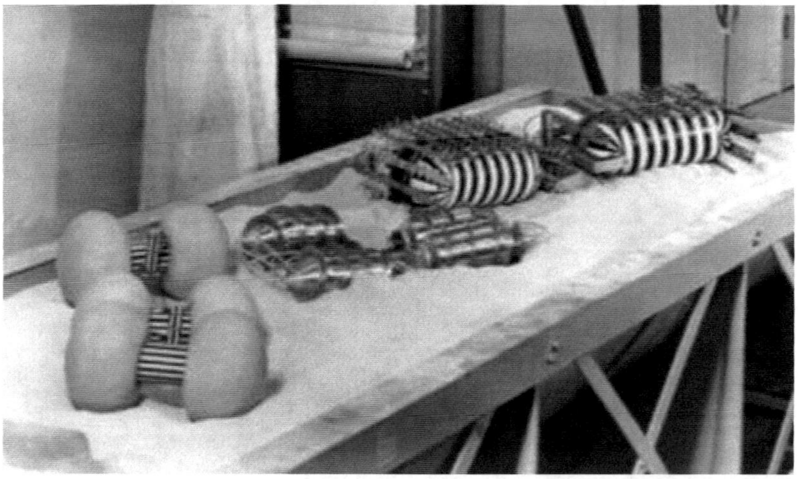

Figure 6. Test of lunar rover models in the Becker soil canal [35]

IOP Conf. Series: Materials Science and Engineering 1086 (2021) 012036 doi:10.1088/1757-899X/1086/1/012036

5. Research stand

In this paper, at the first stages of the experiments, a scaled ground channel was used to physically model the interaction of MTV models with weakly-bearing soil. The scale of the tested MTV models has been chosen equal to 1:43 due to the compactness of the testing installation, low cost, wide range of the models of this scale presented on the market.

The experimental installation was made, including the soil channel on the basis of a small wooden box, towed in it the MTV model (Figure 7), microwind on the basis of the collector electric motor (Figure 8). Measuring the traction force on the microwind was carried out by indirect method on the size of current strength, measured with milliameter, programmed in grams. The towed model of automatic telephone exchange was equipped with a cargo platform, allowing to place on it cargoes of the attachment, bringing weight of model in proportional conformity with prototype. A specially selected mixture based on starch was used as a soil simulator.

Figure 7. Model in the soil canal

Figure 8. Microwindlass

The use of the first version of the stand showed the fundamental possibility of using the ground channel to simulate the MTV movement in the snow with the measurement of resistance from the ground simulator. Its main disadvantage was low force measurement accuracy. Replacing the soil simulator was complicated by the fact that the mixture had to be poured into the box and poured out of it.

At the second stage of the experiment, the second modification of the stand was made, which eliminated these drawbacks. Trays with soil imitator were made replaceable, installed in an aluminum frame mounted on a wooden shield. The micro winch was mounted hinged (Figure 9), the force from it was transferred to the grammeter, which directly measured the towing force. The resulting single traction-dynamometer module significantly improved the measurement accuracy.

The main disadvantage of the stand version 2 was the ability to measure only instantaneous force values without fixing its change over time. Also, due to the rather small scale of the models being tested, the spread of values even when testing one model exceeded 50%. There was also a lack of detail of the models on the market, which reliably simulated only the top part of the body without a detailed study of the bottom.

The third modification of the stand was designed to improve measurement accuracy. It was planned to reduce friction in the joint of the traction-dynamometer module by installing rolling bearings instead of sliding bearings, as well as add a sensor that removes the instantaneous value of the force with the transfer to a PC to record and plot the force dependence on time. This modification was not carried out, as it would not give a noticeable result because of the small scale of the models. The supposed transition to the use of larger scale models (1:35 or even 1:24) was canceled due to the extremely high cost of these models both in case of their purchase and 3d-modeling with subsequent 3D printing.

Figure 9. Traction and dynamometer module of the second version of the stand

Despite the fact that the ground channel method has its limitations (it allows to remove only one parameter from the scale model - the towing force), it can be used for rough comparative analysis of the drag value of different structures when moving on weakly-bearing soil. If there are sufficient automated computational capabilities, its use is unprofitable as it requires the cost of producing physical models and selecting a soil simulator for each real soil.

6. Virtual simulation

The obvious solution in this situation is to use virtual simulation of soil interaction with the PBX. Until recently, the finite element method (FEM) remained actually the only method for high-precision modeling of terramechanics problems [36]. As applied to vehicle movement, it allows, for example, to calculate ground stresses during wheel rolling [37]. The method allows considering the ground as a solid body consisting of a finite number of elements. When the destruction condition is specified, the elements with stresses exceeding the specified value are excluded from the task space.

Therefore, FEM is not applicable for solving problems related to soil transfer after its destruction - for example, optimization of bulldozer knife design to reduce energy consumption. With the development of computational power and software products, it became possible to use the discrete element method (DEM)[38]. In it, the soil is represented as an array of particles connected to each other by ties. By changing the mass, shape (it may be different from a sphere), size and character of bonds between particles, it is possible to simulate almost any soil. The soil may consist of different types of particles, in which case the nature of the relationship is defined individually for each particle pair.

DEM is very demanding for computational resources, because the number of interactions, and, consequently, the resource intensity of the calculation increases with the number of particles. This method supports specialized calculation packages such as AMBER, Chute Maven (Hustrulid Technologies Inc.), PFC2D and PFC3D, EDEM (DEM Solutions Ltd.) GROMOS 96, ELFEN, MIMES, PASSAGE®/DEM, Bulk Flow Analyst, Rocky. Discrete elemental modeling is also possible in finite element packages such as LS-Dyna [39].

When using DEM the soil (in this case snow) is represented as an array of separate interrelated elements, in which an absolutely rigid body moves with some speed (Figure 10). In this case, each virtual element of the ground is a material point having a mass but not a size. Each element is connected with other links (Figure 11), the nature and force of which determine the type of material [40]. These can be both elastic and inelastic connections. In addition, electrostatic, thermal, friction force, etc. can be introduced. Dimensional correspondence of the virtual element of the soil to the real one is set by specifying the so-called contact radius - the distance between material points at which the bonds begin to operate. Under the influence of external growing force, the virtual elements can approach or stretch to a certain distance, which is influenced by the nature and force of the connection, and then the connection breaks.

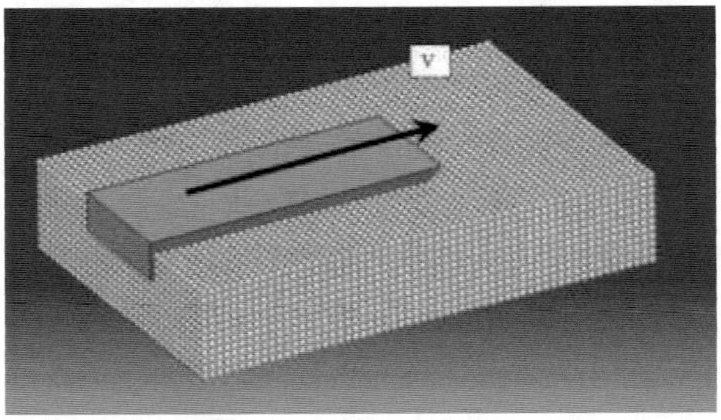

Figure 10. Setting the task of solid body movement in snow in DEM

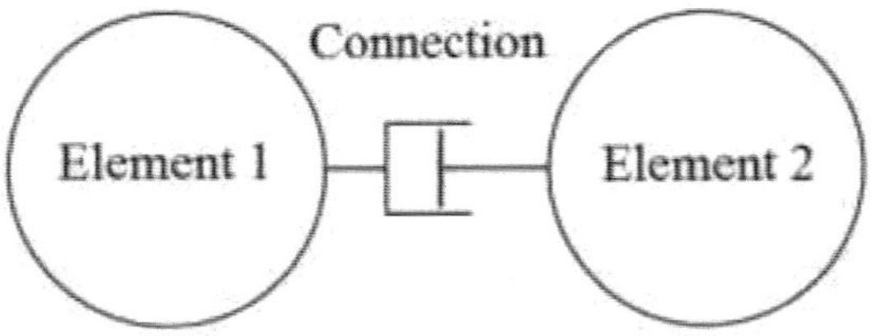

Figure 11. Interaction of two separate soil elements

Let us consider the elementary area of interaction between a moving body and snow elements (Figure 12): On the body side, each element is affected by a pressure force F_t,, which is balanced by an equal resistance force from the neighboring R elements. Also, the element is affected by the force of gravity G and the force of normal reaction of the support *N*, which is played by the underlying element. After the force under the action of force from the side of the body, the connection of the element with the neighboring elements is torn, the element is detached from the snow mass and starts its own movement. In this case, it continues to be affected by the force of friction on the side of the body, the force of pressure on the side of the body, the force of friction on the side of other elements, the force of normal support reaction. In this case, the relationship with the neighboring elements becomes different, episodic.

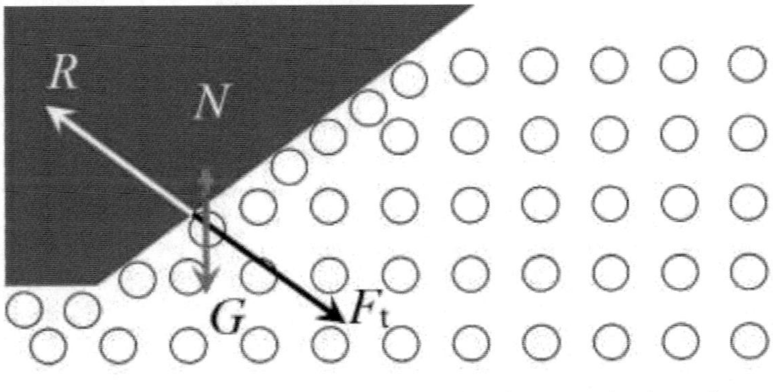

Figure 12. Elementary area of interaction between body and snow

In order to solve the task of modeling the MTV movement in the snow, the EDEM system was chosen, because at present it is actively developing, has a visual interface and is equipped with training materials from the manufacturer. It is important that DEM Solutions Ltd. provides an academic license, which was obtained by the author specifically for this research.

Discrete-element at the moment seems to be the most adequate to the task of all considered, at the same time, it is not without drawbacks. The main one is that DEM-packages, including EDEM, are not originally intended for solving terra-mechanics problems, and body movement in them can be set only with a certain fixed or variable speed.

But to fully simulate the movement of the MTV in the snow requires a different approach - to set the engine power, then virtually "bring" it to the wheels to assess the adequacy of power to overcome the obstacle. For this purpose specialized MBD-packages (Multi body Dynamics) such as MSC Adams, LMS VL Motion are used.

The solution is a connection of MBD and DEM packages. In this case DEM supplies realistic values of ground loads to the common solution space, and MBD controls kinematics (Figure 13). Such approach will allow to model the terra-mechanical system in full, because the modeling of suspension, transmission, etc. is performed in MBD - package, and the soil model in the resulting calculation provides discrete-element modeling software. MSC Adams was chosen as MBD - package, as its licenses are available in the Center of Engineering Competences of NSTU n.a. R.E. Alekseev.

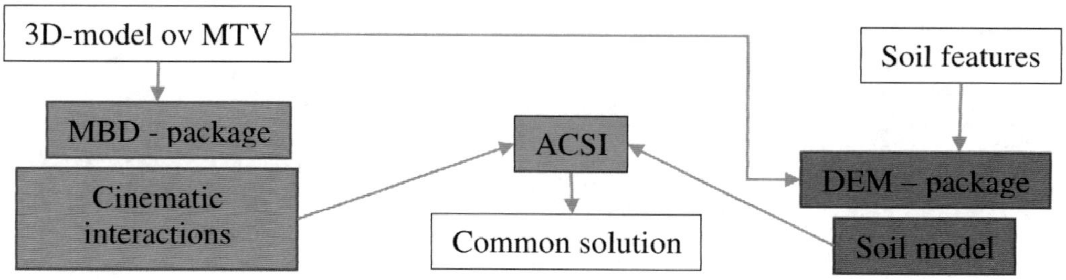

Figure 13. Structure of software interaction for solving terra-mechanics problems.

For joint calculations in EDEM the Coupling Server unit is used, and Adams has a special Adams/Co-simulation Interface (ACSI) module. The sequence of actions to create a joint calculation is as follows: first, prepare a model in EDEM, which includes both 3D model of the MTV, and the soil with all its interactions.

Then, the same model as used in EDEM is loaded into Adams, but instead of soil, kinematic interactions between the parts are defined on it. A *.adm file describing all fixations and interactions within the model is generated as well as a *.asf file describing the duration of the simulation and its discreteness. After that, in the ACSI interface the interactions between the corresponding parts of EDEM and Adams - models are alternately established, and on the basis of these data a *.cosim file is formed (Figure 14). After starting the EDEM Coupling Server, it is possible to run a joint simulation in the ACSI window. In this case, the result of the simulation will be displayed in the EDEM window (Figure 15).

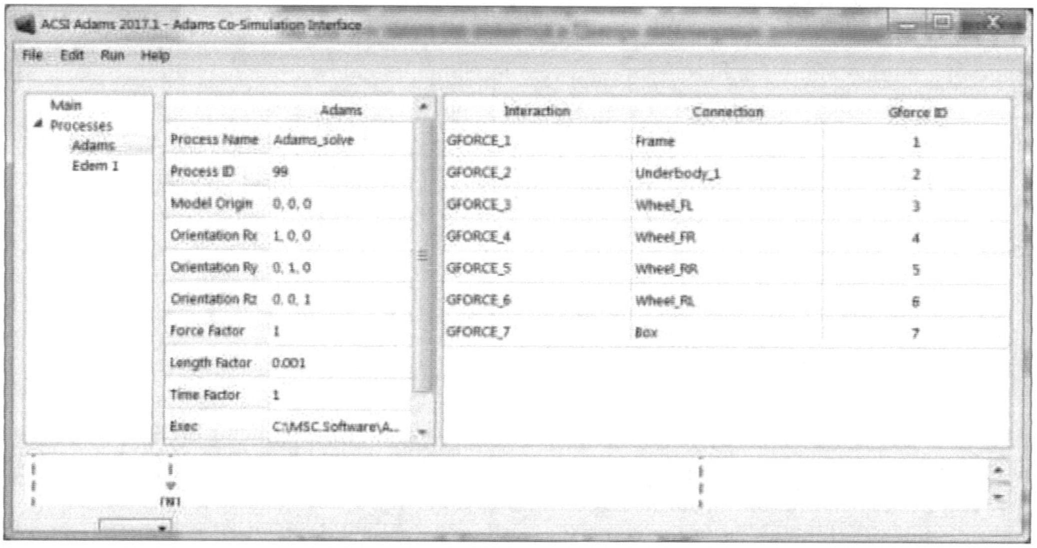

Figure 14. The ACSI interface window.

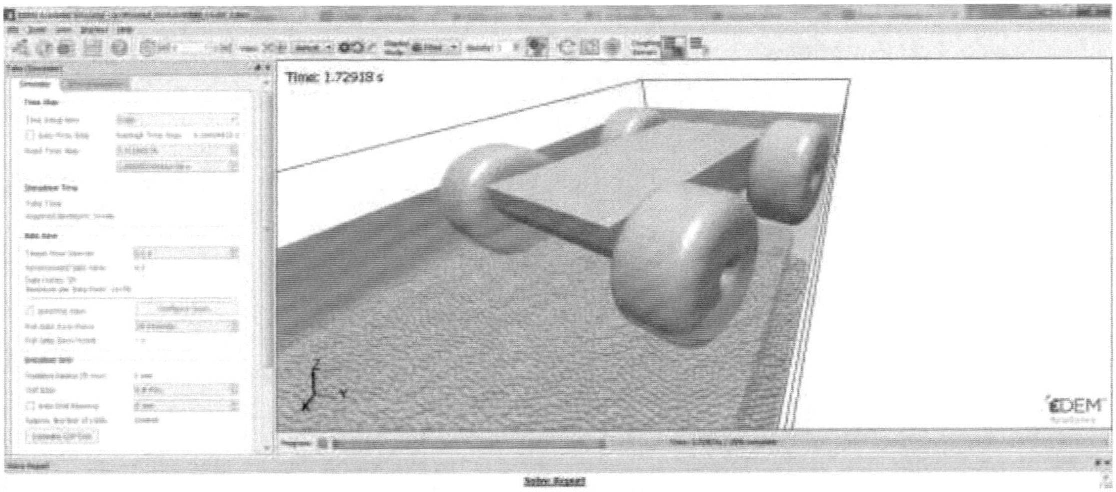

Figure 15. EDEM result window

7. Results and future work

Thanks to the use of the DEM+MBD bond, it is possible to simulate terramechanical interactions relatively cheaply and fairly accurately with minimal use of natural experiments. In the future publications it is planned to give the results of such tests on the basis of the test stand created by the students of Moscow Politech within the framework of the subject "Design activity".

Acknowledgment

This study was conducted in continuation of the research conducted by "Nizhny Novgorod scientific and practical school of snow vehicles" and with financial support of the grant of the President of the Russian Federation No. MD-226.2020.8.

References

[1] Belyakov V V et al. 2013 Belyakov V.V., et al. 2013 The concept of mobility of ground transport and technological vehicles *Transactions of NNSTU n.a. R.E. Alekseev* pp 145-175

[2] Belyakov V V 1991 Methodology for calculating and analyzing ways to increase the cross-country ability of multi-axle wheeled vehicles on snow *Diss. Cand. tech. sciences N. Novgorod* p 307

[3] Belyakov V V, Goncharov K O and Makarov V S 2010 Experimental studies of a multi-wheeled vehicle *Science and education: scientific publication of the Moscow State Technical University n.a. N.E. Bauman* **12**

[4] Makarov V, Zeziulin D and Belyakov V 2014 Prediction of all-terrain vehicles mobility in snowscape scenes *18th International Conference of the ISTVS*

[5] Goncharov K O, Belyakov V V and Makarov V S 2018 Estimating method of efficiency of wheel chassis driving on snow *IOP Conference Series: Earth and Environmental Science* **194(3)** 032005 doi: 10.1088/1755-1315/194/3/032005

[6] Kurkin A et al. 2017 Ground vehicle for ice conditions monitoring *13th International MEDCOAST Congress on Coastal and Marine Sciences, Engineering, Management and Conservation, MEDCOAST 2017* **2** pp 827-838

[7] Porubov D et al. 2017 Study of efficiency of using all-terrain vehicles during the winter period *19th International and 14th European-African Regional Conference of the ISTVS*

[8] Anikin A et al. 2017 Calculation of traction capabilities of wheeled vehicles on low-pressure tires on snow *19th International and 14th European-African Regional Conference of the ISTVS*

[9] Manianin S et al. 2019 Experimental Research of the Effect of Track Chains Pre-static Tension on the Mobility of a Tracked Vehicle *ICMT 2019 - 7th International Conference on Military Technologies* 8870025 doi: 10.1109/MILTECHS.2019.8870025
[10] Rukavishnikov S V et al. 1986 *Snowmobile cars.* Gorky: Volgo-Vyatka book publishing house p 191
[11] Anikin A A, Belyakov V V and Donato I O 2006 *Cross-country ability of wheeled vehicles in snow* Moscow: MSTU n.a. N.E.Bauman p 240
[12] Lavrentyev V B 1974 *Driving cross-country vehicles. To help builders BAM* Moscow: Transport p 96
[13] Lazebnikov M G 1958 *On the cross-country ability of vehicles on virgin soil and snow* Moscow: Military Publishing House of the USSR Ministry of Defense p158
[14] Anikin A A and Barakhtanov L V 2010 Analysis of resistance to the movement of tracked vehicles in the snow *Science and Education (electronic edition)* **8**
[15] Belyakov V V 1999 *Interaction with the snow cover of elastic propellers of special transport vehicles* Nizhny Novgorod. Moscow State Technical University. N.E. Bauman p 486
[16] Teslenko D S et al. 2017 Express-method for comparative assessment of the influence of body elements and other units on the mobility of wheeled vehicles when driving on snow *Proceedings of NAMI* **2** pp 28-36
[17] Bouguelia M et al. 2017 Unsupervised Detection of Soil Embedding Events for Planetary Exploration Rovers *Journal of Terramechanics* **73** pp 95-106 DOI: 10.1016/j.jterra.2017.09.001
[18] Korchak V Y, Lapshov V S and Rubtsov I V 2015 Prospects for the development of ground-based robotic systems for military purposes *Izvestia SFedU. Engineering Science* pp 83-95
[19] Becker M G 1973 *Entry into the theory of terrain-machine systems* Moscow/ Mechanical Engineering p 520
[20] *Russian combat robots conquer snowy battlefield in drill outside Moscow* URL: https://www.rt.com/news/372304-russia-combat-robots-test/

[21] Voitkovsky K F 1977 *Mechanical properties of snow* Moscow. Nauka pp126
[22] Shalman D A 1973 *Snow blowers* Leningrad: Mechanical Engineering p 215
[23] Evenov I M 1954 *Snow blowers* Leningrad. Mashgiz pp 144
[24] Voskresensky G G 2011 *Scientific foundations for the design of working equipment for the destruction of compacted snow on highways* Ph.D. Khabarovsk p325
[25] Merdanov Sh M 2010 *Scientific bases for the creation of complexes of machines for the construction of temporary winter roads in the regions of the North and Siberia* Avtoref. dis. Dr. tech. sciences. Tyumen pp 38
[26] Sokolov A M 2007 *Improving the efficiency of the operation of snow removal equipment taking into account the layering of the snow cover* Avtoref. dis .. cand. tech. sciences. N. Novgorod pp 19
[27] Zakirov M F and Baranchik V P 2011 A new approach to the selection of snow destruction modes by the working bodies of machines *Intelligent systems in production* **2 (18)** pp 107-110
[28] Balovnev V I 1994 *Modeling the processes of interaction with the environment of the working bodies of road construction machines* Moscow. Mechanical Engineering p 432
[29] Baranchik V P, Subbotina A S and Zakirov M F 2008 Determination of resistance to cutting using the theory of similarity *Interstroymekh-2008* pp 265-268
[30] Ivanov O I 2014 On the question of modeling the process of chip formation when cutting metals *Vektor nauki TSU* **3 (29)** pp 57-61
[31] Zelenin A N, Karasev G N and Krasilnikov L V 1969 *Laboratory workshop on soil cutting. Tutorial* Moscow: Higher School p310
[32] Kutenyov V F 1984 *The NAMI auto test site is a testing center for the automotive industry of the USSR* Dmitrov NAMI
[33] Ageikin Ya S 1981 *Passage of cars* Moscow. Mechanical Engineering p 232

[34] Bekker M G 1964 Mechanics of Locomotion and Lunar Surface Vehicle Concepts *SAE Technical Paper*
[35] Bekker M G 1963 Mechanics of locomotion and lunar surface vehicle *Defense Research Laboratories at General Motors paper* 632K
[36] Teslenko D S et al. 2014 The use of the finite element method for solving problems of terramechanics. *Transactions of NNSTU n.a. R.E. Alekseev* **5(107)** pp 52-58
[37] Madsen J et al. 2012 A Physics-Based Vehicle-Terrain Interaction Model for Soft Soil Off-Road Vehicle *SAE International Journal of Commercial Vehicles* **№5(1)** pp 280–290
[38] O'Connor R M et al.1997 Discrete Element Modeling of Sand Production *International Journal of Rock Mechanics & Mining Science* **34 (3-4)** p 373
[39] *Injection Analysis* http://www.dynaexamples.com/dem/injection-analysis
[40] Teslenko D S et al. 2017 Modeling the workflow of a plow snowblower using the discrete element method *Building and road machines* **10** pp 29-33

IOP Conf. Series: Materials Science and Engineering 1086 (2021) 012037 doi:10.1088/1757-899X/1086/1/012037

Track snow and swamp-going vehicle with 16 t GVW

S E Manianin[1], E V Lisagov, P E Dmitriev, A A Eremin, U Sh Vakhidov and V S Makarov

Nizhny Novgorod State Technical University n.a. R.E. Alekseev, Minin str. 24, 603950, Nizhny Novgorod, Russia.

E-mail: [1]sergmanian@yandex.ru

Abstract. The paper handles the relevance problem of all-terrain vehicle utilization to provide for transportation in hard-to-access territories of the Russian Federation. The most advisable is the utilization of track snow and swamp-going vehicle. It has been demonstrated that, for Russian market, the most relevant is the development of vehicles with 4-6 t load capacity. Analyzed was 5901 track snow and swamp-going vehicle developed in Nizhny Novgorod State Technical University n.a. R.E. Alekseev and "Zavolzhsky Crawler Vehicle Plant" CJSC Demonstrated were better load carrying capacity, soil pressure and motor power performance values compared to the existing similar vehicles. The paper presents the basic specifications.

1. Introduction

Development regions of deposits of mineral resources in Russia are characterized by absence of normal access ways. In addition, these territories have specific climate, with average monthly air temperatures -10° and below. Similar regions occupy up to 2/3 of the territory of this country. No less significant is the solid snow blanket duration from 4 to 9 months annually. 3 to 5 months annually, most of the territory of the country is covered by a snow blanket deeper than 0.5 m [1-3], seriously impeding, in most cases, the economic activities of the regions. The severities encountered in the course of the transportation development of such regions can be illustrated by the experience gained in the West Siberia, where almost complete absence of roads is combined with thick snow blanket, tundra, vast swamps and a vast annual interval from minus 50°C to plus 40°C. In such conditions, the climate does not allow for road construction making it impossible to organize manufacturing of foods, construction materials and other essential goods. All these goods must be imported from other regions. For solving of transportation problems in these territories, wheel and truck all-terrain vehicles are widely used. From the point of view of the transportation efficiency [4-6], track snow and swamp-going vehicles are most suitable. In the Russian market, track snow and swamp-going vehicles are represented mostly by domestic-made vehicles.

2. Overview of tracked vehicles

The Russian market of track snow and swamp-going vehicles has a capacity of max. 1000 vehicles annually. Thereby, the number of the manufacturers of vehicles with 4-6 t load carrying capacity is small:

- SEMMESH Machine-Building Factory of Semipalatinsk – GT-TBU Track Vehicle;
- RuMZ Machine-Building Plant of Rubtsovsk OJSC - MT-LBu-N Track Vehicle;
- Chetra – TM-140 Track Vehicle;

- TTM – TTM-4902 Track Vehicle;
- SKBM OJSC - Truck Chassis Vehicle on BMP-3 Infantry Fighting Vehicle basis.

Thereby, there are almost now newly designed vehicle with 4-6 tons load carrying capacity in the market. Up to 80% are civil versions of MTLBu vehicles for armed forces. That is why, Nizhny Novgorod State Technical University n.a. R.E. Alekseev and "Zavolzhsky Crawler Vehicle Plant" CJSC have jointly developed 5901 Track Snow and Swamp-Going Vehicle see Fig. 1 for the general appearance of the vehicle.

Figure 1. 5901 Track Snow and Swamp-Going Vehicle

The vehicle has a comfortable passenger compartment and a capacious cargo compartment. A distinctive feature of the vehicle is its high load carrying capacity (up to 5 t) along with better all-road capability than similar vehicles, due to wider tracks. In addition, 5901 Snow and Swamp-Going Vehicle 5901 is manufactured 100% of Russian-made components.

3. Analysis of the main parameters of the tracked vehicle

The performance of the vehicle are in-line with the best specifications of existing comparable track vehicles. Figures 2-5 demonstrate average statistical parameters of track snow and swamp-going vehicles dependent on the vehicle's weight [7, 8], as well as 5901 vehicle.

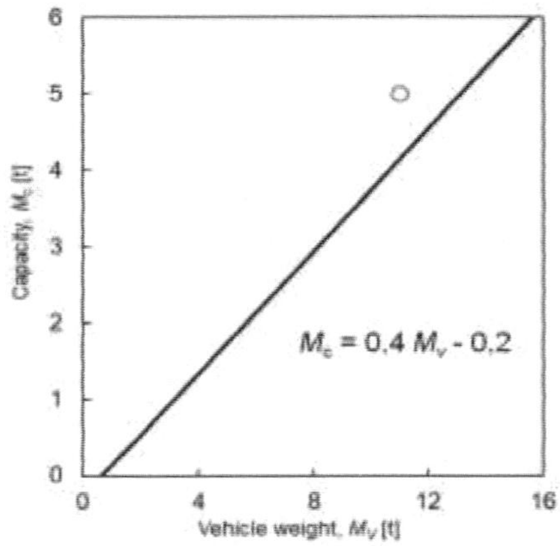

Figure 2. Capacity - Vehicle weight

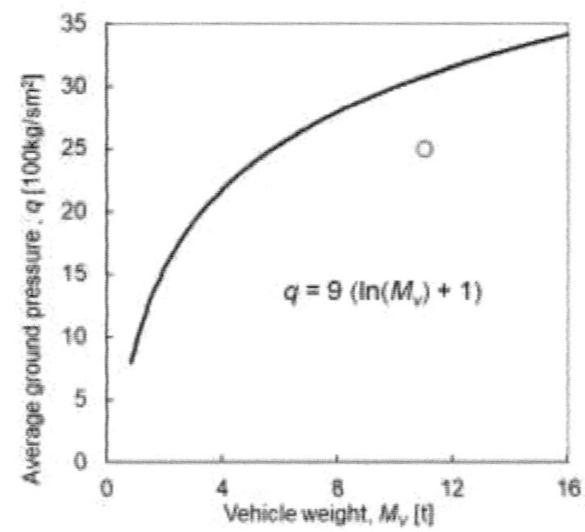

Figure 3. Average ground pressure – Vehicle weight

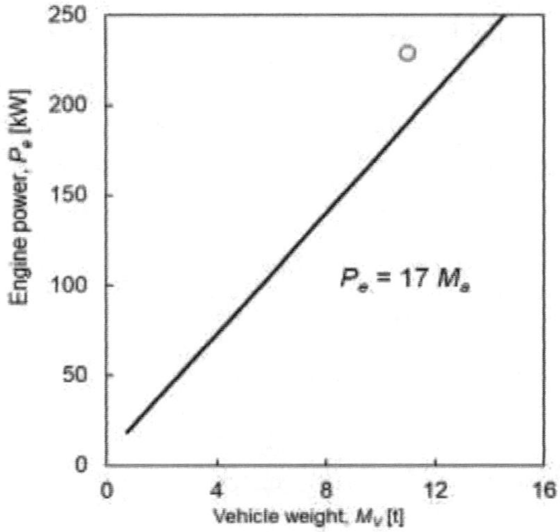

Figure 4. Engine power – Vehicle weight

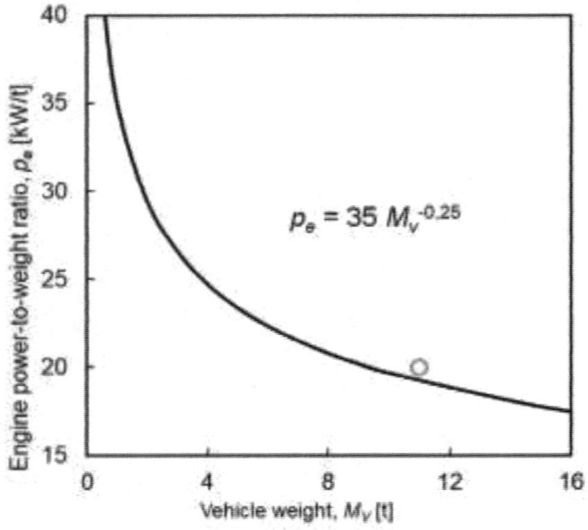

Figure 5. Engine power-to-weight ratio - Vehicle weight

The analysis of the provided data demonstrates the following. Figures 2-5 show that the weight of the transported goods is higher, the specific ground pressure is lower, the power and the specific power of the engine are higher that those of a statistically average vehicle, that is why the designed vehicle has better performance specifications than existing similar vehicles. Table 1 shows the basic specifications of the vehicle.

Table 1. Basic specifications of 5901 Track Snow and Swamp-Going Vehicle

Specifications	Value
Gross vehicle weight (GVW), t	16
Load-carrying capacity, t	5
Configuration diagram:	1) Passenger version; 2) Cargo version
cabin	front section
engine compartment	middle section
Number of seats	4
Average ground pressure with full load, MPa (kgf/cm^2)	0.025 (0.25)*
Engine	YaMZ-536 diesel engine
Rest surface length of the tracks, mm	4440
Overall width (by tracks), mm	3200
Track width, mm	2490
Weight of towed trailer, t	5
Travel speed with full load, w/o trailer, km/h	
on-road mode	55
swim mode;	5
Road clearance, mm	420
Fuel endurance, on-road mode, km	600
Obstacles bridged:	
ascend (descend)	35°
ascend (descend) with a trailer	25°
flank slope	25°
Bridged log diameter, m	0.7

Minimum turning radius, m	3
Maximum angles (unflooded):	
water entry	20°
water exit	15°
Transmission:	
gearbox	1) Mechanical; 2) Automatic
axle gear	Single-stage
side transmissions	Planetary gears
Turning mechanism	side clutches with brakes
Turning control	steering wheel
Chassis / suspension:	independent, torsion bar suspension
Propulsion unit:	
type	track-type with front drive sprockets
track rollers	single-ended, 7 pc./side, dia. 670 mm
support rollers	single-ended, 3 pc./side
idler sprockets	rear, suspended
caterpillar chains	steel, with metal-rubber joints, with elastic wideners, 720 mm wide
tension mechanisms	screw type, inside the body.
Body:	self-carrying, steel, all-welded

Conclusions

The paper describes the design of a promising snow and swamp-going vehicle with 16 GVW designed by NNSTU n.a. R.E. Alekseev, to be manufactured by "ZZGT" CJSC. The authors of the paper are designers of the vehicle. It was demonstrated that the performance specifications of the track vehicle, as well as its being manufactured completely of Russian-made components, allows an assumption that the snow and swamp-going vehicle will be in demand in the Russian market.

Acknowledgment

This study was conducted in continuation of the research conducted by "Nizhny Novgorod scientific and practical school of snow vehicles" in the framework of cooperation between the Nizhny Novgorod State Technical University n.a. R.E. Alekseyev and Zavolzhsky Crawler Vehicle Plant under Contract № 28/19/DPR/117/19/2468.

References

[1] Makarov V, Zeziulin D and Belyakov V 2014 Prediction of all-terrain vehicles mobility in snowscape scenes *18th International Conference of the ISTVS*

[2] Goncharov K O, Belyakov V V and Makarov V S 2018 Estimating method of efficiency of wheel chassis driving on snow *IOP Conference Series: Earth and Environmental Science* 194(3) 032005 doi: 10.1088/1755-1315/194/3/032005

[3] Kurkin A et al. 2017 Ground vehicle for ice conditions monitoring *13th International MEDCOAST Congress on Coastal and Marine Sciences, Engineering, Management and Conservation, MEDCOAST 2017* **2** pp 827-838

[4] Porubov D 2017 Study of efficiency of using all-terrain vehicles during the winter period *19th International and 14th European-African Regional Conference of the ISTVS*

[5] Anikin A et al. 2017 Calculation of traction capabilities of wheeled vehicles on low-pressure tires on snow *19th International and 14th European-African Regional Conference of the ISTVS*

[6] Manianin S et al. 2019 Experimental Research of the Effect of Track Chains Pre-static Tension on the Mobility of a Tracked Vehicle *ICMT 2019 - 7th International Conference on Military Technologies* 8870025 doi: 10.1109/MILTECHS.2019.8870025

[7] Belyakov V et al. 2020 Analysis of the diagrams and main parameters of the existing structures of the articulated tracked vehicles *IOP Conference Series: Materials Science and Engineering* **820(1)** 012001 doi: 10.1088/1757-899X/820/1/012001
[8] Belyaev A et al. 2019 Development of 8×8 all-terrain vehicle with individual wheel drive *VEHITS 2019 - Proceedings of the 5th International Conference on Vehicle Technology and Intelligent Transport Systems* pp 556-561

IOP Conf. Series: Materials Science and Engineering 1086 (2021) 012038 doi:10.1088/1757-899X/1086/1/012038

Use of active related transportation systems in the extractive industries of the national economy and protection complex

E E Bazhenov[1], L A Novopashin[1] and D O Chernyshev[2]

[1] Ural State Agrarian University, Karl Liebknecht street 42, Ekaterinburg, Russia
[2] Ural State Forest Engineering University, Siberian highway 37, Ekaterinburg, Russia

E-mail: [2] den_is-best@mail.ru

Abstract. The range of practical applications of active articulated transport systems is considered (AATS). AATS have a lower specific pressure of the propellers on the supporting surface, all other things being equal with single machines. This allows you to increase the payload on the transport system, increase the off-road capability of vehicles and on supporting surfaces with low bearing properties. The presence of separate power and technological sections makes it possible to create a range of technological systems with different functional properties. It is most expedient to use AATS in the raw materials and extractive industries and in the defense complex.

1. Introduction

The work of transport and technological machines in some sectors of the economy is specific. This is due to their operation on temporary roads or off-road conditions. In the timber industry, the operation of transport systems takes place on simplified temporary roads, which make up about 40% of all types of timber haul roads [1-3].

The emergence of new energy-rich multi-operational machines allows mechanizing most of the work in the forestry, agriculture, oil and gas and mining complexes, the defense industry and other areas of the national economy. [4, 5]

2. Experimental results

The use of all-wheel drive transport systems based on the activation of the trailer stock and the use of one of the promising tasks in the implementation of many tasks that perform during the operation of transport and technological complexes in the conditions of winter roads, dirt roads during the thaw period and in other specific conditions.

The concept of an articulated transport system (ATS) includes a whole range of machines with two or more sections, connected by a hinge with one or more degrees of freedom. Some ATS have a specific steering design. ATS is often considered as road trains with active trailers (TAT). [6-8]

The use of TAT and ATS in the sectors of agriculture, forestry, oil and gas complexes and other raw material sectors of the economy makes it possible to create a wide range of technological and transport systems.

The most general concept of all-wheel drive ATS should be considered active articulated transport systems - AATS. [9, 10]

ATS are technological and transport. This division is determined by the purpose of the transport system.

Technological AATS consists of two or more sections, one of which plays the role of a mobile power plant (energy module), and the second has a narrow utilitarian function (technological module). Similar AATS schemes can be widely used in the forestry, oil and gas industries, and municipal services.

A transport AATS is a car or tractor train with a different distance from the energy module to the trailer link. The transport AATS includes wheeled and tracked transport systems with a "breaking frame". [11, 12]

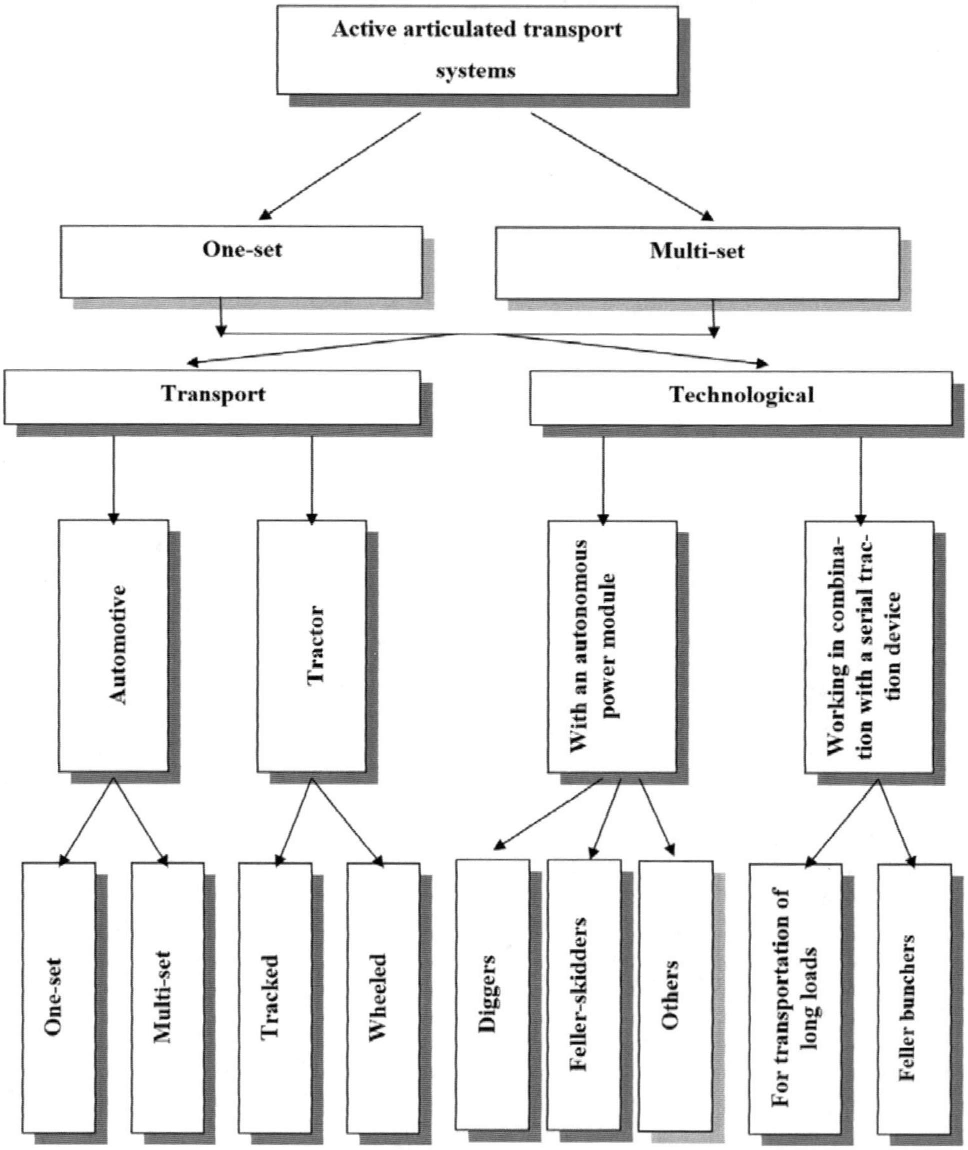

Figure 1. Articulated transport system classification

As an example, consider the use of AATS in agriculture, mining, and the defense industry.

In agriculture, the share of transport work in the annual employment of wheeled tractors exceeds 50%, and most of them fall on temporary roads and field conditions. Under unfavorable weather conditions, especially in the zones of "risky farming", such roads become a significant obstacle in the performance of technological operations and transportation of products.

Most of agricultural machinery performing specific technological operations, high power of the internal combustion engine. This technique is used no more than one month a year. At the same time, the equipment must be stored in special conditions, undergo maintenance during operation and during storage. Maintenance of equipment with internal combustion engines must be carried out by highly qualified personnel in special conditions. [13]

The use of AATS in agriculture implies the following scheme: one energy module (a serial tractor can be used), depending on the technological operation being performed, is connected to a technological module (section) having drive wheels to increase cross-country ability (Figure 2).

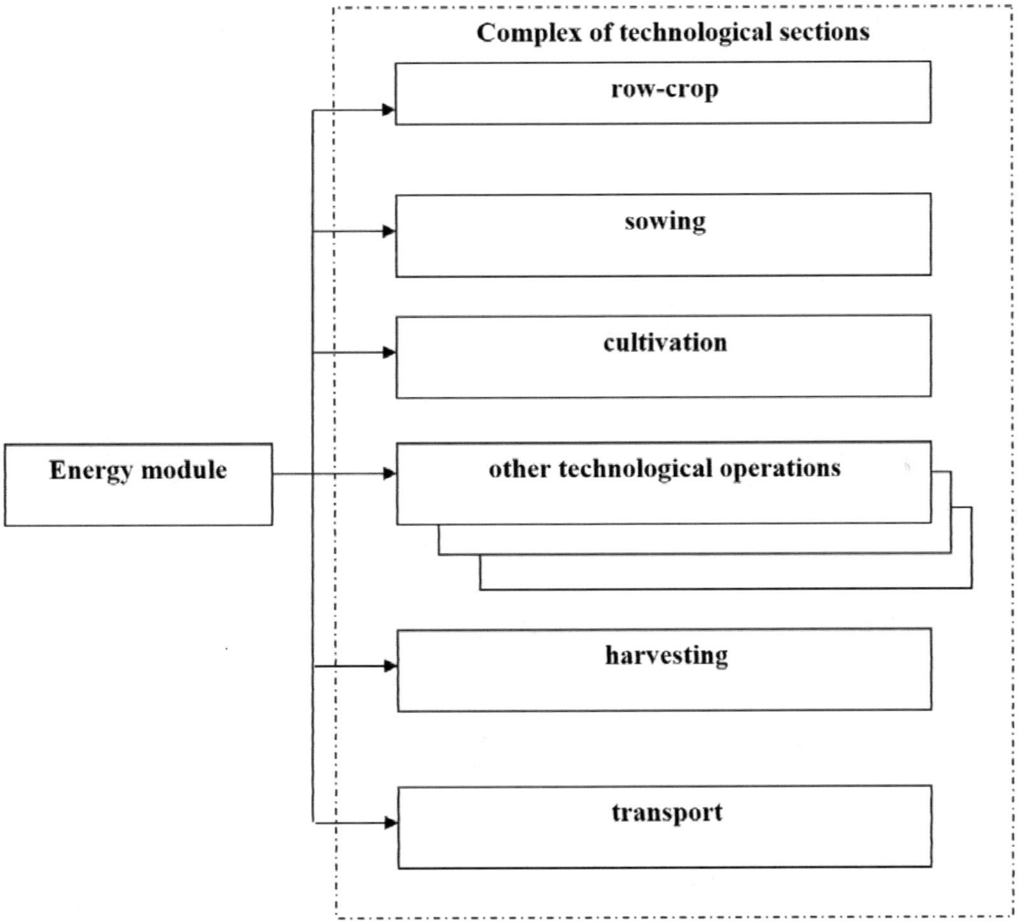

Figure 2. The scheme of using AATS in agriculture

In the mining industry, the use of AATS allows you to abandon heavy-duty dump trucks with a lifting capacity of 1.5 - 1.7, and using an energy module as part of a trailer link to reduce the total weight of the transport system by at least 25% with an increase in the lifting capacity to 3 - 3,5. [14]

In the defense complex, the use of AATS makes it possible to create fundamentally new platforms for the deployment of advanced types of weapons. The appearance of new propellant charges makes it possible to significantly increase the rate of fire and the range of fire in self-propelled artillery. Self-propelled artillery pieces (SPAP) are designed to conduct short-term (time of a safe stay at one firing position up to 1 minute) high-intensity fire with a change in firing position. SPAP move into position, conduct high-intensity fire at distant targets, then move to another position before the enemy can return fire. With the growth of automation of fire control of modern artillery and the widespread introduction of radar reconnaissance means for detecting firing positions, the reaction time of enemy artillery as part of reconnaissance and firing systems operating in real time, and, accordingly, the time for the safe stay of the SPAP in a firing position is significantly reduced. Therefore, a means for overcoming these problems is the creation of a SPAP, with fundamentally new, unconventional structural and layout schemes, providing a significant increase in mobility and firepower (and, accordingly, a decrease in the time spent on a firing position of the SPAP). [15, 16]

In this case, an artillery installation with a loader is considered the basic layout. For example, the experimental vehicle "Coalition - SV" based on the SPAP 2S19 "MSTA-S", as part of the design Uraltransmash (Yekaterinburg). This basic layout of an articulated self-propelled artillery unit (ASPAP) consists of two separate machines designed to perform specific tasks. Each of the vehicles has a power plant and a crew, and can operate both as part of the ASPAP and independently.

A fundamentally new platform does not have to have two propulsion systems and two crews. The most rational and promising is the use of a platform with one power plant and one crew. The loading process is controlled by the crew of the artillery gun. This solution involves the creation of a separate structure (Figure 3). [17, 18]

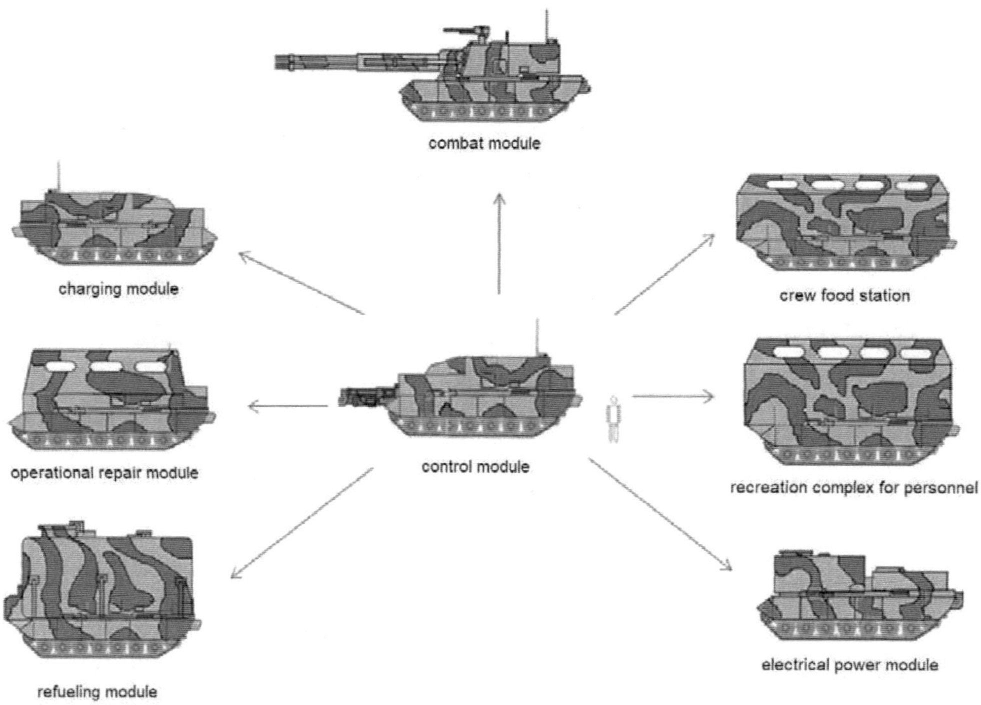

Figure 3. The modular principle of building an ensemble of machines

The installation, for performing any task, is located in the combat (technological, transport) module - a section that does not have a power plant and has a drive for propellers. At the same time, there is a control (energy, power) section - a module where the power plant and the crew are located. Modules have the ability to connect using a articulation unit. The articulation unit allows the power flow to be transferred from the power module to the technological one, changes the relative position of the modules to rotate the system and provides mutual movement of the modules in three degrees of freedom. The modular principle of building the platform is not limited to the use of two sections - the combat module and the control module. [19-21]

3. Conclusion

The use of a multi-set or multi-section principle creates complexes with a large ammunition load, which has great survivability and mobility.

To increase the firepower and increase the tactical and operational-tactical mobility of the CAO, it is necessary to create a fundamentally new look of the platform. To the greatest extent, the requirements are met by multi-set articulated transport systems, which make it possible to create an ensemble of transport systems that ensure the combat capability of an artillery unit. [22]

Thus, one of the most important reserves for increasing the efficiency of transport and transport-technological operations is the increase in cross-country ability and maneuverability, achieved by the creation of structurally different chassis. The requirements are met by multi-set articulated systems that create an ensemble of transport and transport-technological machines, and ensure high efficiency in performing production tasks.

References

[1] Klubnichkin E E et al. 2018 Interaction between elements of the track ground contacting area with the soil at curvilinear motion of the timber harvesting machine IOP Conference Series: Materials Science and Engineering 386(1) 012016 doi: 10.1088/1757-899X/386/1/012016

[2] Klubnichkin V et al. 2019 Researching rolling resistance of the wheeled forestry tractor at skidding IOP Conference Series: Materials Science and Engineering 695(1) 012003 doi: 10.1088/1757-899X/695/1/012003

[3] Makarov V S et al. 2019 Study of uneven surfaces distribution on forestry roads Journal of Physics: Conference Series 1177(1) 012041 doi: 10.1088/1742-6596/1177/1/012041

[4] Platonov V F 1989 All-wheel drive cars (Moscow: Mechanical Engineering) p 312

[5] Ageykin Ya S, Volskaya NS 2008 Theory of Automobiles: A Tutorial (Moscow: MGIU) p 318

[6] Polungyan A A 2009 Design of all-wheel drive vehicles (Moscow: MSTU them. N.E. Bauman) p 488

[7] Bazhenov E E 2009 Articulated transport and technological systems (Ekaterinburg: USTU – UPI) p 174

[8] Belyakov V at al. 2020 Analysis of the diagrams and main parameters of the existing structures of the articulated tracked vehicles IOP Conference Series: Materials Science and Engineering 820(1) 012001 doi: 10.1088/1757-899X/820/1/012001

[9] Bazhenov E E and Vlyukhin A V 2010 Application of articulated transport systems in the extractive industries of the national economy *Truck: Scientific and Technical. magazine* **4** p 34

[10] Dhir A and Sankar S 1997Analitical model for dynamic simulation of off-road Tracked Vehicles *Vehicle system dynamics* **27** p 37

[11] Bazhenov E E, Golomidov S I and Yushkov A I 1985 Determination of the drive power of the active axle of trailers *Exploitation of timber transport rolling stock: interuniversity* Sat. sci. tr., Sverdlovsk p 72-75

[12] Watanabe K and Kitano M 1986 Study on streerability of articulated tracked vehicles *Journal of Terramechanics* **23** p 69

[13] Veits V L 1969 *Dynamics of engine aggregates with internal combustion engines* Leinigrad p 370

[14] Bazhenov E E 2010 Modular principle of modeling of articulated transport-technological systems *Tractors and agricultural machinery* p 20

[15] Setinc M, Gradisar M and Tomat L 2015 *Optimization of a highway project planning using a modified genetic algorithm* p 687-707
[16] Buynachev S K and Bazhenov E E 2008 *Modeling of the structure of mechanisms* Ekaterinburg p 27-29
[17] Chelomey V N 1980 *Vibration in Engineering: A Handbook* Moscow p 544
[18] Bazhenov E E 2010 *Fundamentals of the theory of articulated transport systems* Ekaterinburg: UrFU p 257
[19] Bazhenov E E, Buynachev S K and Kruchinin I N 2010 *Fundamentals of the theory of articulated transport systems* Yekaterinburg. UrFU p 257
[20] Bazhenov EE 2010 Modular principle of the synthesis of transport and technological systems *Izvestiya VUZov. The Volga region. Technical sciences* **2** p 79
[21] Tarasik V P 1998 *Mathematical Modeling of Technical Systems* Moscow p 204-640
[22] Bazhenov E E 2009 *Articulated Transport and Technological Systems Monograph* Ekaterinburg: USTU – UPI p 174

Ways to reduce fuel consumption of a light commercial vehicle for WLTC and NEDC cycles

A A Kolin, S E Silantyev and P S Rogov[1]

[1]Nizhny Novgorod State Technical University n.a. R.E. Alekseev, Minin st., 24, 603950, Nizhny Novgorod, Russia.

E-mail: [1] petr.rogov@nntu.ru

Abstract: The paper presents the results of the simulation model employed for fuel consumption analysis of a light commercial vehicle in road traffic cycles; virtual tests are performed. The motor vehicle main parameters are analysed for impact on fuel consumption in NEDC and WLTC cycles. Changes of average fuel consumption are quantified for varying main parameters of the vehicle structure. The distribution of energy consumption for the vehicle in motion is demonstrated. On the basis of the regression analysis impact coefficients of individual gear ratios on the average fuel consumption of a light commercial vehicle are obtained.

Introduction

For many consumers the fuel consumption is one of the main criteria for a vehicle model selection.

Higher prices for hydrocarbon fuel and deterioration of the environment associated with high-intensity traffic brings about the development and implementation of single-approach methods for calculation and evaluation of fuel, economic and environmental properties based on standard driving cycles.

The most common standard cycles: New European Driving Cycle (NEDC) [1], American Driving Cycle (FTP-75) and Japanese Driving Cycle (JC-08). In addition to the above, the World Harmonized Driving Cycle (WLTC) [2] was introduced in 2017; it is based on the world static study of driving modes and features high accelerations and lack of steady-motion intervals.

The objective of the study is to quantify the impact of various design parameters on fuel efficiency of a light commercial vehicle with gross vehicle weight of 3500 kg in NEDC and WLTC driving cycles.

Main methods of improving the fuel efficiency of a vehicle are specified in earlier studies [3-7]. These include the reduction of aerodynamic drag, tire rolling resistance and optimization of the transmission performance. Everything stated above shall have an impact on result of the vehicle motion simulation in driving cycles.

Methods of the analysis

NEDC cycle consists of one urban driving cycle (or 4 simple urban cycles) and one suburban driving cycle. Cycle driving is performed using a detailed operational map. The map details the the gear box speeds to be used for every segment of the vehicle motion as well as accelerations of the vehicle.

The operational map and requirements for the vehicle motion in NEDC driving cycle are included in the standard [1]. The standard applies to transport vehicles of category M1, M2, N1 and N2 with reference mass not exceeding 2 610 kg.

The vehicle driving diagrams based on NEDC cycle are shown in figure 1.

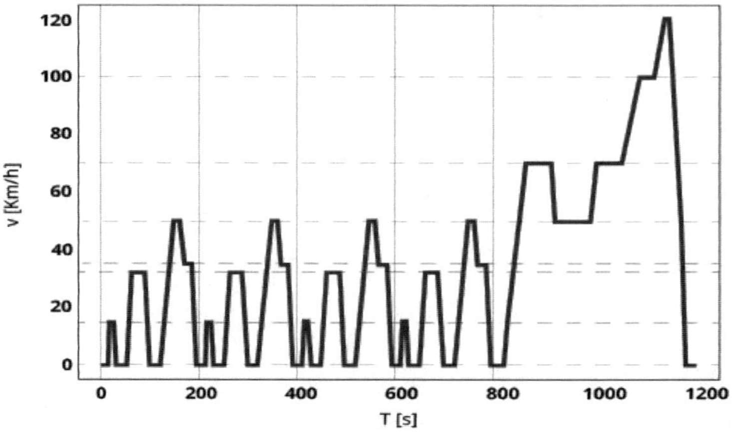

Figure 1. Vehicle motion diagram based on NEDC driving cycle

As shown in figure 2, WLTC cycle is separated into four phases by short-duration stops: Low-speed phase when the vehicle accelerates to a maximum of 56.5 km/h; medium-speed phase (76.6 km/h), high-speed phase (97.4 km/h) and extra-high speed phase (131.6 km/h).

The vehicle motion for each driving cycle is based on the operational map. The phase is selected depending on the power-to-weight ratio of a vehicle. Extra High-speed phase is not included in the study which is acceptable for the vehicle class under review. WLTC driving cycle is performed according to rules specified in [2].

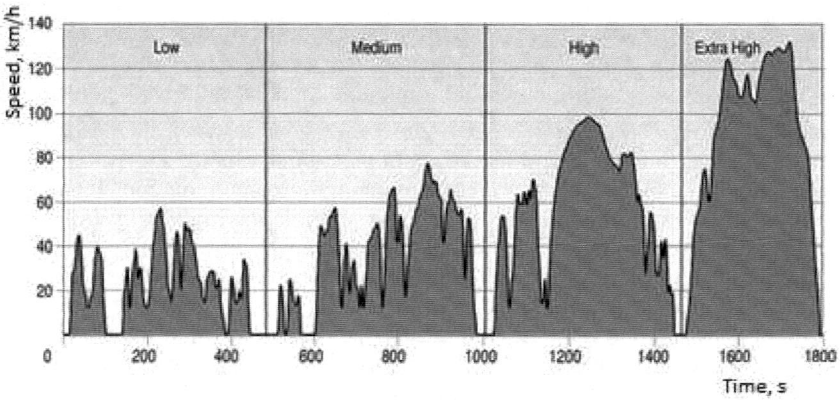

Figure 2. Vehicle motion diagram based on WLTC driving cycle

The simulation model built is used for modelling of the vehicle motion. Well-known approaches to the mathematical representation of the vehicle motion are used, as specified in papers on the vehicle theory [8].

In case of WLTC cycle simulation, the vehicle acceleration is unknown.

The acceleration of the vehicle is computed using the following formula:

$$a_a = \frac{v_{j+1} - v_j}{3{,}6 \cdot (t_{j+1} - t_j)}, \tag{1}$$

where a_a (m/s) – acceleration of the vehicle; V_j, V_{j+1} (km/h) – initial and final speeds over the selected interval, respectively; t_j, t_{j+1} (s) – start and end times, respectively.

Gears are shifted as required in [1-2]. After gear shifting ICE starting speed of rotation is determined subject to correlation of the vehicle speed and ICE speed of rotation:

$$V = \omega_e \, r_e / u_{\text{тр}} \tag{2}$$

The traction force is computed by formula:

$$F_{\text{т}} = T_e \, u_{\text{тр}} \, \eta_{\text{тр}} / r_{\text{к}} \tag{3}$$

where T_e – ICE torque, $u_{\text{тр}}$ – transmission gear ratio, $\eta_{\text{тр}}$ – transmission efficiency, $r_{\text{к}}$ – wheel rolling radius.

T_e is normally a function of the accelerator pedal position and ICE speed, or $T_e = f(d, \omega_e)$. In the mathematical model T_e represents data array $T_{e\,i,j}$ $i = 1 \ldots 17, j = 1 \ldots 11$.

Alternatively, the traction force can be determined based on the following dependency.

$$F_m = F_f + F_w + \delta m_a a_a, \tag{4}$$

F_m – traction force of drive wheels; F_f – rolling resistance force; F_w – air resistance force; δ – rotational inertia coefficient, m_a – vehicle mass, a_a – vehicle acceleration.

ICE torque T_e is calculated using formula (3), and subject to the engaged transmission gear that is determined using formula (2). The specific fuel consumption in the model is represented by data array $g_{e\,ij}$; it is a function of ICE torque and speed of ration $g_e = f(T_e, \omega_e)$. T_e and ω_e values are computed using expressions (2-4). For calculation of the total fuel consumption per driving cycle, the current fuel consumption values are multiplied by validity times of T_e and ω_e, and then added up. The unknowns are computed by formulae (5-7),

The rolling resistance force is determined as:

$$F_f = f m_a g \tag{5}$$

where f – rolling resistance factor; g – gravity acceleration.

The air resistance is determined as follows:

$$F_w = 0{,}5 \cdot Cx \cdot p_в \cdot A_в V^2, \tag{6}$$

where C_x – aerodynamic drag coefficient; $\rho_в$ – air density; $A_в$ – transverse projection area of the vehicle; V – vehicle speed.

The rotational inertia coefficient is determined as:

$$\delta = 1 + \frac{I_e u_{TP}^2 \eta_{TP}}{m_a r_к^2} + \frac{I_{кn} u_{ГП}^2 \eta_{ГП}}{m_a r_к^2} + \frac{\sum I_к}{m_a r_к^2}, \tag{7}$$

where I_e – ICE moment of inertia, $I_{кn}$ – gear drive moment of inertia, $u_{ГП}$ – final drive gear ratio, $\eta_{ГП}$ – final drive efficiency, $I_к$ – moment of inertia of a half-axle wheel (if any).

Results of the study

The developed simulation model shall reliably respond to the variation of input data, i.e. the reduction of the aerodynamic drag component will contribute to lower fuel consumption. Simulation results for a light commercial vehicle are specified in tables 1 and 2.

The gear box efficiency changed by 1% per shift.

It is to be noted that the results obtained conclusively prove that these parameters affect the average fuel consumption rate. Increasing the cardan drive efficiency by 5% brings down the average fuel consumption rate by 1.2% for NEDC cycle and 1.02% for WLTC cycle

Increasing the final drive and the gear box efficiency by 1% in NEDC cycle brings down the average fuel rate by 1.3%, and 0.74%, respectively. For WLTC cycle the change of these parameters results in the reduction of the average fuel consumption rate by 1.4% and 0.93%, respectively.

If the aerodynamic drag coefficient is reduced by 5%, the average fuel rate goes down by 2% for NEDC cycle, and 1.8% for WLTC cycle.

The rolling resistance factor will have a lower impact as compared to other indices reviewed, since its variation to the extent of 5% causes the average fuel rate to change within the limits of 0.7% for NEDC cycle, and 0.65% for WLTC cycle.

The energy expended on the light commercial vehicle motion is specified in table 3.

Table 1. Impact of specified parameters for NEDC and WLTC cycles

Index	Variation, %	Fuel rate, l/100 km	Variation, l
Aerodynamic drag coefficient	5	11.02	0.21
	0	10.81	0
	-5	10.66	-0.21
Rolling resistance factor	5	10.88	0.07
	0	10.81	0
	-5	10.74	-0.07
Ring and pinion set efficiency	1	10.67	-0.14
	0	10.81	0
	-1	10.95	0.14
Cardan drive efficiency	0.5	10.68	-0.13
	0	10.81	0
	-0.5	10.94	0.13
Gear box efficiency	1	10.73	-0.08
	0	10.81	0
	1	10.89	0.08

Table 2. Impact of specified parameters for WLTC and WLTC cycles

Index	Variation, %	Fuel rate, l/100 km	Variation, l
Aerodynamic drag coefficient	5	10.98	0.19
	0	10.79	0
	-5	10.6	-0.19
Rolling resistance factor	5	10.86	0.07
	0	10.79	0
	-5	10.72	-0.07
Ring and pinion set efficiency	1	10.64	-0.15
	0	10.79	0
	-1	10.94	0.15
Cardan drive efficiency	0.5	10.68	-0.11
	0	10.79	0
	-0.5	10.9	0.11
Gear box efficiency	1	10.69	-0.1
	0	10.79	0
	1	10.89	0.1

Table 3. Energy consumption of the vehicle motion for NEDC and WLTC cycles

Driving cycle	Total resistance of motion energy			Energy of mechanical losses in transmission			
	Aerodynamic drag energy, %	Roling resistance energy, %	Speed up resistance energy, %	Half-axles and hubs, %	Gear box, %	Final drive, %	Cardan drive, %
NEDC	40.16	18.47	18.58	14.75	4.04	3	1.01
WLTC	31.54	18.89	28.51	13.02	4.04	3	1.01

It is to be noted that the energy of mechanical losses in the transmission can be as high as 23% of the total energy expended on the transport vehicle motion, as specified in table 3. The aerodynamic drag energy can be as high as 30 to 40% of the total energy used for NEDC and WLTC cycles, respectively.

Furthermore, the impact assessment of the current 6-speed transmission gear ratio optimization on fuel consumption is performed.

The Plackett-Burman plan is employed to reduce the number of simulations and, thereby shorten the time of the analysis, as specified in [9]. Transmission gear ratios are used as the factors affecting the average fuel consumption, specifically: X1 – gear ratio of 1st gear; X2 – gear ratio of 2nd gear; X3 – gear ratio 3rd gear; X4 – gear ratio of 4th gear; X5 – gear ration of 5th gear; X6 – gear ration of 6th gear.

The range of the gear ratio variability is 10%; the changed gear ratios are specified in table 4.

Table 4. Levels and ranges of variability of factors

Level of factor	X1	X2	X3	X4	X5	X6
Basic level	5.065	2.780	1.591	1.000	0.807	0.643
Bottom level	4.5585	2.502	1.4319	0.9	0.7263	0.5787
Top level	5.5715	3.058	1.7501	1.1	0.8877	0.7073

Further, regression equations are composed.

Coefficients of equation Y = b0 + b1·X1 + b2·X2 + b3·X3 +…+ b4·X4 + b5·X5 +… are determined using formulae:

$$b_0 = \frac{1}{N}\sum_{j=1}^{N} Y_j \tag{8}$$

$$b_i = \frac{1}{N}\sum_{j=1}^{N} X_{ij} Y_j \tag{9}$$

where $i = 1, 2, \ldots, n$ – factor number; N – number of experiments; j - experiment reference number.

Further, coefficients of interest and results of the experiments are included in table 5.

Table 5. Target and calculated values of average fuel consumption for NEDC cycle.

Experiment	X1	X2	X3	X4	X5	X6	S	σ_j^2	S^P	$\left(S_{кр} - S_{кр}^P\right)^2$
1	1	1	1	-1	1	-1	11.124	0.111	11.12	0.0000
2	-1	1	1	1	-1	1	11.400	0.117	11.40	0.0000
3	-1	-1	1	1	1	-1	10.625	0.102	10.62	0.0000
4	1	-1	-1	1	1	1	10.721	0.103	10.72	0.0000
5	-1	1	-1	-1	1	1	11.250	0.114	11.26	0.0001
6	1	-1	1	-1	-1	1	10.774	0.104	10.78	0.0000
7	1	1	-1	1	-1	-1	10.947	0.108	10.94	0.0000
8	-1	-1	-1	-1	-1	-1	10.351	0.096	10.36	0.0001
	b1	b2	b3	b4	b5	b6	b0			
	-0.01	0.28	0.08	0.02	0.03	0.14	10.9	0.856		0.0003

The problem of interest is not resolved by calculation of regression coefficients; the problem solution also requires the model quality (accuracy) to be found. According to [9], each experiment is subjected to dispersion analysis, with correction for the design model inaccuracy applied to the obtained results that amounts to 3%. Having calculated the error correction for the experiment, the model validity can be found, and the model efficiency checked based on Fisher's criterion [9].

Tabulated value F of Fisher's criterion, at $P=0.95$, $f_1= f_{eff}=6$, $f_2= f_y=8$; $F=3.58$. Calculated value $F_c=0.003$.

Since $F_c < F$, the model adequately simulates the dependence of interest with a confidence level of 0.95.

The degree of impact of this factor is dependent on the value of the regression coefficient. The 'plus' sign indicates that the complex optimization parameter goes up when the factor increases; the 'minus' sign indicates the complex optimization parameter goes down.

According to table 5, regression coefficients b1-b6 indicate that the reduction of the first speed gear ratio causes the average fuel consumption to increase. The reduction of gear ratios for 2nd gear to 6th gear inclusive brings down the average fuel consumption. In terms of the impact on the average fuel consumption of a light commercial vehicle, it is mostly affected by the reduction in the 2nd speed gear ratio, with the regression coefficient 0.28; then 6th speed gear ratio, with the regression coefficient 0.14, and 3rd speed gear ratio with the regression coefficient 0.08.

It should be noted that reducing the transmission gear ratio for higher fuel efficiency has a negative impact on traction, speed and environmental characteristics of the vehicle. Keeping the above in view, a compromise solution is needed in the future.

Summary

According to tables 1 and 2, the following results are obtained:

Increasing the cardan drive efficiency by 5% brings down the average fuel consumption rate by 1.2% for NEDC cycle and 1.02% for WLTC cycle.

Increasing the final drive and the gear box efficiency by 1% in NEDC cycle brings down the average fuel rate by 1.3%, and 0.74%, respectively. For WLTC cycle, changing these parameters results in the reduction of the average fuel consumption rate by 1.4% and 0.93%, respectively.

If the aerodynamic drag coefficient is reduced by 5%, the average fuel rate goes down by 2% for NEDC cycle, and 1.8% for WLTC cycle.

The rolling resistance factor will have a lower impact as compared to other indices reviewed, since its variation to the extent of 5% causes the average fuel rate to change within the limits of 0.7% for NEDC cycle, and 0.65% for WLTC cycle.

It is to be noted that according to table 3 the energy of mechanical losses in the transmission can be as high as 15% of the total energy used for the transport vehicle motion. The aerodynamic drag energy can be as high as 30 to 40% of the total energy used for NEDC and WLTC cycles, respectively.

Given the simulation results, the highest potential for reducing the average fuel rate of a light commercial vehicle is associated with its improved aerodynamic shape. Higher transmission efficiency can have a substantial impact on the average fuel consumption rate, however, the potential for improvement of this index is limited. The rolling resistance factor has the least impact on the parameter of interest. All the parameters listed above can be optimized together that can help to achieve the desired fuel efficiency index of a vehicle.

According to table 5, regression coefficients b1-b6 indicate that reducing the first speed gear ratio brings up the average fuel consumption. Lower gear ratios for 2nd gear to 6th gear inclusive bring down the average fuel consumption. In terms of the impact on the average fuel consumption of a light commercial vehicle, it is mostly affected by the reduction in the 2nd speed gear ratio, with the regression coefficient 0.28; then 6th speed gear ratio, with the regression coefficient 0.14, and 3rd speed gear ratio with the regression coefficient 0.08.

In the future, the developed simulation model will be used to analyse the impact of transmission gear ratios on fuel efficiency and environmental friendliness of a light commercial vehicle.

The studies are conducted under the sponsorship of the Ministry of Education and Science of the Russian Federation as part of the project High-end production of the product range of GAZelle Next vehicle fitted with new electronic systems architecture under Agreement No. 075-11-2019-027 dated 29.11.2019 (Decree No.218 dated April 09, 2010 of the Government of the Russian Federation).

References

[1] *Uniform provisions concerning the approval of vehicles with regard to the emission of pollutants according to engine fuel requirements* 2015 UNECE **No. 83** url:https://www.unece.org/ fileadmin/DAM/trans/main/ wp29/wp29regs/R083r5e.pdf

IOP Conf. Series: Materials Science and Engineering 1086 (2021) 012039 doi:10.1088/1757-899X/1086/1/012039

[2] *Proposal for new global technical regulations concerning worldwide harmonized light vehicle emissions test procedure* 2014 UNECE url: https:// www.unece.org/fileadmin/ DAM/trans/doc/ 2014/ wp29/ ECE- TRANS-WP29-2014-027r.pdf

[3] Ogorodnov S M, Tikhomirov A N and Maleev S I 2015 Evaluation of potential application of analytical methods of research of automobile fuel efficiency *Proceedings of universities. Mechanical engineering* pp 53-61

[4] Beresnev P et al. 2017 Simulation of road impacts for forecasting of durability of VE-hcle's suspension elements and supporting system *19th International and 14th European-African Regional Conference of the ISTVS*

[5] Zezyulin D et al. 2014 Modeling of roads impacts for life prediction of light commercial vehicles parts *FISITA 2014 World Automotive Congress*

[6] Blokhin A N 2006 *Development of search methods for rational transmission gear ratios taking into account performance characteristics and designated purpose of the vehicle*, PhD thesis p 256

[7] Valeev D Kh and Karabtsev V S 2014 Ways to reduce fuel consumption of freight vehicles *Mechanics of machines, mechanisms and materials* **4** pp 33-39

[8] Kravets V N 2013. *Automobile theory* Nizhny Novgorod State Technical University n.a. R.E. Alekseev p 413

[9] Lunev V A 2012 *Mathematical simulation and planning of the experiment* Saint-Petersburg State Polytechnic University p 153

AUTHOR INDEX

Alyabiev, A. 80
Amirseyidov, Sh A 36, 211
Bazhenov, E E 192, 206, 253
Bazhenov, M. 140
Bazhenov, Y 140
Belyaev, A M. 116
Belyakov, V V 1, 235
Berdiev, A L 30
Berdnikov, L A 74
Birykov, A L 86
Bushmanov, D 229
Busikov, Sh V 86
Butin, D A 121
Chegurov, M K 128
Cherevastov, M.G. 6
Chernyshev, D O 192, 253
Chikaev, A O 161
Denezhko, L V 192
Dmitriev, P E 248
Eliseev, M E 161
Eremin, A A 248
Erofeeva, A V 202
Evdokimova, T S 11, 100
Fedosova, L O 65, 100
Filatov, V I 183
Glazunov, D V 147
Glushkov, M N 94
Gnenik, M E 121, 183
Golubev, M 167
Goncharov, K O 161
Goncharov, K. 167
Gots, A N 36, 211
Gulin, M 167
Kalyashina, A V 100
Kalyonov, V. 140
Kantor, P Ya 106
Kapustin, A G 116
Kartashevich, A N 94, 106
Kartashov, A B 44
Kartoshkin, A P 86
Khozerov, V. 134, 167
Kindeyev, E A 197
Klubnichkin, E E 20, 44
Klubnichkin, V E 44, 53
Kocherov, A. 229
Kolesnichenko, D A 202
Kolin, A A 259
Kolobov, M 229
Korzhachkin, M G. 74
Kotiev, G O 20, 53
Kuklina, I G 218
Kulagin, A. 134, 167
Kustovsky, A N 206
Kuzmichev, V 167
Laschuk, K V 161
Latysheva, V 80
Lazarevich, S V 223
Leliovsky, K Y 177
Lepeshkin, A A 74
Lisagov, E V 248
Makarov, V S 53, 248
Maleev, S I 153
Mambetalin, K T 30
Manianin, S E 248
Mantserov, S 229
Markovnina, A I 1
Mazunova, L N 1
Mokerov, D S 59
Molev, Y I 6, 59
Moshkov, P S 183
Musarsky, R A 153
Novopashin, L A 253
Novoselov, A A 223
Ogorodnov, S M 128, 153
Okunev, A 229
Orlov, L N 128
Panov, V A 74
Papunin, A V 116
Pechalinov, D N 202
Plotnikov, S A 94, 106
Plotnikov, S.A. 86
Proshin, D N 147
Rogachev, D. 80
Rogov, P S 259
Saidullozoda, S S 30
Samokhvalov, E E 223
Sergeev, D G 86
Silantyev, S E 259
Simonov, M V 94
Sinodkin, A A 11, 100
Sizov, A Y 65
Stepanov, E V 128, 153
Teslenko, D S 235
Tiurikov, M I 11, 100
Toropov, Ye I 183
Trusov, Yu P 183

Tumanov, A A 65
Tumasov, A V 121
Umirzokov, A M 30
Usupov, M L 206
Vakhidov, U Sh 59, 248
Vashurin, A S 121, 183
Vasilyeva, K V 44
Vasin, A V 223
Vtyurina, M N 106
Vyatkin, R V 202
Yerasov, I A 59, 147
Zolotov, A V 65

Institute of Physics
Dirac House, Temple Back
Bristol BS1 6BE UK

ISSN: 1757-8981
ISBN 978-1-7138-3349-9

9 781713 833499

CURRAN ASSOCIATES INC.
proceedings.com